Bacterial Biofilms

Edited by Sadik Dincer, Melis Sümengen Özdenefe and Afet Arkut

Published in London, United Kingdom

IntechOpen

Supporting open minds since 2005

Bacterial Biofilms
http://dx.doi.org/10.5772/intechopen.82929
Edited by Sadik Dincer, Melis Sümengen Özdenefe and Afet Arkut

Contributors
Marián Maďar, Jana Kačírová, Eva Styková, Radomíra Nemcová, Michaela Maďarová, Gabriela Štrkolcová, Aladár Maďari, Arwa Abu Khweek, Amal Amer, Shuai Wang, Vasan Sivalingam, Sudeep Parajuli, Rune Bakke, Wei Chen, Ting-ting Xie, Hong Zeng, Cynthia Amaning Danquah, Samuel Osei-Djarbeng, Theresa Appiah, Yaw Duah Boakye, Francis Adu, Sadik Dincer, Fatıma Masume Uslu, Anıl Delik, John A Glaser, Mohsin Khurshid, Aqsa Shahid, Maria Rasool, Naheed Akhter, Bilal Aslam, Ali Hassan, Sadia Sana, Muhammad Hidayat Rasool, Andrea Stašková, Andrej Jenča, Stanislav Lauko, Juliana Cabrini Carmello, Sarah Raquel De Annunzio, Carla Raquel Fontana, Maria Betânia Melo De Oliveira, Lívia Caroline, Angela Di Somma, Carolina Canè, Antonio Moretta, Angela Duilio, Arianna Cirillo, Tsiry Rasamiravaka, Mondher El Jaziri, Julie Carette, Amandine Nachtergael, Pierre Duez, Sangeeta Singh, Princy Choudhary, Vishnu Agarwal, Jixiang Yang, Victor Alves Carneiro, Ramaiana Soares Melo, Águida Maria Albuquerque Azevedo, Antonio Mateus Gomes Pereira, Rafaela Mesquita Bastos Cavalcante, Renan Rhonalty Rocha, Vinícius Queiroz Albuquerque, Jesús Alberto Pérez Guerrero, Maria Nágila Carneiro Matos, Anca Nicolau, Elena-Alexandra Alexa, Leontina Grigore-Gurgu, Florentina Bucur, Daniela Borda, Corina Neagu, Andrew Spiers, Robyn Jerdan, Olga Iungin, Olena Moshynets, Geert Potters, Francisco Eduardo Aragão Catunda Junior

First published in London, United Kingdom, 2020 by IntechOpen
IntechOpen is the global imprint of INTECHOPEN LIMITED, registered in England and Wales, registration number: 11086078, 5 Princes Gate Court, London, SW7 2QJ, United Kingdom
Printed in Croatia

British Library Cataloguing-in-Publication Data
A catalogue record for this book is available from the British Library

Additional hard and PDF copies can be obtained from orders@intechopen.com

Bacterial Biofilms
Edited by Sadik Dincer, Melis Sümengen Özdenefe and Afet Arkut
p. cm.
Print ISBN 978-1-78985-899-0
Online ISBN 978-1-78985-900-3
eBook (PDF) ISBN 978-1-83968-819-5

For EU product safety concerns:
IN TECH d.o.o., Prolaz Marije Krucifikse Kozulić 3, 51000 Rijeka, Croatia, info@intechopen.com or visit our website at intechopen.com.

Meet the editors

For the past 32 years, Prof. Sadık Dincer has been involved in teaching, research, and academic work in numerous distinguished universities in Turkey. Currently, he is working at Cukurova University, Biology and Biotechnology Departments, Adana, Turkey. His manuscripts and book chapters have been published in national and international journals and his works has been cited 897 times. To date he has trained twenty-five MSc and eleven PhD students. He received the Technology Development Award from the Scientific and Technological Research Council of Turkey (TÜBİTAK) in 2013 and a national study patent in 2019. His research is focused on bacteriology, microbial ecology, industrial biotechnology, and microbial genetics.

Melis Sümengen Özdenefe received her BSc in Biology from Cukurova University, Turkey, in 2009. She obtained an MSc and PhD in Biology from Cukurova University in 2011 and 2014, respectively. During her MSc studies she attended Anhalt University, Germany, for six months as an international exchange student (Erasmus) in 2010–2011. In 2014, she began work as a doctor in the Department of Biomedical Engineering at Near East University, North Cyprus. In August 2016, she became an assistant professor. Her teaching and research interests include industrial microbiology, bacteriology, biotechnology, enzymology, and microbial ecology. Her research areas include enzymes and biosurfactants, which are produced from various bacteria and fungus for industrial applications, the production and characterization of bacterial enzymes and bacteriocins, antimicrobial and antioxidant activity of various plant structures, multiple antibiotic resistance, and heavy metal resistance in Gram-negative bacteria.

Afet Arkut Özyapici received a BSc and PhD in Biology from Cukurova University, Turkey, in 2009 and 2013, respectively. During her postgraduate studies, she attended Anhalt University, Germany, for six months as an international exchange student (Erasmus) in 2010–2011. In 2013, she began working as a doctor in the Health Sciences Faculty at Cyprus International University. In March 2014, she became an assistant professor. Dr. Özyapici has been working as head of the Department of Social Work and vice dean of the Health Sciences Faculty at Cyprus International University since 2016 and 2017, respectively. Her teaching and research interests include microbiology, bacteriology, microbial ecology, hospital infection, hygiene, and sanitation.

Contents

Preface

A biofilm is a microbial form of life formed by the organic exopolysaccharide structures produced by microorganisms, irreversibly attached to a living or inanimate interface. Today, biofilms are widespread in nature and produced by many bacteria.

Biofilms protect bacteria from changes in environmental conditions and the harmful effects of ultraviolet light, making it easier for the bacteria to store food and remove waste materials. Due to biofilms, it is difficult for microorganisms to undergo phagocytosis by the immune system cells. When favorable conditions occur, biofilms can be created by pathogenic microorganisms. Bacterial attachment levels, surface properties, temperature, type and number of bacteria, pH of the environment, cell-wall structure of the bacteria, mobility, amount and content of the nutrients in the environment, and ion concentration all affect biofilm formation. Microorganisms create biofilms for reasons such as protection from the harmful effects of the environment as well as obtainment of nutrients and new genetic features. Biofilm formation is not a random event, and many microorganisms signal each other using small diffuse molecules to coordinate activities.

The intercellular communication required for biofilm formation is provided by a system called "quorum sensing." This system provides many advantages to bacteria, but it also controls the formation of biofilms through signals transmitted from cell to cell. While the quorum sensing system detects the population density around the bacteria, the bacteria provide gene control. In addition to developing coordination between food sources and bacteria, quorum sensing also increases communication with the bacteria during infection to develop resistance to the immune response in the host.

Studies have shown that bacteria living in biofilms are more resistant to antibacterial agents, iodine, the iodine polyvinylpyrrolidone complex, chlorine, monochloramine, pyrogens and biocides, and heat.

Biofilms are known to cause economic losses due to their negative effects in many areas, especially in health and food. New methods incorporating enzymes, detergents, ultrasound, and electricity are being used to control and prevent bacteria via mechanical cleaning, use of biocides, preventing biofilm development, preventing microbial adhesions, and supporting biomass extraction. However, the effectiveness of each methods varies according to the applied surface, the type of bacteria that forms the biofilm, and the application methods.

Biofilms can cause great industrial losses and threaten public health. Therefore, more research is needed to understand biofilms and develop efficient ways of preventing their formation.

Sadık Dincer
Cukurova University,
Adana, Turkey

Melis Sümengen Özdenefe
Near East University,
Nicosia, Turkish Republic of Northern Cyprus

Afet Arkut
Cyprus International University,
Nicosia, Turkish Republic of Northern Cyprus

Section 1

Biofilm in Wastewater Treatment

Chapter 1

Biofilm in Moving Bed Biofilm Process for Wastewater Treatment

Shuai Wang, Sudeep Parajuli, Vasan Sivalingam and Rune Bakke

Abstract

A brief introduction of the long history of biofilm-based wastewater treatment is given together with basics of biofilm behavior and mechanisms in removal and transformation of pollutants. Moving bed biofilm reactor (MBBR) principles and applications of such are presented. Advantages and limitations of such solutions are given together with evaluations of emerging MBBR applications. The basis of biofilm processes and biofilm layer classification based on dissolved oxygen gradient is discussed. Organisms grow at the protected surface of the biocarrier where oxygen gradients create aerobic, anoxic, and anaerobic layers allowing simultaneous nitrification and denitrification in one MBBR (nitrification, nitritation, autotrophic, and heterotrophic denitrification). Combination of MBBR with activated sludge, continuous flow intermittent cleaning (CFIC®), and integration with anaerobic digestion increases the potential usage of MBBR for enhanced efficiency and energy recovery and is partly discussed as case studies (COD, ammonium, and solid removal). Biofilm thickness and scaling control can be crucial for MBBR performance. The type of carriers, filling degree, and operational conditions play a major role for process performance; hence, the effect of those parameters is presented.

Keywords: moving bed biofilm, TN removal, scaling on biofilm, biocarriers

1. Introduction

The use of biofilm systems in wastewater treatment is being increased rapidly because of its tempting approach of pollutant removal from wastewater, which has been proved to be effective in terms of both cost and environmental perspectives [1, 2]. Biofilm can have both positive and negative effects in treatment processes depending on the type of treatment concept applied. Processes such as a moving bed biofilm reactor (MBBR) depend on biofilm development, while it can cause problems in membrane bioreactor (MBR) through membrane biofouling. Those processes taking advantage of biofilms have been widely used for the removal of organic and inorganic matters from different wastewaters [1], by mechanisms such as biodegradation, bioaccumulation, biosorption, biomineralization, and bio-immobilization [1, 3, 4].

There are several benefits of using biofilm system in wastewater treatment, as compared to suspended growth system (activated sludge for example), such as flexible procedures, smaller space demand, lower hydraulic retention time, increased resilience to changes in the environment, higher biomass retaining period, high active biomass concentration, as well as low sludge production [3, 5, 6]. The use of biofilm systems also enhances the control of reaction rate and population mechanisms [5, 7].

Microorganisms tend to form clusters/colonies to expedite the organism's growth and facilitate access to food, etc., by forming biofilm [8]. In biofilm or attached growth systems, the growth of the biomass responsible for the conversion of organic material and/or nutrient occurs on the surface of support packing material [9]. Biofilm formation is enhanced by substratum provided to retain and grow microorganisms. The support medium can be rocks, stones, gravels, sand, soil, wood, rubber, plastic, and agglomerates of the biomass itself (granules) or any other synthetic materials [3, 8]. The packing material provides a large surface area per unit volume for biofilm development in high-rate processes; thus, substratum material selection is important to maintain a high quantity of active biomass and to uphold different varieties of microbial populations [10]. The large surface area of the biofilm enables the media to efficiently adsorb a high amount of substrates from the influent wastewater. As the biofilm develops on the media, it provides diverse habitats so that different constituents such as carbon and nitrogen components of the wastewater are transformed and mineralized, thus increasing the removal efficiency of the organic substances from influent wastewater.

There are generally three steps involved in biofilm formation, including biofilm attachment, growth, and detachment (**Figure 1**). Microorganisms attach on to the substratum, such as the surface of carriers in MBBR processes, by adhesion, and the attachment is reversible at the early stage. Tight connections between organisms and the substratum can be gradually established by extracellular polymetric substances (EPS) produced by the organisms. EPS is a mixture of polysaccharides, proteins, and extracellular DNA, and it is recognized to also be important for the communication between biofilm cells, biofilm 3D structure formation, and multicellular living [11]. Biofilm detachment from the surface is a natural mechanism where biomass (individual cells or lumps of cells) is released into bulk liquid. It can be influenced by hydrodynamic shear forces and other environmental conditions such as toxic chemical exposure. Detachment process limits biofilm accumulation and thickness and thus balances the attached biofilm quantity at steady-state conditions [11].

Different species can be found in the same biofilm clusters. They can vary from rapidly growing to inactive organisms, from heterotrophic to autotrophic organisms depending on substrate gradients, mutation, genetic regulatory switches, and signaling pathways [11]. Due to oxygen transfer limitations in an aerated system, the biofilm can contain aerobic, anoxic, or anaerobic organisms at the same time [5, 8]. A well-established biofilm can have any thickness, but around 0.1 mm is considered

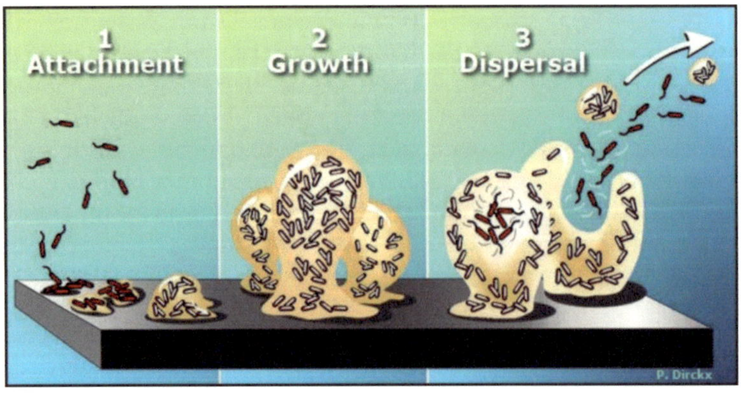

Figure 1.
Biofilm life cycle. Adapted from the Center for Biofilm Engineering, Montana State University [11].

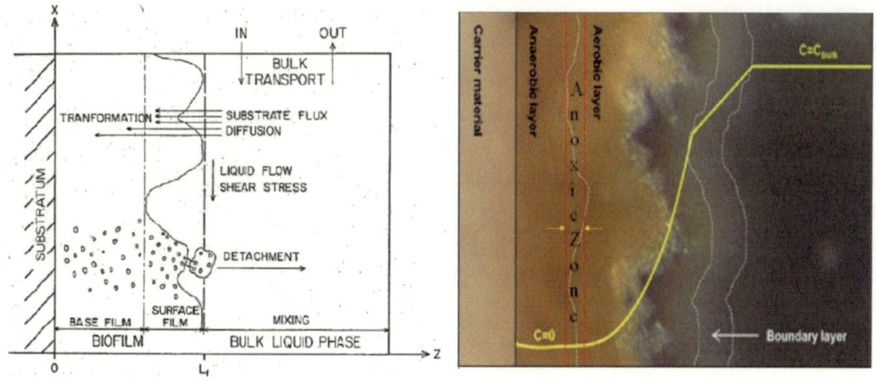

Figure 2.
Left, graphical illustration of biofilm processes [15]; right, biofilm picture and layer classification based on dissolved oxygen gradient [12].

suitable in an efficient MBBR, where mass transfer in the biofilm structure and between the interphase of biofilm and liquid is critical for efficient mass transfer [12, 13]. Both diffusion and convention can occur [11, 14] in the biofilm mass transfer (**Figure 2**), while substrate diffusion is considered to usually be the rate-limiting process within the biofilm structure. Substrate accessing of biofilm can be enhanced by, for example, enhanced aeration/mechanical mixing to enhance mass transfer from the bulk liquid to the biofilm surface. The internal biofilm growth (**Figure 1**) and external forces such as abrasion are important factors for biofilm morphology, development, and effectiveness of biofilm processes.

Biofilm processes applied for wastewater treatment have a long history. Trickling filters (TF) and rotating biological contactors (RBCs) utilizing biofilm growing on the packing medias are biofilm processes being widely applied of low-cost and low maintenance comparing to activated sludge process [1, 3, 16]. Moving bed biofilm process, which was developed in the 1980s [14, 17], has been widely applied for organic and inorganic wastewater treatment of high efficiency, low maintenance, and low operation cost [8, 17, 18]. A membrane-aerated biofilm reactor (MABR) has been developed recently for organic and ammonia removal, based on an aerated membrane where biofilm attaches on the fiber [10]. Biofilm in the form of granular sludge for energy recovery (methane) from wastewater organics, such as by upflow anaerobic sludge blanket (UASB) [8, 19], expended granular sludge bed (EGSB) [20, 21], and internal recycle reactor (IC) [22], and aerobic granular sludge reactors for shortcut ammonia removal, such as anammox process [23], and for simultaneously organic, phosphor, and ammonia removal, Nereda [24] has been developed and is increasingly used in both industrial and municipal wastewater treatments.

Biofilm applied in MBBR processes are focused in this book chapter. Commonly applied MBBR and its derivatives processes are introduced and compared. A case study based on MBBR concept for municipal wastewater treatment is also provided.

2. Moving bed biofilm reactor

Moving bed biofilm reactor (MBBR) is an advanced wastewater treatment technology, which employs the benefits of both biofilm and activated sludge processes for highly efficient wastewater treatment [14]. Developed in the 1980s, MBBR has been established in the last two decades worldwide as a simple, robust, flexible, and compact wastewater technology for both municipal and industrial wastewater treatment [14].

2.1 Biocarriers

Plastic carriers of different shapes and surface areas have been developed and applied in the MBBR systems as biofilm substratum. The carriers' shape, density, protected areas, and void volume are important factors that affect the performance of MBBR processes. Carriers can be made of different shapes such as squares, round, and sphere. The shape can affect the carrier's strength, shearing, and colliding conditions. The carrier density is normally lower than water at around 0.98 kg/L, so that it can be suspended in wastewater with biofilm attachment without introducing strong mixing. The carriers protected areas range from 300 to over 1000 m^2/m^3 depending on the shapes and internal structure. Large carrier protected areas normally mean high complexity of the carrier structure and higher production cost. Carriers of protected areas of 500–1000 m^2/m^3 are normally applied in full-scale wastewater treatment plants due to their costs and process benefits. **Figure 3** shows two different types of plastic carriers that are made of high-density polyethylene (HDPE) with respective protected surface area of 650 (BWTX®) and 828 (BWT15®) m^2/m^3. The biofilm on carriers develops as illustrated in **Figure 1** and maintains active organisms in thin layers. A well-designed carrier enables stable biofilm in the MBBR process, so that the void is not easily blocked by wastewater particles or excessive biofilm accumulation. Effective mixing/aeration combining a good carrier design leads to good system performance and low-maintenance requirements.

2.2 Carrier filling degree

A typical MBBR process can have a biocarrier filling ratio lower than 70%, where carriers are continuously mixed in the reactor and the whole reactor has homogenous conditions. Due to shear forces from mixing/aeration, biofilm growth and detachment processes are balanced to maintain a relatively constant biofilm thickness at steady-state condition. The limitation of filling degree is related to energy consumption and mixing effects for mass transfer purposes [26]. Higher filling degrees will result in higher energy requirement for sufficient mixing of the suspended carriers. It is especially challenging for aerobic systems where aeration energy consumption can account for more than 70% of the complete treatment energy demand [8, 13]. While different carrier filling degrees have been attempted, a different setup based on over 90% filling degree has been developed by biowater technology. The process is named continuously flow intermittent cleaning (CFIC®) which constitutes of two individual modes, a normal operation mode and a washing mode. In the normal operation, over 90% filling degree leads to an almost stagnant

Figure 3.
Biocarrier BWTX® (left) and BWT15® (right) (biowater technology AS, Norway) with biofilm growth [25].

carrier bed. Oxygen field transfer efficiency (OTE) has been documented to be 1.5 times higher than in a normal MBBR with lower filling degree by applying coarse bubble aeration. Big bubbles are cut through the carrier bed with better utilization. Due to high filling degree, sludge will accumulate on carriers which also work as a filter bed media for wastewater treatment. High sludge accumulation will lead to effluent solid increase after certain times depending on load situations. A washing mode is therefore introduced by increasing the water level in the reactor which resembles a normal MBBR operation washed-out accumulated excess sludge (similar to backwashing of sand filters). Wastewater can be fed continuously to the reactor without stops during the washing cycle. This high filling degree process has been applied in full scale for petrochemical wastewater treatment [27] and municipal wastewater treatment both in China and Brazil for organic and ammonia removal, confirming high efficiency and compactness. Carrier filling degree around 30% is also applied for systems to remove dissolved oxygen before feeding to a denitrification system, for example.

2.3 MBBR treatment process

In a MBBR biofilm system, the process effectiveness depends on the active organism's concentration, mass transfer efficiency, and system setup, for example, feed distribution and mixing. Organisms' concentration is relatively constant in a stable process, depending on feed substrates and biofilm mass on carriers, which is on average below 20 g/m^2. The carrier mass value can be higher in a system with scaling, for example, while the active organisms are mainly on the outer surface of the scaling mass. For processes like nitrification or anammox, the mass per area can be lower due to the slow growth rates. The organic loading rate in MBBR is generally based on the protected surface areas, such as $gCOD/m^2/d$. The organic loading rate can be as high as 100 $gCOD/m^2/d$ depending on the biofilm condition and loading history. A reduced removal efficiency is expected in such high load system where oxygen supply can be a limiting factor. Comparing to activated sludge system, a MBBR can sustain higher sludge concentration per reactor volume. With an on average 20 g/m^2 biofilm on carrier surface and a filling degree of 70%, the sludge content is about 7 g/L for a surface area 500 m^2/m^3 carriers. This is achieved without sludge return and thus reduces the operation complexity and equipment for sludge return is avoided.

MBBR process has also been developed for ammonia removal through both traditional nitrification and denitrification processes and anammox (**Figure 4**) [13]. In conventional nitrogen removal process, ammonium ion is oxidized to nitrate by complete nitrification, and subsequently nitrate is reduced to nitrogen gas by pre-/post-denitrification. Such nitrogen removal is usually carried out in two different reactors. Inorganic carbon as alkalinity is normally supplied to perform ammonium oxidation. Denitrification requires easily degradable organic such as methanol as electro acceptor. Partial nitrification, called nitritation, and anaerobic ammonium oxidation can also be achieved to remove nitrogen from wastewater in one reactor by manipulating dissolved oxygen concentration into the biofilm. That means oxidation of nitrite to nitrate is suppressed, and denitrification can occur according to "shortcut" in **Figure 4** [13].

Ammonium removal by nitrite is performed by a group of autotropic bacteria, named anammox bacteria [28, 29]. The anammox process requires 40% less energy and generates 88% less CO_2 emission comparing to traditional nitrogen removal process [10, 24]. Due to low growth rate (0.06 g VSS/g VSS d), a doubling time being ~10–14 days at relatively high temperature (30–35°C) [30], anammox requires long start-up period. The biofilm attached to the MBBR carrier, being protected from the environment, maintaining long sludge retention time,

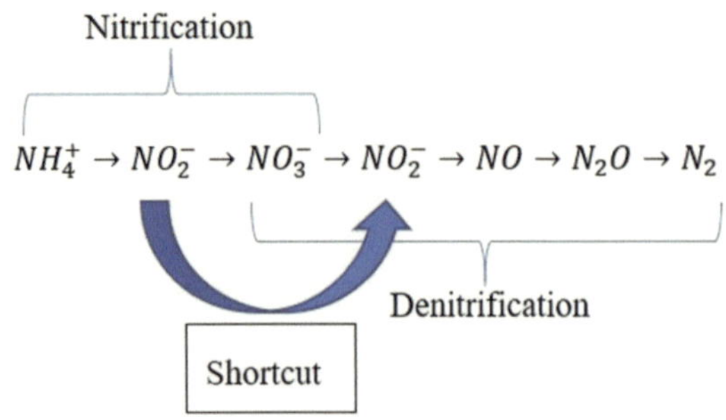

Figure 4.
Nitrification and denitrification with shortcut mechanism illustrated [13].

and thereby preventing the slow-growing organisms from being washed out of the system, is suitable for slow-growing anammox biomass. Limited research on anammox process in MBBR is documented, but it has been observed that removal rates of up to 1.2 kg N/m^3.d can be achieved using MBBR for side-stream reject wastewater treatment in municipal application [31]. Nitrite formation is a limiting step, and dissolved oxygen needs to be well controlled, so advanced process control is required for efficient MBBR anammox solutions.

MBBR has also been applied for biological phosphor removal in Norway by physically moving carriers with biofilm from anaerobic stage to aerobic stage and back to anaerobic stage so that the P-accumulating organisms undergo the same cycles as in activated sludge "Bio-P" processes [32].

2.4 Different MBBR reactor setups

Due to MBBR's compact nature, high effectiveness, and reliability, the MBBR process is also integrated with other processes (summarized in **Table 1**), such as with activated sludge for enhancing ammonia removal, with anaerobic granular sludge process to form a hybrid system, such as HyVAB® [13, 27], for combined anaerobic and aerobic wastewater polishing, and with membrane bioreactor (MBR) for high strength and stricter wastewater treatment requirements [10].

Based on the MBBR technology, there are several commercially proven technologies available in the market [25], such as:

- CFAS®—Combined fixed film activated sludge
- CFIC®—Continuous flow intermittent cleaning reactor
- HyVAB®—Hybrid vertical anaerobic biofilm reactor

The typical layout for the above processes for organic removal is shown in **Figure 5**. **Table 1** briefly compares the abovementioned technologies and key benefits. Most of the technologies only focus on COD and nutrient removal from wastewater except HyVAB®. HyVAB® is the technology that focuses on both COD, nutrient removal together with energy recovery as biogas [13, 25]. Biogas production from the HyVAB® reactor makes the treatment process partially or fully energy self-sufficient.

2.5 MBBR operational issues

Depending on MBBR process operational knowledge and full-scale design experience, several problems can be encountered for a full-scale MBBR process, such as

Technology	Process	Benefits
MBBR [12, 13, 25]	Freely moving plastic carriers with attached biofilm removing both organic and inorganic nitrogen	• High effective surface area in carrier gives large protected growth area, hence less space requirement • Self-regulating biofilm on carriers requires less monitoring and ensures stable treatment process
CFAS®/IFAS [33]	Uses the existing activated sludge process together with MBBR carriers, by introducing plastic carriers into the activated sludge process	• Suitable for retrofitting existing activated sludge plant to enhance nitrification and BOD removal • Small foot print • BOD, P, and N removal can be achieved together • Achieve low SVI, ensures efficient sludge removal
CFIC® [34]	High carrier filling degree of over 90–99% that allows high substrate transfer efficiency. Operates in normal and washing modes with continuously wastewater feeding. Excess biomass removal is needed	• Very compact and energy-efficient process (20% smaller footprint and 50% less energy demand than MBBR) • Higher oxygen (1.5 times to MBBR) and substrate transfer efficiency • Very low SVI enables fast sludge settlement, 80% less effluent sludge than MBBR in normal mode
HyVAB® [27]	Hybrid system integrates both anaerobic and aerobic high-rate processes. Anaerobic stage recovers energy (methane) from wastewater and the aerobic part with biocarrier removes the remaining organics and nutrients	• Ultra-high rate and compact process • Suitable for wide range of application; reject water treatment and industrial wastewater treatment • Very low sludge production • High COD removal and generate high methane content biogas

Table 1.
MBBR integrated technologies with other biological treatment process.

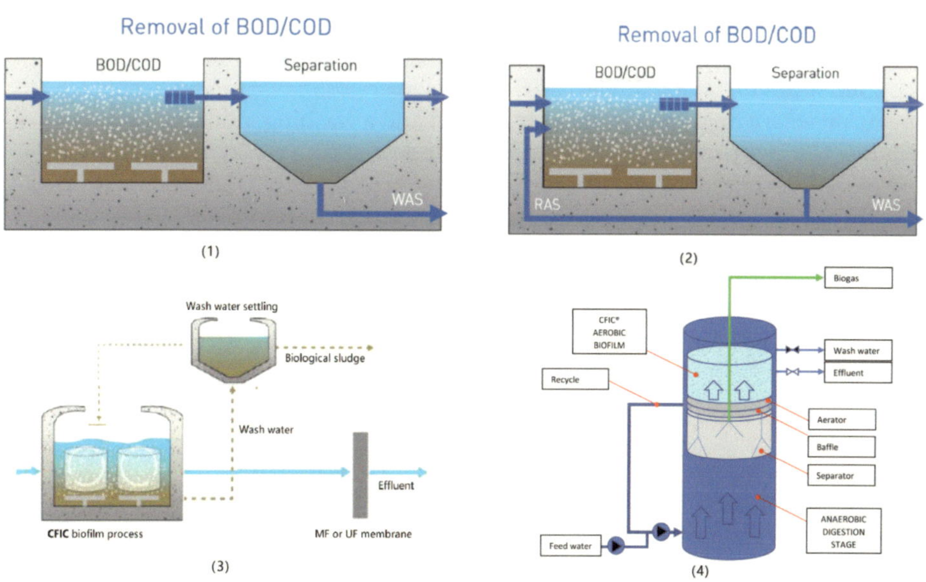

Figure 5.
Typical layout of (1) MBBR, (2) CFAS, (3) CFIC, and (4) HyVAB for organic (BOD/COD) removal [25].

feed pipe/effluent sieve blocking, nonhomogeneous mixing, carrier voids blocking, destroyed carriers, carrier accumulating at the effluent sieves, and carrier overflow out of reactor. These can be all prevented through skilled design, based on accumulated project knowledge and operation experience.

Depending on wastewater characteristics, problems such as chemical scaling on carries can happen, especially for wastewater that contains high calcium, ammonia, and other minerals, such as anaerobic digestion reject water and diary wastewater [35]. Mineral precipitation can occur when wastewater is supersaturated with relevant ion concentration [35]. The composition of mineral scaling varies and can contain struvite, hematite, hydroxyapatite, maghemite, etc. [8, 35]. Scaling on biofilm carriers creates negative effects on the reactor's performance by reducing effective surface area, hindering the mass transfer, and demanding more energy to keep the carriers in suspension. Carriers with excess scaling become heavier and settled down at the reactor bottom and need to be replaced [35]. The pH and concentration of the ions are the main factors influencing chemical precipitates on carriers. Minerals tend to precipitate more at higher pH; thus, pH control can alleviate scaling. Buffer dosing, reduced air stripping of CO_2, and alkalinity removal could help to hinder scaling rates. Pure oxygen aeration is an option to avoid air stripping of CO_2 to avoid pH increase. Pretreatment by chemical precipitation such as adding lime to remove calcium and magnesium could also be an option.

Feed wastewater composition changes can cause disturbances such as increased organic load in nitrification or anammox processes that will lead to a shift in competition between heterotrophic to autotrophic bacteria. In such cases, the heterotrophic bacteria that have higher growth rate can gradually dominate the MBBR biofilm, leading to unfavorable condition for ammonia removal.

Unwanted biofilm detachment caused by toxic chemicals or abrupt operational condition changes, such as sudden increase of aeration can lead to process problems and even failure in extreme cases, but inner layers of biofilms are protected by the outer layers, making biofilms quite resilient to such disturbances.

3. MBBR case study

This chapter provides a case study where the novel CFIC biofilm process has been studied for municipal wastewater treatment, including for organic, ammonia, and total nitrogen removal. The CFIC process operates in two modes, a normal operation where high carrier filling is applied and a washing mode for extra sludge removal (**Figure 6**). Detailed process concept description can be referred to [34], and more information is given in the following presentation of a three-stage CFIC pilot for municipal organic and nitrogen removal. The first full-scale three-stage CFIC process has also been accomplished for a 30,000 m^3/d municipal wastewater treatment in Guiyang, China, in 2017.

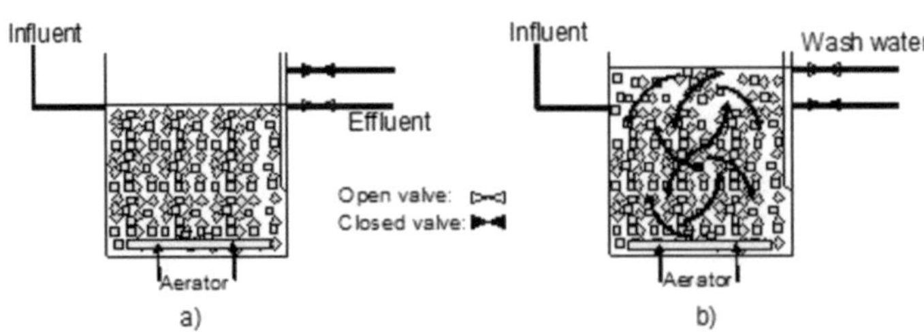

Figure 6.
The CFIC® during (a) normal operation and during (b) the cleaning cycle.

3.1 Pilot layout

A pilot CFIC plant with a maximum feeding capacity of 6 m^3/h has been constructed for municipal wastewater treatment study at NRA, Norway. The pilot plant constitutes of a pre-denitrification (R1), two aerobic CFIC stages (R2 and R3), and a sludge settler for sludge removal and supernatant return to biological stages (**Figure 7**). The three biological stages are 8.7, 8.3, and 8.3 m^3, respectively, in volume. Biocarriers of BWT15® and BWTX® (**Figure 3**) were filled in the first and the other two stages separately. During normal CFIC operation, a filling ratio of 62, 86, and 83%, respectively, is applied. The filling degree of the pre-denitrification was kept constant at 62% while reduced to 71 and 69% when intermittent washing cycle was performed in the other two aerobic stages.

The pilot was fed with municipal wastewater directly pumped from the full-scale primary clarifier onsite (**Figure 7**), and the wastewater characteristics are given in **Table 2**. The wastewater temperature was around 15°C in the whole year. The wastewater was fed at 3–6 m^3/h to the system with a recycle ratio of 1–1.5 during the study. To facilitate biofilm growth on carriers, washing mode was applied at the beginning of the test until stable biofilm growth was observed. The pre-denitrification stage was washed daily, and the two aerobic stages were washed together in every 1 or 2 days after the first reactor washing cycle finished. The washing cycle for each stage

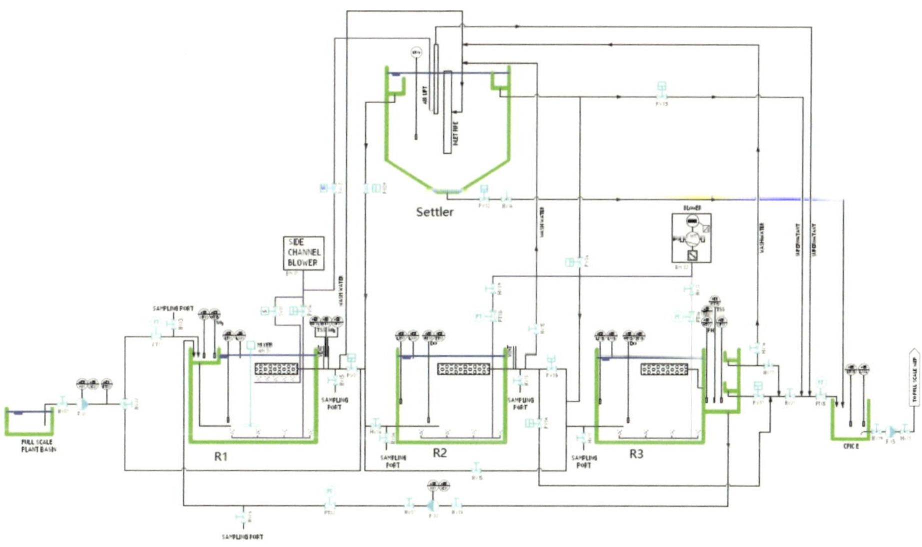

Figure 7.
Pilot system PID layout.

Average feed condition	Period 1 (15.05–29.06)	Period 2 (24.10–01.12)
TCOD (mg/L)	392	214
TSS (mg/L)	264	123
NH_4-N (mg/L)	20.4	14.5
TN (mg/L)	45	29
pH	7.2	7.0

Table 2.
Feed wastewater characteristics in the two test periods.

is normally 1 h. Wastewater samples were taken for analysis before and in the washing cycle to record parameters such as COD and suspended solid.

3.2 COD and ammonium removal

Pilot performance in period 1 (**Table 2**) is presented below. Feed wastewater characteristics show that during this period more than 80% of the feed COD was particles. The influent total COD was mostly removed/retained in the denitrification reactor (R1), and the effluent TCOD in R2 and R3 is identical (**Figure 8**). Soluble COD removal was about 30%, with 16% removed in R1 and the rest was removed after R3. The feed ammonium concentration was around 20 mg/L (**Figure 9**) after combining with recycle wastewater from R3, the ammonium content was diluted to about half of initial value, and it can be seen that significant NH_4-N is removed in the first (80%) aerobic reactor (R2) (**Figure 9**) to an average concentration of 1.5 mg/L. After aerobic stage 2, the NH_4-N concentration was on average 0.6 mg/L. Due to very low available organic for denitrification (C to NOx-N ratio of on average 1.7), the total nitrogen removal was about 36%. Limited flow capacity of the pilot giving the TN and NH_4-N loading rate about 0.4 g N/m_2/d, which was much lower than previously tested (>2 g N/m^2/d in a small-scale reactor).

3.3 Solid removal

Comparing to a traditional MBBR, CFIC process has good capacity to retain particles inside the carrier filter bed (instead of being continuously washed out of the system in conventional MBBR). The pilot study shows that during the MBBR mode (CFIC washing), the total suspended solid (TSS) content in the three stages was similar at around 250 mg/L, which was slightly lower than the feed TSS of about 300–400 mg/L. While in the CFIC normal operation model, the TSS was lower than 100 mg/L in all three stages with an average value of 50 mg/L. This is five times lower than a MBBR effluent TSS content, which indicates that solids were retained in the CFIC process.

CFIC washing cycle can normally bring out a large quantity of solid attached or accumulated in a short period. The average solid content for the washing water

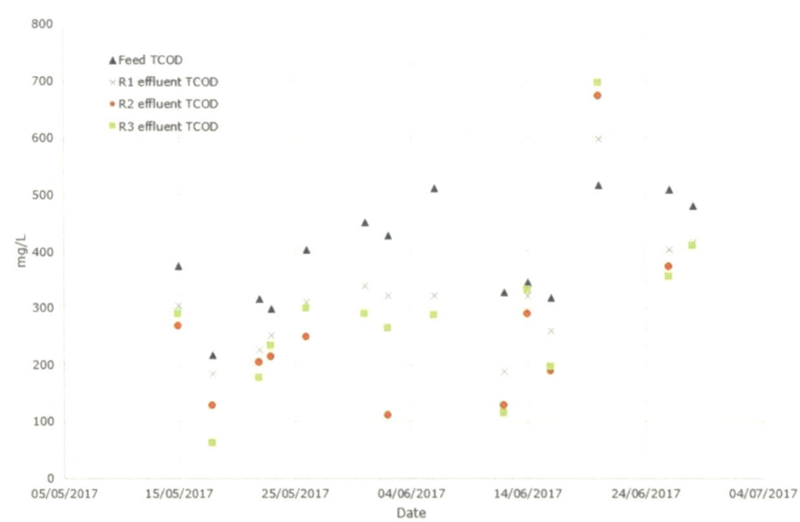

Figure 8.
COD removal by the CFIC pilot, R1, pre-denitrification; R2 aerobic stage 1 and R3, aerobic stage 2.

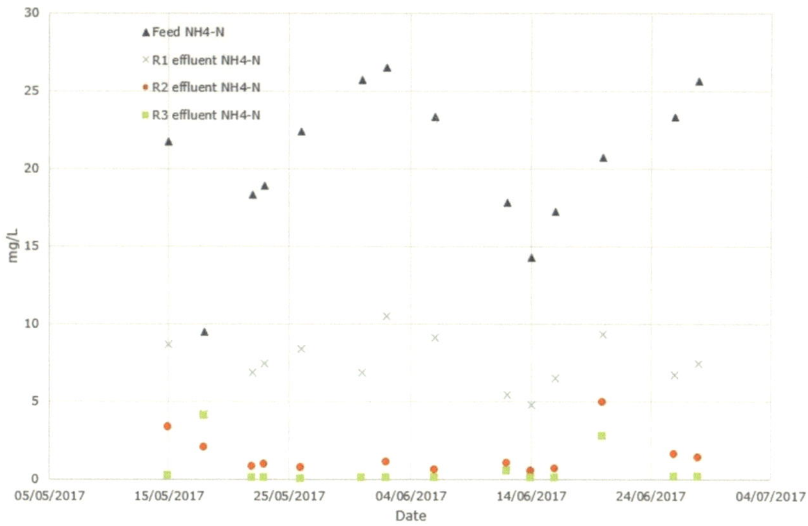

Figure 9.
NH_4-N removal by the CFIC pilot, R1, pre-denitrification; R2 aerobic stage 1 and R3, aerobic stage 2.

is 1.3–6.4 times of the influent wastewater TSS in this study depending on washing frequency and accumulation of solids. The washed-out sludge has low sludge volume index (SVI) of around 70 mg/L and can settle quickly in a fast sludge settler. This feature enables at least two times smaller clarifier for sludge settlement comparing to the one needed for conventional MBBR processes. The solids washed out of the system can be from 3 to 12 g TSS/m^2 carrier surface, accounting for 3–14% of the total attached TSS [36]. The washing water peak TSS content can reach over 2000 mg/L and gradually reduced with continuous wastewater feeding after washing stops [36, 37]. Over 50% of the washed-out particles are larger than 60 μm, which is larger than normal influent and effluent values [36], explaining the low SVI level. It may take 1–4 h until the effluent solid content reaches a stable condition after each washing.

4. Conclusions

Wastewater treatment by applying biofilm has been developed over the years, and various biofilm processes are playing important roles at different stages of wastewater treatment industries. MBBR concept based on biofilm is widely used for organic and inorganic removal in both industrial and municipal wastewater remediation. It is approved to be a compact, energy-efficient, and robust solution comparing to a traditional activated sludge process. Due to biofilm growth on a protected area, different organism species coexist in the MBBR biofilm clusters which enhances their resilience to the environmental condition variations. The development based on MBBR to even compact process such as CFIC and HyVAB and the integration of MBBR with other high-rate and efficient processes could potentially reduce the footprint and complexity of wastewater treatment. Future studies to improve MBBR system for high mineral content wastewater treatment, optimize carrier designs and understand the correlation between protected area and organism species in different environmental condition, the biofilm growth, and detachment mechanisms induced by external forces and improving the energy efficiency for enhanced mass transfer are interesting topics to be explored.

Acknowledgements

The authors would like to thank the funding provided by Oslofjordfondet No. 269014 and the University of South-Eastern Norway.

Author details

Shuai Wang[1*], Sudeep Parajuli[1], Vasan Sivalingam[2] and Rune Bakke[2]

1 Biowater Technology AS, Tønsberg, Norway nology AS, Tønsberg, Norway

2 University of South-Eastern Norway, Porsgrunn, Norway

*Address all correspondence to: sw@biowatertechnology.com

References

[1] Asri M, Elabed S, Ibnsouda S, El Ghachtouli N. Biofilm-based systems for industrial wastewater treatment. In: Hussain C, editor. Handbook of Environmental Materials Management. Cham: Springer; 2018. pp. 1-21. DOI: 10.1007/978-3-319-58538-3_1371

[2] Das N, Basak LVG, Salam JA, Abigail EA. Application of biofilms on remediation of pollutants—An overview. Journal of Microbiology and Biotechnology Research. 2012;**2**:783-790

[3] Sehar S, Naz I. Role of the biofilms in wastewater treatment. In: Dhanasekaran D, Thajuddin N, editors. Microbial Biofilms—Importance and Applications. IntechOpen; 2016. DOI: 10.5772/63499

[4] Singh R, Paul D, Jain RK. Biofilms: Implications in bioremediation. Trends in Microbiology. 2006;**14**:389-397. DOI: 10.1016/j.tim.2006.07.001

[5] Shahot A, Idris A, Omar R, Yusoff HM. Review on biofilm processes for wastewater treatment. Life Science Journal. 2014;**11**:1-13. DOI: 10.7537/marslsj111114.01

[6] Sofia A. Characterization of bacterial biofilm for wastewater treatment [thesis]. Stockholm: Kungliga Tekniska Högskolan; 2009

[7] Borkar R, Gulhane M, Kotangale A. Moving bed biofilm reactor—A new perspective in wastewater treatment. IOSR Journal of Environmental Science, Toxicology and Food Technology. 2013;**6**:15-21. DOI: 10.9790/2402-0661521

[8] Tchobanoglous G, Metcalf, Eddy, Aecom. Wastewater Engineering: Treatment and Resource Recovery: Volume 2. 5th international ed. New York: McGraw-Hill; 2014

[9] Metcalf & Eddy Inc. Wastewater Treatment Disposal. New York: McGraw-Hill; 1986. 553 p

[10] Loupasaki E, Diamadopoulos E. Attached growth systems for wastewater treatment in small and rural communities: A review. Journal of Chemical Technology and Biotechnology. 2013;**88**:190-204. DOI: 10.1002/jctb.3967

[11] Montana State University. Essential biofilm concepts and phenomena [Online]. 2010. Available from: http://www.biofilm.montana.edu/biofilmbasics/index.html [Accessed: 21 June 2019]

[12] Piculell M. New dimensions of moving bed biofilm carriers: Influence of biofilm thickness and control possibilities [thesis]. Lund: Lund University; 2016

[13] Vasan S. Nitrogen transformation in biofilm for reject water treatment [thesis]. Porsgrunn: University of South-Eastern Norway; 2019

[14] Barwal A, Chaudhary R. To study the performance of biocarriers in moving bed biofilm reactor (MBBR) technology and kinetics of biofilm for retrofitting the existing aerobic treatment systems: A review. Reviews in Environmental Science and Biotechnology. 2014;**13**:285-299. DOI: 10.1007/s11157-014-9333-7

[15] Bakke R. Biofilm detachment [thesis]. Bozeman Montana: Montana State University; 1986

[16] Pipeline. The attached growth process—An old technology takes new forms [Online]. 2004. Available from: http://www.nesc.wvu.edu/pdf/WW/publications/pipline/PL_WI04.pdf [Accessed: 30 June 2019]

[17] Rusten B, Ødegaard H, Lundar A. Treatment of dairy wastewater in a novel moving bed biofilm reactor. Water

Science and Technology. 1992;**26**: 703-711. DOI: 10.2166/wst.1992.0451

[18] Ødegaard H. The moving bed biofilm reactor. Water Environmental Engineering and Reuse of Water. 1999:250-305

[19] Wang S, Ghimire N, Xin G, Janka E, Bakke R. Efficient high strength petrochemical wastewater treatment in a hybrid vertical anaerobic biofilm (HyVAB) reactor: A pilot study. Water Practice Technology. 2017;**12**:501-513. DOI: 10.2166/wpt.2017.051

[20] Lim SJ. Comparisons between the UASB and the EGSB reactor. Iowa State University. 2009:17

[21] Chu LB, Yang FL, Zhang XW. Anaerobic treatment of domestic wastewater in a membrane-coupled expended granular sludge bed (EGSB) reactor under moderate to low temperature. Process Biochemistry. 2005;**40**:1063-1070. DOI: 10.1016/j.procbio.2004.03.010

[22] Mutombo DT. Internal circulation reactor: Pushing the limits of anaerobic industrial effluents treatment technologies. In: Proceedings of the 2004 Water Institute of Southern Africa (WISA) Biennial Conference 2004; Cape Town; 2004. pp. 608-616

[23] Winkler MKH. Magic granules [thesis]. Delft: Delft University of Technology; 2012

[24] Pronk M, De Kreuk MK, De Bruin B, Kamminga P, Kleerebezem R, Van Loosdrecht MCM. Full scale performance of the aerobic granular sludge process for sewage treatment. Water Research. 2015;**84**:207-217. DOI: 10.1016/j.watres.2015.07.011

[25] Biowater Technology As. Technology [Online]. 2019. Available form: http://www.biowatertechnology.com/en/technology/ [Accessed: 30 June 2019]

[26] Barwal A, Chaudhary R. Impact of carrier filling ratio on oxygen uptake & transfer rate, volumetric oxygen transfer coefficient and energy saving potential in a lab-scale MBBR. Journal of Water Process Engineering. 2015;**8**:202-208. DOI: 10.1016/j.jwpe.2015.10.008

[27] Wang S, Savva I, Bakke I. A full-scale hybrid vertical anaerobic and aerobic biofilm wastewater treatment system: Case study. Water Practice Technology. 2019;**14**:189-197. DOI: 10.2166/wpt.2018.123

[28] Mulder A, van de Graaf AA, Robertson LA, Kuenen JG. Anaerobic ammonium oxidation discovered in a denitrifying fluidized bed reactor. FEMS Microbiology Ecology. 1995;**16**:177-183. DOI: 10.111/j.1574-6941.1995.tb00281.x

[29] Li J, Li J, Gao R, Wang M, Yang L, Wang X, et al. A critical review of one-stage anammox processes for treating industrial wastewater: Optimization strategies based on key functional microorganisms. Bioresource Technology. 2018;**265**:498-505. DOI: 10.1016/j.biortech.2018.07.013

[30] Xie H, Ji D, Zang L. Effects of inhibition conditions on anammox process. In: Paper presented at IOP Conference Series: Earth and Environmental Science; 100: 012149; 2017

[31] Lemaire R, Thesing G, Christensson M, Zhao H, Liviano I. Experience from start-up and operation of deammonification MBBR plants and testing of a new deammonification IFAS configuration. In: WEFTEC, the Water Environment Federation's Technical Exhibition and Conference. 2013. DOI: 10.2175/193864713813667857

[32] Rudi K, Goa IA, Saltnes T, Sørensen G, Angell IL, Eikås S.Microbial ecological processes in MBBR biofilms for biological phosphorus removal

from wastewater. Water Science and Technology. 2019;**79**:1467-1473. DOI: 10.2166/wst.2019.149

[33] Sander S, Behnisch J, Wagner M. Energy, cost and design aspects of coarse-and fine-bubble aeration systems in the MBBR IFAS process. Water Science and Technology. 2017;**75**: 890-897. DOI: 10.2166/wst.2016.571

[34] Ghimire N, Wang S. Biological treatment of petrochemical wastewater. In: Petroleum Chemicals—Recent Insight, Mansoor Zoveidavianpoor. IntechOpen; 2018. DOI: 10.5772/intechopen.79655. Available from: https://www.intechopen.com/books/petroleum-chemicals-recent-insight/biological-treatment-of-petrochemical-wastewater

[35] Vasan S, Sergey K, Babafemi O, Osama KMI, Eshetu J, Wang S, et al. Chemical equilibrium model to investigate scaling in moving bed biofilm reactors (MBBR). In: Proceedings of the 60th International Conference of Scandinavian Simulation Society; SIMS; 2019

[36] Rathnaweera SS, Rusten B, Korczyk K, Helland B, Rismyhr E. Novel biofilm reactor for denitrification of municipal wastewater. Water Science and Technology. 2018;**78**(7):1566-1575

[37] Rusten B, Stang P, Rogne E, Siljudalen J, Marcolini L. Development of a compact, cost effective, and energy efficient biofilm reactor for wastewater treatment and effluent reuse. Proceedings of the Water Environment Federation. 2011;**2011**(11):5222-5235

Chapter 2

Effect of Heavy Metals on the Biofilm Formed by Microorganisms from Impacted Aquatic Environments

Lívia Caroline Alexandre de Araújo and Maria Betânia Melo de Oliveira

Abstract

The aquatic environment is highly complex and diverse, consisting of several types of ecosystems that are dynamic products of complex interactions between biotic and abiotic components. Changes in the physical and chemical properties of these ecosystems can significantly affect the balance of life forms present, especially in their microbiota. Among the main pollutants present in these environments are heavy metals. Several studies demonstrate the effects of these minerals on the structure and function of microbial communities, which may develop adaptation mechanisms for survival and permanence in these sites. In addition, the resistance to heavy metals may contribute to the evolution of resistance genes to the different types of antimicrobials due to the increase of the selective pressure in the environment, becoming a public health problem. One of the adaptive mechanisms present in bacteria from impacted environments that has been frequently investigated is the formation of biofilms. Recent studies have reported significant changes in the structure and amount of biofilm formed in the presence of different metals, and consequently, an increase in the tolerance to these pollutants and antimicrobials. This review will discuss the effects of some metals on bacterial biofilms and their consequences for the marine environment.

Keywords: chemical pollution, toxicity, mechanisms of adaptation, metals, antimicrobials

1. Introduction

The aquatic environment is highly complex and diverse, comprising various types of ecosystems that are dynamic products of complex interactions between biological and abiotic components. Changes in physical properties and ecosystems may affect the balance of life forms present there [1, 2].

In recent decades, these ecosystems have been significantly altered due to multiple environmental impacts from the release of large amounts of effluent without adequate prior treatment, resulting in the scarcity of existing natural resources [3, 4]. Among the main pollutants that generate negative impacts on life forms are heavy metals. The presence of these contaminants may cause changes in the structure and function of microbial communities [5], which can develop various resistance mechanisms that enable their survival [6]. In addition, heavy metal resistance may contribute to the evolution of resistance genes to different types of antimicrobials due to increased selective pressure in the environment [7].

Adaptability as well as metabolic and physiological differences are essential characteristics for microorganisms to remain in these locations. One of the adaptive mechanisms present in bacteria that has been frequently investigated is biofilm formation [8]. Biofilms are structures composed mainly of microbial cells and a matrix formed by a cluster of extracellular polymeric substances (EPSs) [9]. Biofilm-grown cells have some distinct properties from planktonic cells, one of which is increased resistance to antimicrobials and heavy metals [10]. In this review, we propose to report the latest findings on the survival strategies of microorganisms in impacted aquatic environments, more precisely on the influence of heavy metals on biofilm formation.

2. Microorganisms in contaminated aquatic environments

Water is an indispensable natural resource for the survival of man and other living beings [11, 12]. According to Raucci and Polette [13], 97% of the planet's water is found in the oceans, and of the remaining 3%, only 0.3% is available for human consumption and is stored in springs, lakes, rivers, and groundwater.

According to the United Nations (UN), access to water supply and sanitation is a human right and vital to the dignity and health of all people. However, there are still about 1.1 billion people without access to clean water and 2.4 billion people without access to basic sanitation services [14].

The decline in water quality has become one of the most serious problems worldwide, a fact that has been intensified by the increase in population and the absence of public policies aimed at the preservation of water resources. According to the World Health Organization—WHO [15], approximately half of the world's developing population will be affected by diseases that are directly related to poor-quality water and/or lack of adequate or even no sanitation.

Contamination of natural waters represents one of the main risks to public health, a fact that is directly related to the discharge of untreated domestic, hospital, and industrial effluents, which cause contamination of aquatic bodies by pathogenic microorganisms such as bacteria, viruses, protozoa, and helminth eggs [16].

Among the bacteria can be highlighted those belonging to the Enterobacteriaceae family, represented by species *Escherichia coli*, *Klebsiella pneumoniae*, *Proteus mirabilis*, *Enterobacter cloacae* and *Providencia rettgeri*. Most of these species are commonly found in the intestinal tract of humans and animals, and their presence in aquatic environments indicates fecal contamination [4, 17].

Another problem found in the aquatic environment is the contamination by resistant bacteria from humans and animals exposed to antimicrobials [18, 19], as well as the disposal of antimicrobial waste from domestic and hospital effluents. Water is not only a means of spreading resistant microorganisms, but also the pathway through which resistance genes are introduced into the ecosystem, altering the environmental microbiota [20].

Studies have shown bacterial resistance in various aquatic environments including rivers and coastal areas, domestic sewage, hospital sewage, sediment, surface water, lakes, oceans, and drinking water [4, 21–24].

Among the main pollutants found in this environment, we highlight the heavy metals that when introduced into the environment can cause changes in

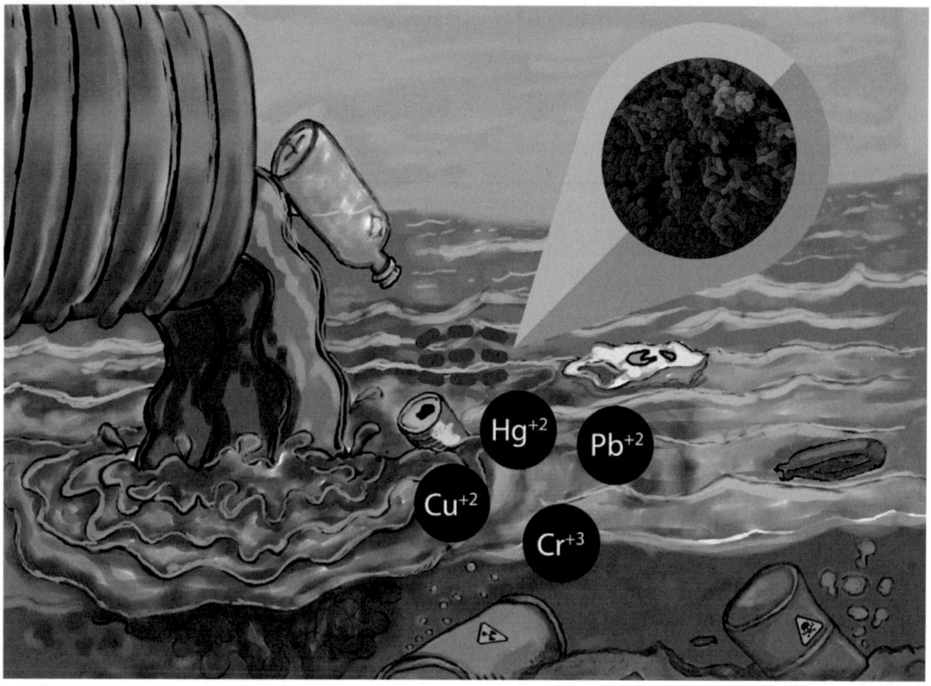

Figure 1.
Impacted aquatic environment and survival strategy of the present microorganisms.

the structure and function of microbial communities [25]. Aquatic systems may be introduced as a result of natural processes such as weathering, erosion, and volcanic eruptions [26]. However, in recent decades, the increase in urbanization and industrialization has contributed to the large increase of these environmental contaminants worldwide [27].

Thus, microorganisms have been developing various resistance mechanisms that allow their survival [6]. Among the various mechanisms, intra and extra-cellular, are bioaccumulation [28], biosorption [29], biomineralization and precipitation [30, 31], oxidation and enzymatic reduction of the metal to the less toxic form [32], production of siderophores [33], and biofilm formation [34]. **Figure 1** shows an impacted aquatic environment and a survival strategy for the microorganisms present there.

3. General characteristics of heavy metals

The term "heavy metals" is used to identify a group of chemical elements that have atomic density greater than 5 g cm^{-3} or have atomic number greater than 20 [35]. Some of these elements, such as sodium (Na), magnesium (Mg), potassium (K), calcium (Ca), zinc (Zn), and copper (Cu), are essential microelements for various life forms, as they are necessary for the functioning of some metabolic pathways [36]. However, the excess or lack of these elements can lead to disturbances in organisms, and in extreme cases, even death [37]. Other elements such as mercury (Hg), lead (Pb), cadmium (Cd), and arsenic (As) are highly toxic even when present in low concentrations, and account for most health problems due to environmental pollution [38].

Heavy metals participate in the global ecobiological cycle, derived from numerous sources and are dynamically transported through the atmosphere, soil, and water; also, because they are not biodegradable, they can remain in the environment for long periods [39].

Among the various metals, mercury, cadmium, and lead stand out for being associated with contamination of the aquatic environment, which can cause problems of poisoning to man and other organisms. These elements are capable of reacting with molecules and ligands present in cell membranes, conferring them with the properties of bioaccumulation, food chain biomagnification, persistence in the environment, and metabolic disturbances of living beings [40].

4. Effects of heavy metals on biofilm

Biofilm is a porous and complex structure formed by one or more species of microorganisms, organized in several layers irreversibly adhered to a biotic or abiotic surface and enclosed in a matrix composed of extracellular polymeric substances (EPS) [9].

They are formed dynamically and gradually, involving several stages. The first is reversible bacterial adhesion that can occur on biotic surfaces mediated by molecular interactions or abiotic surfaces through physicochemical interactions. The second is irreversible adhesion, where the adhesion process is consolidated through the production of EPS. After the establishment and maturation of the protective matrix in the irreversible phase, the cycle ends with the rupture of the biofilm and the release of bacterial cells (**Figure 1**) [9, 41].

Bacteria in the form of free (planktonic) cells are not often found in nature; most of them live in communities or attached to various biotic or abiotic surfaces, such as clinical and industrial equipment. Several factors may contribute to bacterial adhesion such as flagella, fimbriae, adhesin, and polymers, as well as adhesion forces such as electrostatic and hydrophobic attraction, van der Waals interactions, hydrogen bridges, and covalent bond [10].

Biofilm formation is an effective strategy for microbial survival and persistence under stress conditions, such as in the presence of antimicrobials and heavy metals [42]. The biofilm structure may be associated with a protective mechanism that allows the bacteria to survive and persist in environments with high metal concentrations [43]. Studies have shown that subinhibitory heavy metal concentrations can induce biofilm formation [44, 45], like lead [46], cadmium [47], and nickel [48] among others.

Giovanella et al. [46] evidenced the increase in formation by an isolate of *Pseudomonas* sp. in the presence of mercury (Hg^{2+}). Similarly, Araújo et al. [49] verified an increase in biofilm formation in *Klebsiella pneumoniae* isolates obtained from an impacted urban stream. However, other studies show that depending on the metal and its concentration, biofilm formation may be reduced [50, 51]. These differences may be related to the fact that the effects of metals depend on their concentration and speciation [47, 51, 52], growth conditions, and especially the bacterial isolate that is being exposed [53, 54].

Recent studies have shown that metals can affect various stages in biofilm formation and development [55]. Metals can impact cell surface adhesion and/or cell-to-cell aggregation process, promoting biofilm formation and, consequently, its resistance. Harrison et al. [56] verified that the increase in cadmium concentration induces cell adhesion and biofilm formation in *Rhizobium alamii* YAS34. Subinhibitory concentrations of manganese (Mn) and zinc (Zn) affected cell aggregation in *Xylella fastidiosa* isolates. Mn increased the process of biofilm formation in this bacterium, while Zn impaired this process probably by reducing cell adhesion on the surface [50, 57]. Perrin et al. [48] observed that some isolates of *Escherichia coli* K-12 formed biofilm in response to subinhibitory nickel (Ni) concentrations and that cells embedded in the biofilm were less affected by metal exposure than planktonic cells. These studies show that bacterial cells exposed to metals generally respond by inducing adhesion processes, and consequently, biofilm formation and maintenance [55].

In addition to changes in cell adhesion, exposure to heavy metals may cause structural changes in the biofilm extracellular polymeric substance (EPS) matrix. Araújo et al. [49] verified by scanning microscopy, the increase of EPS in *K. pneumoniae* biofilms formed when exposed to subinhibitory mercury concentrations (Hg^{2+}). Sheng et al. [58] also demonstrated that heavy metals stimulate EPS production in *Rhodopseudomonas acidophila*. Schue et al. [59] observed in *R. alamii* isolates the formation of a more condensed biofilm in the presence of subinhibitory concentrations of Cd when compared to isolates not exposed to this metal. The increase of extracellular matrix in *Thiomonas* sp. subinhibitory concentrations of arsenic (III) possibly contributed to biofilm integrity and physiological heterogeneity of immobilized cell subpopulations [60].

In stabilized biofilm, the presence of metals impacts cells via passive processes by the influence of gene expression, resulting in mechanisms of resistance or tolerance to these pollutants [55]. Extracellular polymeric matrix (EPS) acts as a barrier to toxic metals, which can be sequestered, immobilized, mineralized, and precipitated, diminishing their effect on bacteria [61]. In *Pseudomonas putida* ATCC 33015, sugars present in the biofilm matrix exposed to chromium (Cr) probably facilitated the immobilization process of this metal [62]. The biomineralization process was described in *Cupriavidus metallidurans* CH34, which was able to form gold (Au) nanoparticles in biofilm through the reduction and precipitation mechanism of the toxic gold complex (Au III) [63].

5. Heavy metal resistance

Environmental contamination by heavy metals has been increasing in recent years, due to various anthropogenic activities. Heavy metals, because they are not biodegradable, have a tendency for biomagnification and bioaccumulation and are extremely toxic to various biological functions, causing serious impacts on the environment and human health [64].

Microorganisms present in contaminated environments have developed different resistance mechanisms to adapt to stress caused by heavy metals. The ability to survive under these extreme conditions depends on acquired biochemical and physiological attributes, as well as genetic adaptations [65].

Several studies suggest that metal contamination in the natural environment may play an important role in maintaining and proliferating antimicrobial resistance (**Table 1**) [67–69]. In the environment, selective pressure exerted by metals may select resistant isolates similar to antibiotics, since both resistance genes are often located on the same moving elements [70, 71].

Bacteria develop some mechanisms to neutralize mercury toxicity, the most common being enzymatic reduction of the highly toxic mercuric ion (Hg^{2+}) to the volatile and less toxic elemental mercury (Hg^0). This reduction is catalyzed by the cytosolic mercury reductase (MerA) enzyme encoded by a gene belonging to the operon mer. Studies have shown the frequent association between operon mer and antimicrobial resistance [66, 72]. Péres-Valdespino et al. [73] demonstrated that several clinical isolates of *Aeromonas* sp. that presented the *merA* gene were resistant to different antibiotics such as tetracycline, trimethoprim, nalixidic acid, and streptomycin. Araújo et al. [49] verified, when comparing isolates of *K. pneumoniae*, that the isolate that presented the *merA* gene was resistant to the highest number of antimicrobials and presented the minimum inhibitory concentration

Resistance mechanisms	Heavy metals	Antibiotics	References
Reduction in permeability	As, Cu, Zn, Mn, Co, Ag	Cip, Tet, Cholr, β-lactâmicos	[32, 74]
Drug and metal alteration	As, Hg	β-lactâmicos, Chlor	[75, 76]
Drug and metal outflow	Cu, Co, Zn, Cd, Ni, As	Tet, Chlor, β-lactâmicos	[77, 78]
Cell signaling change	Hg, Zn, Cu	Cip, β-lactâmicos, Trim, Rif	[79, 80]

Abbreviations: Cholr, chloramphenicol; Cip, ciprofloxacin; Rif, rifampicin; Tet, tetracycline; Trim, trimetropim. Adapted from Baker-Austin et al. [66].

Table 1.
Examples of characteristics and negative effects on metal and antibiotic resistance mechanisms.

(MIC) value up to four times higher than the others, suggesting a co-resistance mechanism for mercury and antimicrobials tested.

Martins et al. [81] observed that isolates of *P. aeruginosa*, obtained from a contaminated river in southeastern Brazil, had a conjugative plasmid with co-resistance to tetracycline and copper, reinforcing that resistance to antibiotics may be induced by selective pressure of heavy metals in the environment. Caille et al. [82] demonstrated that in *P. aeruginosa*, copper can induce imipenem resistance by the CopR-CopS two-component regulatory system mechanism. Ghosh et al. [83] verified resistance to ampicillin, arsenic, chromium, cadmium, and mercury in *Salmonella abortus equi* isolates and observed that after removal of the plasmids, isolates became sensitive to these compounds.

In order to corroborate the evidence of co-resistance of metals and antibiotics, some studies compared the resistance profiles of bacteria collected in contaminated and uncontaminated environments. Rasmussen and Sørensen [84] demonstrated an increase in the occurrence of conjugative plasmids at contaminated sites and found that the mercury and tetracycline resistance genes were located on the same plasmid. Mcarthur and Tuckfield [85] examined metal and antibiotic resistance profiles in contaminated and uncontaminated stream sediments and found that isolates obtained from the contaminated sediment were more resistant to kanamycin and streptomycin than the others.

Thus, not only the indiscriminate use of antibiotics but also environmental contamination by heavy metals can pose risks and harm to human health, as resistance genes can be transferred horizontally from environmental microorganisms to human diners [66].

6. Conclusions

Increased urbanization and industrialization have contributed to heavy metal contamination in aquatic ecosystems, modifying the structure and function of microbial communities. The ability of microorganisms to survive under stress conditions, such as in the presence of heavy metals, depends on structural and biochemical attributes, as well as physiological and/or genetic adaptations. The studies cited demonstrated that the presence of heavy metals influences at different stages of biofilm formation. Additionally, the correlation between resistance to metals and antimicrobials was demonstrated, showing the environmental impact that these contaminants can cause in aquatic environments.

Author details

Lívia Caroline Alexandre de Araújo[1] and Maria Betânia Melo de Oliveira[2]*

1 Federal University of Pernambuco, Recife, Brazil

2 Department of Biochemistry of the Federal University of Pernambuco, Recife, Brazil

*Address all correspondence to: mbetaniam2008@gmail.com

References

[1] Rand GM, Wells PG, Mccarty LS. Fundamentals of Aquatic Toxicology: Effects, Environmental Fate, and Risk Assessment. 2nd ed. Washington: Taylor & Francis; 1995

[2] Zhang X-X, Zhang T, Fang HHP. Antibiotic resistance genes in water environment. Applied Microbiology and Biotechnology. 2009;**82**(3):397-414

[3] Freitas JHES, Santana KV, Nascimento ACC, et al. Evaluation of using aluminum sulfate and water-soluble *Moringa oleifera* seed lectin to reduce turbidity and toxicity of polluted stream water. Chemosphere. 2016;**163**:133-141

[4] Purificação-Júnior AF, Araújo LCA, Lopes ACS, et al. Microbiota sampled from a polluted stream in Recife-PE, Brazil and its importance to public health. African Journal of Microbiology Research. 2017, 2017;**11**:1142-1149

[5] Dixit G, Singh AP, Kumar A, Mishra S, Dwivedi S, Kumar S, et al. Reduced arsenic accumulation in rice (*Oryza sativa L.*) shoot involves sulfur mediated improved thiol metabolism, antioxidant system and altered arsenic transporters. Plant Physiology and Biochemistry. 2015;**99**:86-96

[6] Adarsh VK, Mishra M, Chowdhyry S, Sudarshan M, Thakur AR, Ray CS. Studies on metal microbe interaction of three bacterial isolates from east Calcutta wetland. Journal of Biological Sciences. 2007;**7**:80-88

[7] Sarma B, Axharya C, Joshi SR. Pseudomonas: A versatile bacterial group exhibiting dual resistance to metals and antibiotics. African Journal of Microbiology Research. 2010;**4**:2828-2835

[8] Chouduri AU, Wadud A. Twitching motility, biofilm communities in cephalosporin resistant *Proteus* spp and the best in vitro amoxicillin susceptibility to isolates. American Journal of Microbiological Research. 2014;**2**(1):8-15

[9] Trentin DS, Giordani RB, Macedo AJ. Biofilmes bacterianos patogênicos: Aspectos gerais, importância clínica e estratégias de combate. Revista Liberato. 2013;**14**(22):113-238

[10] Flemming HC, Wingender J. The biofilm matrix. Nature Reviews Microbiology. 2010;**8**:623-633

[11] Agência Nacional das Águas. Panorama da qualidade das águas superficiais do Brasil, 2012; Agência Nacional de Águas. Brasília: ANA; 2012

[12] Paz VPS, Teodoro REF, Mendonça FF. Recursos hídricos, agricultura irrigada e meio ambiente. Revista Brasileira de Engenharia Agrícola e Ambiental. 2000;**4**(3):465-473

[13] Raucci GD, Polette M. Subsídios para análise da capacidade de suporte da praia central de balneário Camboriú–SC. Perfil do Usuário. In: XIV Semana Nacional de Oceanografia Rio Grande-RS Livro de síntese do evento citado: Furg. 2001. pp. 117-118

[14] WHO. Water Sanitation and Health. 2014. Available from: http://www.who.int/water_sanitation_health/diseases [Accessed: 24 September 2017]

[15] WHO. Sanitation Safety Planning: Manual for Safe Use and Disposal of Wastewater, Greywater and Excreta. Geneva: World Health Organization; 2016

[16] Geldreich EE. The bacteriology of water. In: Microbiology and Microbial Infections. London: Arnold; 1998

[17] Liu SY, Zhangs SN, Geng TY, Li CM, Yed D, Zhang F, et al. High diversity of extended-spectrum beta-lactamase-producing bacteria in an urban river sediment habitat. Applied and Environmental Microbiology. 2010;**76**:5972-5976

[18] Shakibaie MR, Jalilzadeh KA, Yamakanamardi SM. Horizontal transfer of antibiotic resistance genes among Gram negative bacteria in sewage and lake water and influence of some physico-chemical parameters of water on conjugation process. Journal of Environmental Biology. 2009;**30**(1):45-49

[19] Al-Bahry SN, Mahmood IY, Al-Khaifi A, Elshafie AE, Al-Harthy A. Hability of multiple antibiotic resistant bacteria in distribuition lines of treated sewage effluent used for irrigation. Water Science and Technology. 2009;**60**(11):2939-2948

[20] Caumo KS, Duarte M, Cargin ST, Ribeiro VB, Tasca T, Macedo AJ. Revista Liberato: Revista de divulgação de educação, ciência e tecnologia. Novo Hamburgo, RS. 2010;**11**(16):89-188

[21] Maal-Bared R, Bartlett KH, Bowie WR, Hall ER. Phenotypic antibiotic resistance of *Escherichia coli* and *E. coli* O157 isolated from water, sediment and biofilms in an agricultural watershed in British Columbia. Science of the Total Environment. 2013;**443**:315-323

[22] Middleton JH, Salierno JD. Antibiotic resistance in triclosan tolerant fecal coliforms isolated from surface waters near wastewater treatment plant outflows (Morris County, NJ, USA). Ecotoxicology and Environmental Safety. 2013;**88**:79-88

[23] Zhang X, Li Y, Liu B, Wang J, Feng C, Gao M, et al. Prevalence of veterinary antibiotics and antibiotic-resistant *Escherichia coli* in the surface water of a livestock production region in northern China. PLoS One. 2014;**9**:e111026

[24] Chen Z, Yu D, He S, Ye H, Zhang L, Wen Y, et al. Prevalence of antibiotic-resistant *Escherichia* coli in drinking water sources in Hangzhou City. Frontiers in Microbiology. 2017;**8**:1133

[25] Doelman P, Jansen E, Michels M, Van TM. Effects of heavy metals in soil on microbial diversity and activity as shown by sensitivity-resistance index, an ecologically relevant parameter. Biology and Fertility of Soils. 1994;**17**:177-184

[26] Foster IDL, Charlesworth SM. Heavy metals in the hydrological cycle: Trends and explanation. Hydrological Processes. 1996;**10**:227-261

[27] Fashola MO, Ngole-Jeme VM, Babalola OO. Heavy metal pollution from gold mines: Environmental effects and bacterial strategies for resistance. International Journal of Environmental Research and Public Health. 2016;**13**:1047

[28] Blindauer CA, Harrison MD, Robinson AK, et al. Multiple bacteria encode metallothioneins and Smt A-like zinc fingers. Molecular Microbiology. 2002;**45**:1421-1432

[29] Quintelas C, Rocha Z, Silva B, Fonseca B, Figueiredo H, Tavares T. Biosorptive performance of *Escherichia coli* biofilm supported on zeolite NaY for the removal of Cr(VI), Cd(II), Fe(III) and Ni(II). Chemical Engineering Journal. 2009;**152**:110-115

[30] Podda FP, Zuddas A, Minacci M, Baldi F. Heavy metal coprecipitation with hydrozincite $[Zn_5(CO_3)_2(OH)_6]$ from mine waters caused by photosynthetic micro organisms. Applied and Environmental Microbiology. 2000;**66**:5092-5098

[31] Mire CE, Tourjee JA, O'brien WF, Ramanujachary KV, Hecht GB. Lead precipitation by *Vibrio harveyi*: Evidence for novel quorum-sensing interactions. Applied and Environmental Microbiology. 2004;**70**(2):855-864

[32] Silver S. Bacterial heavy metal resistance: New surprises. Annual Review of Microbiology. 1996;**50**(69):753-789

[33] Schalk IJ, Hannauer M, Braud A. New roles for bacterial siderophores in metal transport and tolerance. Environmental Microbiology. 2011;**13**:2844-2854

[34] Taga ME, Bassler BL. Chemical communication among bacteria. Proceedings of the National Academy of Sciences of the United States of America. 2003;**100**(Suppl):14549-14554

[35] Marques JJGSM, Curi N, Schulze DG. Trace elements in Cerrado soils. In: Alvarez VH, Schaffer CEGR, et al., editors. Tópicos em Ciência do Solo (Topics in Soil Scince). Viçosa: Sociedade Brasileira de Ciência do Solo; 2002. pp. 103-142

[36] Aguiar MRMP, Novaes AC, Guarino AWS. Remoção de metais pesados de efluentes industriais por aluminossilicatos. Química Nova. 2002;**25**(6b):1145-1154

[37] Virga RSP, Geraldo LP, Santos FH. Avaliação de contaminação por metais pesados em amostras de siris azuis. Ciência Tecnologia e Alimentação, Campinas. 2007;**27**(4):779-785

[38] World Health Organization. Trace Elements in Human Nutrition and Health. Geneva: WHO; 1996

[39] Linde AR, Arribas P, Sanchez-Galan S, Garcia-Vazquez E. Eel (*Anguilla anguilla*) and Brown Trout (*Salmo trutta*) target speecies to assess the biological impact of trace metal pollution in freshwater ecosystems. Archives of Enviromental Contamination and Toxicology. 1996;**31**:297-302

[40] Tavares TM, Carvalho FM. Avaliação de exposição de populações humanas a metais pesados no ambiente: Exemplos do Recôncavo Baiano. Revista Química Nova. 1992;**15**(2):147-154

[41] Monroe D. Looking for chinks in the armor of bacterial biofilms. PLoS Biology. 2007;**5**(11):307

[42] Azevedo NF, Cerca N. Biofilmes: Na Saúde, no Ambiente, na Indústria. 1st ed. Portugal: Publindústria Edições Técnicas; 2012

[43] Muller D, Médigue C, Koechler SA. A tale of two oxidation states: Bacterial colonization of arsenic-rich environments. PLoS Genetics. 2007;**3**:e53

[44] Kaplan JB, Izano EA, Gopal P, et al. Low levels of β-lactam antibiotics induce extracellular DNA release and biofilm formation in *Staphylococcus aureus*. MBio. 2012;**3**(4):1-14

[45] Hennequin C, Aumeran C, Robin F, Traore O, Forestier C. Antibiotic resistance and plasmid transfer capacity in biofilm formed with a CTX-M-15-producing *Klebsiella pneumoniae* isolate. The Journal of Antimicrobial Chemotherapy. 2012;**67**(9):2123-2130

[46] Giovanella P, Cabral L, Costa AP, et al. Metal resistance mechanisms in Gram-negative bactéria and their potential to remove Hg in the presence of outher metals. Ecotoxicology and Environmental Safety. 2017;**140**:162-169

[47] Wu X, Santos RS, Fink-Gremmels J. Cadmium modulates biofilm formation by *Staphylococcus epidermidis*. International Journal of Enviromental

Research and Public Health. 2015;**12**:2878-2894

[48] Perrin C, Briandet R, Jubelin G, Lejeune P. Nickel promotes biofilm formation by *Escherichia coli* K-12 strains that produce curli. Applied and Environmental Microbiology. 2009;**75**:1723-1733

[49] Araújo LCA, Purificação-Júnior AF, Silva SM, Lopes ACS, et al. In vitro evaluation of mercury (Hg^{2+}) effects on biofilm formation by clinical and environmental isolates of *Klebsiella pneumoniae*. Ecotoxicology and Environmental Safety. 2019;**169**:669-677

[50] Navarrete F, De La Fuente L. Response of *Xylella fastidiosa* to zin: Decreased culturability, increased exopolusaccharide production, and formation of resilient biofilms under flow conditions. Applied and Environmental Microbiology. 2014;**8010**:97-107

[51] Jomova K, Valko M. Advances in metal-induces oxidative stress and human disease. Toxicology. 2011;**283**:65-87

[52] Lemire JA, Harrison JJ, Turner RJ. Antimicrobial activity of metals: Mechanisms, molecular targets and apllications. Nature Reviews Microbiology. 2013;**11**:371-384

[53] Tremarolli V, Fedi S, Turner RJ, Ceri H, Zannoni D. *Pseudomonas pseudoalcaligenes* KF707 upon biofilm formation on a polystyrene surfasse acquire a strong antibiotic resistance with miner changes in their tolerance to metal cátions and metalloid oxyanions. Archives of Microbiology. 2008;**190**:29-39

[54] Booth SC, George IFS, Zannoni D, Cappelletti M, Duggan GE, Ceri H, et al. Effect of aluminium and copper on biofilm development of *Pseudomonas pseudoalcaligenes* KF707 and *P. fluorescens* as a function of diferente media compositions. Metallomics. 2013;**5**:723-735

[55] Koechler S, Farasin J, Cleiss-Arnold J, Arsene-Ploetze F. Toxic metal resistance in biofilms: Diversity of microbial responses and their evolution. Research in Microbiology. 2015;**166**:764-773

[56] Harrison JJ, Ceri H, Stremick C, Turner RJ. Differences in biofilm and planktonic cell mediates reduction of metalloid oxyanions. FEMS Microbiology Letters. 2004;**62**:235-357

[57] Conibe PA, Cruz LF, Navarrete F, Ducan D, Tygart M, De La Fuente L. *Xylellafastidiosa* differentiallyaccumulates mineral elements in biofilm na planktonic cells. PLoS One. 2013;**8**: e54936

[58] Sheng GP, Yu HQ, Yu Z. Extraction of extracellular polymeric substances from the photosynthetic bacterium *Rhodopseudomonas acidophila*. Applied Microbiology and Biotechnology. 2005;**67**(1):125-130

[59] Schue M, Fekete A, Ortet P, Brutesco C, Heulin T, Schmitt-Kopplin P, et al. Modulation of metabolism and switching to biofilm prevail over exopolysaccharide production in the response of *Rhizobium alamii* to cadmium. PLoS One. 2011;**6**:e26771

[60] Marchal M, Briandet R, Halter D, Koechler S, Dubow MS, Lett MC, et al. Subinhibitory arsenite concentrations lead to population dispersal in *Thiomonas* sp. PLoS One. 2011;**6**(8): e23181

[61] Kazy SK, Sar P, Songh SP, Sem Asish K, D'Souza SF. Extracellular polysaccharides of a copper-sensitive and copper-resistant *Pseudomonas aeruginosa* strain: Synthesis, chemical nature and copper binding. World

Journal of Microbiology and Biotechnology. 2002;**18**:583-588

[62] Priester JH, Olson SG, Webb SM, Neu MP, Hersman LE, Holden PA. Enhanced exopolymer production and chromium stabilization in *Pseudomonas putid*a unsaturated biofilms. Applied and Environmental Microbiology. 2006;**72**:1988-1996

[63] Reith F, Etschmann B, Grosse C, Moors H, Benotmane MA, Monsieurs P, et al. Mechanisms of gold biomineralization in the bacterium *Cupriavidus metallidurans*. PNAS. 2009;**106**:17757-17762

[64] Li B, Zhao Y, Liu C, et al. Molecular pathogenesis of *Klebsiella pneumoniae*. Future Microbiology. 2013;**9**:1071-1081

[65] Abou-Shanab RAI, Van Berkum P, Angle JS. Heavy metal resistance and genotypic analysis of metal resistance genes in gram-positive and gram-negative bactria present in Ni-rich serpentine soil and the rhizosphere of *Alyssum murale*. Chemosphere. 2007;**68**:360-367

[66] Baker-Austin C, Wright MS, Stepanauskas R, Mcarthur JV. Co-selection of antibiotic and metal resistance. Trends in Microbiology. 2006;**14**(4):176-182

[67] Summers AO et al. Mercury released from dental silver fillings provokes an increase in mercury-resistant and antibiotic-resistant bacteria in oral and intestinal floras of primates. Antimicrobial Agents and Chemotherapy. 1993;**37**:825-834

[68] Summers AO. Generally overlooked fundamentals of bacterial genetics and ecology. Clinical Infectious Diseases. 2002;**34**:S85-S92

[69] Alonso A et al. Environmental selection of antibiotic resistance genes. Environmental Microbiology. 2001;**3**:1-9

[70] Fugimore H, Kiyono M, Nobuhara K, Pan-Hou H. Possible involvement of red pigments in defense against mercury in Pseudomonas K-62. FEMS Microbiology Letters. 1996;**135**(2-3):317-321

[71] Mcintosh D, Cunningham M, Ji B, Fekete FA, et al. Transferable, multiple antibiotic and mercury resistance in Atlantic Canadian isolates of *Aeromonas salmonicida* subsp. *salmonicida* is associated with carriage of an IncA/C plasmid similar to the Salmonella enterica plasmid pSN254. The Journal of Antimicrobial Chemotherapy. 2008;**61**(6):1221-1228

[72] Mathema VB, Thakuri BC, Sillanpã M. Bacterial mer operon-mediated detoxification of mercurial compounds: A short review. Archives of Microbiology. 2011;**193**:837-844

[73] Pérez-Valdespino A, Celestino-Mancera M, Villegas-Rodriguez VL, Curiel-Quesada E. Characterization of mercury-resistant clinical *Aeromonas* species. Brazilian Journal of Microbiology. 2013;**44**(4):1279-1283

[74] Ruiz N. The role of Serratia marcescens porins in antibiotic resistance. Microbial Drug Resistance. 2003;**9**:257-264

[75] Mukhopadhyay R, Rosen BP. Arsenate reductases in prokaryotes and eukaryotes. Environmental Health Perspectives. 2002;**110**:745-748

[76] Wright JW, Natan MJ, Macdonnell FM, Ralston DM, O'Halloran TV. Mercury(II) Thiolate Chemistry and the Mechanism of the Heavy Metal Biosensor MerR. In: Lippard SJ, editor. Progress in Inorganic Chemistry. Lippard SJ, editor. 38 ed. New York: John Wiley & Sons; 1990. pp. 323-412

[77] Nies DH. Efflux-mediated heavy metal resistance in prokar-yotes. FEMS Microbiology Reviews. 2003;**27**:33-39

[78] Levy SB. Active efflux, a common mechanism for biocide and antibiotic resistance. Journal of Applied Microbiology. 2002;**92**:65-71

[79] Barkay T, Miller SM, Summers AO. Bacterial mercury resistance from atoms to ecosystems. FEMS Microbiol Review. 2003;**27**:355-384

[80] Roberts MC. Update on acquired tetracycline resistance genes. FEMS Microbiology Letters. 2005;**245**:195-203

[81] Martins VV, Zanetti MOB, Pitondo-Silva A, Stehling EG. Aquativ environments polluted with antibiotics and heavy metals: A human health hazard. Environmental Science and Pollution Research. 2014;**21**:5873-5878

[82] Caille O, Rossier C, Perron K. A copper activated two-component systems interacts with zinc and imipenem resistance in *Pseudomonas aeruginosa*. Journal of Bacteriology. 2007;**189**:4561-4568

[83] Ghosh A et al. Characterization of large plasmids encoding resistance to toxic heavy metals in *Salmonella abortus equi*. Biochemical and Biophysical Research Communications. 2000;**272**:6-11

[84] Rasmussen LD, Sørensen SJ. Effects of mercury contamination on the culturable heterotrophic, functional and genetic diversity of the bacterial community in soil. FEMS Microbiology Ecology. 2001;**36**(1):1-9

[85] Mcarthur JV, Tuckfield RC. Spatial patterns in antibiotic resistance among stream bacteria: Effects of industrial pollution. Applied and Environmental Microbiology. 2000;**66**(9):3722-3726

Section 2

Inhibition of Biofilm Formation

Chapter 3

Natural Compounds Inhibiting *Pseudomonas aeruginosa* Biofilm Formation by Targeting Quorum Sensing Circuitry

Julie Carette, Amandine Nachtergael, Pierre Duez, Mondher El Jaziri and Tsiry Rasamiravaka

Abstract

The biofilm lifestyle mode certainly represents one of the most successful behaviors to facilitate bacterial survival in diverse inhospitable environments. Conversely, the ability of bacteria to develop effective biofilms represents one of the major obstacles in the fight against bacterial infections. In *Pseudomonas aeruginosa*, the biofilm formation is intimately connected to the quorum sensing (QS) mechanisms, a mode of cell-to-cell communication that allows many bacteria to detect their population density in order to coordinate common actions. In this chapter, we propose an overview (i) on *P. aeruginosa* QS mechanisms and their implication in biofilm formation, and (ii) on natural products that are known to interfere with these QS mechanisms, subsequently disrupting biofilm formation. The concluding remarks focus on perspectives of these compounds as possible antibiotherapy adjuvants.

Keywords: biofilm, *las*, natural products, PQS, *pseudomonas*, quorum sensing, *rhl*

1. Introduction

Bacterial infections are mainly related to the ability of bacteria to invade and disseminate through their hosts by using different types of motility, by releasing a myriad of virulence factors, by building structured biofilm which lead to host cell and tissue damage but also allow bacteria to evade the immune system and conventional antimicrobial agents [1]. For decades, antibiotics, although less effective in biofilm-growing bacteria [2], have represented our best weapon against bacterial diseases. However, the on-going emergence and worldwide spreading of resistant bacteria is considerably reducing the antibiotic pallet available for the treatment of bacterial infections [3]. This alarming situation forces researchers to consider other strategies to combat bacterial infections, notably the use of phages [4] or the use of alternative agents, such as essential oils [5], silver nanoparticles [6], bacteriocins [7], and antimicrobial peptides [8]. Some interesting strategies propose original compounds that disrupt biofilm formation without affecting the viability of invading bacteria; this strategy is expected (i) to reduce the bacterial aptitude to build protective barriers, but without exerting a selective pressure *per se* [4]; (ii) to allow sufficient time for the immune defenses to effectively destroy invaders; and (iii) to minimize the use of effective antibiotics.

In most bacteria, the expressions of virulence factors are coordinated by quorum sensing (QS) mechanisms, a cell-to-cell communication which allows bacteria to detect their population density by producing and perceiving diffusible signal molecules to synchronize common actions [9]. This cell-to-cell communication has been largely investigated in *Pseudomonas aeruginosa,* an opportunistic pathogen which mainly affects people who are severely immunocompromised, in part due to its ability to evade from both innate and acquired immune defenses through adhesion, colonization, and biofilm forming and to produce various virulence factors that cause significant tissue damage [10, 11]. In this bacterium, QS regulates virulence factors production, motilities and, in particular, biofilm formation for which QS is one of the relevant key actors. Interestingly, within the two past decades, study papers reporting natural and synthetic compounds that interfere with QS and/or biofilm formation are regularly published; QS circuitry and biofilm formation control mechanisms indeed constitute promising targets to struggle against *P. aeruginosa* infection with potential huge clinical interests [12]. The present chapter covers the scope of natural compounds from both prokaryote and eukaryote organisms that have been identified to disrupt the biofilm lifestyle cycle in *P. aeruginosa* via modulation of QS mechanisms. An overview of the entanglement between QS circuitry and biofilm formation is reported as a prerequisite for a better understanding of the mechanisms of action proposed for some of the identified compounds. The concluding remarks focus on the perspectives of these compounds as possible antibiotherapy adjuvants for possible eradication of resistant infections caused by *P. aeruginosa*.

2. *P. aeruginosa* biofilm lifestyle

Like most bacteria, *P. aeruginosa* can develop two distinct lifestyles, planktonic and sessile cells. The planktonic state is encountered when *P. aeruginosa* evolves freely in a liquid suspension, whereas on natural or synthetic surfaces, *P. aeruginosa* can form sticky clusters in permanent rearrangements characterized by the secretion of an adhesive and protective matrix [13]. Defined as "biofilm," this set of bacterial community adherent to a surface appears as an adaptive response to an environment more or less unsuited to growth in planktonic form [14].

The biofilm formation can be delimited in five main stages (**Figure 1**, image A). A first reversible phase corresponds to the initial adhesion of bacteria to surfaces; this adhesion becomes irreversible in the second stage (image B). Then, thanks to a proliferation period corresponding to the third stage, microcolonies are built concomitantly with the production of extracellular matrix (image C), leading to the fourth stage of biofilm structuration and organization in which the growth of three dimensional communities is observed with amplified extracellular matrix production (image D). This biofilm cycle is completed by a dispersion step (image E) [12].

The secreted extracellular matrix mainly consists of proteins, nucleic acids, lipids, and exopolysaccharides (EPS). These account for 50–90% of total organic matter [16]. *P. aeruginosa* produces at least three types of EPS that are required for biofilm formation and architecture [17]. (i) Alginate a linear polysaccharide composed of L-guluronic and D-mannuronic acids linked by β-1,4 bonds [18], (ii) Pel polysaccharide, a glucose-rich matrix material, with unclarified composition, and (iii) Psl polysaccharide, a repeating pentasaccharide consisting of D-mannose, L-rhamnose, and D-glucose. In mucoid strains, EPS are predominantly characterized by the presence of alginate. The alginate participates in the structuring of the

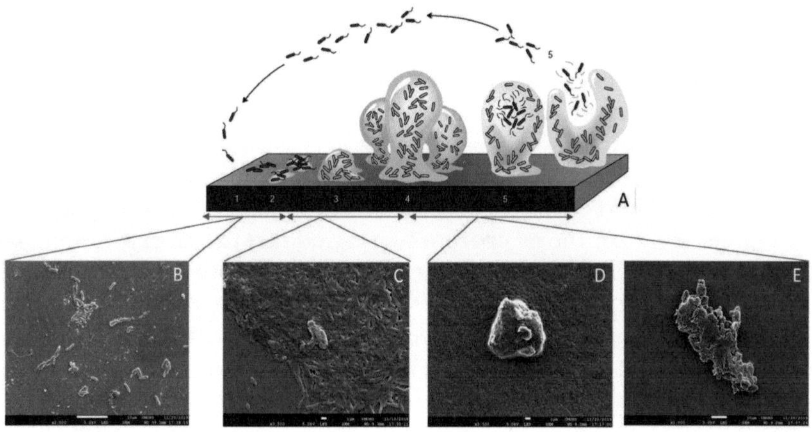

Figure 1.
Sketch of the different steps of a biofilm development (A) [15]. Several representative scanning electron microscopy (SEM-JEOL JSM-7200F) images of the P. aeruginosa *biofilm at different steps of development and with different magnifications (B = reversible and irreversible stages at 8 h growth, C = microcolonies stage at 30 h growth, D = mature biofilm stage at 120 h growth, and E = dispersion stage at 144 h growth).* P. aeruginosa *PAO1 colonies were grown at 37°C with Centers for Disease Control and Prevention (CDC) biofilm reactor (biosurface technologies, MT) on tryptone soy broth (TSB).*

biofilm [19], but its real importance is still controversial since some authors claim that it is not essential; indeed architecture and antibiotic resistance profiles of wild-type and alginate-deficient biofilms are identical [20, 21]. Nevertheless, the overexpression of alginate was shown to protect *P. aeruginosa* from phagocytosis and host responses [22]. In "nonmucoid" *P. aeruginosa* strains, such as the PAO1 strain isolated from an infected wound [23], alginate is even considered poorly produced at the expense of exopolysaccharides rich in glucose and mannose [24], Pel and Psl, which have been described as being more important in the formation and maintenance of the biofilm [25].

Extracellular DNA (eDNA) is an important component of *P. aeruginosa* biofilm matrix, which particularly intervenes in the establishment, maintenance, and perpetuation of structured biofilms [26]. Its importance has been demonstrated since *P. aeruginosa* biofilm formation is prevented by exposition to DNase I [27] and biofilms that are deficient in eDNA have been shown to be more sensitive to the detergent sodium dodecyl sulfate [28]. It has been established that eDNA plays roles in bacterial adhesion and in the structural stability of biofilms by maintaining coherent cell alignments [29]; interestingly, its contribution to antimicrobial resistance has also been proposed as eDNA, a highly anionic polymer, is believed to bind cationic antibiotics, such as aminoglycosides and antimicrobial peptides [30].

3. QS mechanisms and their implication in biofilm formation

The complex regulation of biofilm formation involves multiple bacterial machineries including the QS systems. In *P. aeruginosa*, this mechanism is involved in the development of various common bacterial behaviors, including virulence factors expression and biofilm formation, which are mostly implicated in infection success. Three QS systems have been clearly characterized: (i) the *las* system and the *rhl* system, two LuxI/R type systems using the signal molecules of the family of acyl-homoserine lactones (AHLs); and (ii) the PQS (pseudomonas quinolone signal) system based on molecules of the 2-alkyl-4-quinolone class [10, 31]. The mechanisms of QS in *P. aeruginosa* are summarized in **Figure 2** while the main functions regulated by QS systems and involved in the pathogenesis of *P. aeruginosa* are presented in **Figure 3**.

Evidence that the *las* system is implicated in biofilm formation has been firstly established when Davies et al. [32] demonstrated that the biofilm formed by *lasI* mutant appears flat, undifferentiated, and quickly dispersed from the surface upon exposure to sodium dodecyl sulfate, compared to wild type biofilms.

Furthermore, Gilbert et al. [33] observed the binding of the QS regulator LasR to the promoter region of the *psl* operon, suggesting that the *psl* expression may be regulated by the QS. Considering that the *psl* operon is implied in biofilm modulation, the QS then plays a role in the biofilm formation and architecture. The transcription of the *pel* operon seems to be reduced in *rhlI* mutant, suggesting that the *rhl* system plays a biofilm formation role in *P. aeruginosa* by modulating the biosynthesis of the Pel polysaccharide [34]. The *pqsA* mutant produces a biofilm with less eDNA than the wild type biofilm, suggesting that the PQS system also plays a role in biofilm formation, more particularly in the eDNA releasing [34].

Notably, the production of rhamnolipids and lectins is under QS control, indicating a further indirect link between biofilm formation/degradation and QS.

Indeed, the *rhl* system controls the production of rhamnolipids [35], that play multiple roles in *P. aeruginosa* biofilm formation: (i) as biosurfactant and virulence factor [36]; (ii) in the formation of microcolonies [37]; (iii) in the maintenance of open channel structures necessary for nutrient circulation [38]; (iv) in the development of biofilm mushroom-shaped structures [37]; and (v) in cell dispersion from the biofilm [39]. Indeed, a hyper-detaching property has been observed in the *P. aeruginosa* mutants that produce more rhamnolipids compared to wild type strains [40]. Moreover, the *rhl* system also controls the expression of the cytotoxic virulence factors LecA and LecB. Data obtained on mutant strains indicate that these galactophilic lectins probably contribute to the biofilm development [41, 42]. Similarly, two types of *P. aeruginosa* motilities implicated in biofilm formation are also QS-regulated. The first movement, swarming motility, accomplishes an organized surface translocation, dependent on cell-to-cell

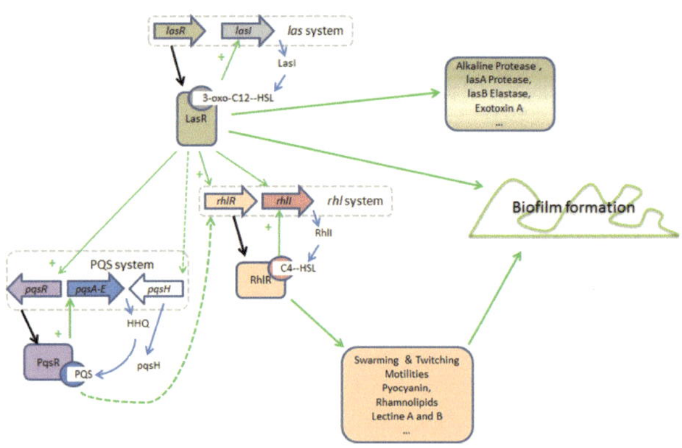

Figure 2.
Systems involved in P. aeruginosa *QS circuitry. The main QS systems in* P. aeruginosa *are the* las, rhl, *and PQS systems. The* las *system consists of a* lasR *regulatory gene coding for the LasR protein, a* lasI *gene coding for a LasI synthase involved in the synthesis of a signal molecule of the acyl-homoserine lactone (AHL) family, the 3-oxo-C12-HSL. The LasR/3-oxo-C12-HSL complex is a transcriptional activator of virulence genes (protease, elastase, and exotoxin) and* lasI *gene. According to the same model, the* rhl *system consists of* rhlR, rhlI *genes, and another AHL, the C4-HSL. This system activates genes in common with the* las *system and also specific genes, such as those coding for the synthesis of rhamnolipids, pyocyanin, and swarming/twitching motilities. The* las *system controls the* rhl *system. The third PQS system is interposed between the two main systems. The PqsABCDE operon produces the precursor 2-heptyl-4-quinolone (HHQ), and PqsH catalyzes conversion of HHQ to 2-heptyl-3-hydroxy-4-quinolone (PQS), detected by the receptor PqsR [10, 31].*

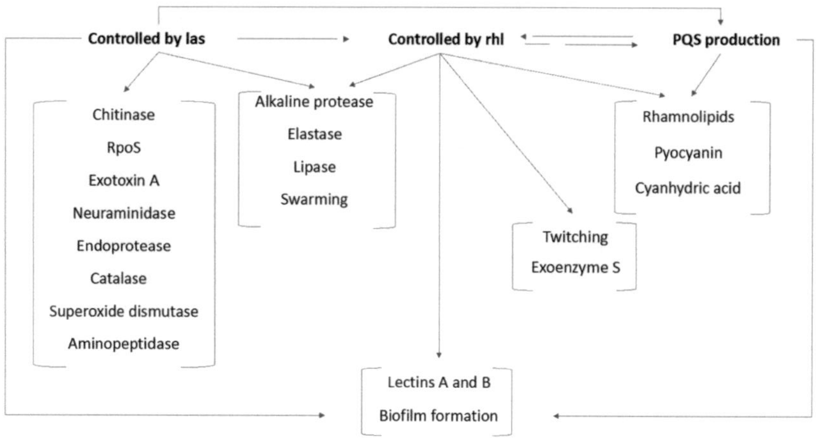

Figure 3.
Functions positively regulated by QS in P. aeruginosa *[10, 31].*

contacts and extensive flagellation [43]; this has been observed during the first stage of *P. aeruginosa* biofilm development and seems to be regulated by the *rhl* system [44]. Flat and uniform biofilms are formed when the strains grow under conditions promoting swarming motility, for example, a growth medium with glutamate or succinate as carbon sources; by contrast, a biofilm without confluent cell aggregates is formed by strains with limited swarming motility [45]. The second movement, a flagella-independent form of translocation, is described as a successive extension and retraction of polar type IV pili [46]. This kind of movement, regulated by the *rhl* system on a Fe-limited minimal medium [47], is necessary to assemble bacteria in monolayers that form microcolonies [38].

4. Other mechanisms implied in biofilm formation

The QS systems are not the sole key actors intervening in biofilm formation by *P. aeruginosa*. Indeed, the complex regulation of biofilm formation involves multiple bacterial machineries that also include the membrane-bound sensor kinase GacS, the transcriptional response regulator GacA (GacS/GacA two-component regulatory system), and the intracellular second messenger bis-(3′-5′)-cyclic dimeric guanosine monophosphate (c-di-GMP). Briefly, the GacS/GacA system acts as a super-regulator of the *las* and *rhl* systems [48], whereas c-di-GMP is important for the biosynthesis of alginate and Pel polysaccharides and for the switch from planktonic to biofilm lifestyle [49].

5. Natural products that affect QS and biofilm formation by *Pseudomonas aeruginosa*

5.1 From prokaryotes

5.1.1 Enzymes

Microorganisms known to have the ability to produce anti-QS enzymes are still limited to a few bacteria from the families of (i) *Actinobacteria (Rhodococcus* and *Streptomyces)*; (ii) *Firmicutes-Arthrobacter (Bacillus* and *Oceanobacillus)*; (iii) *Cyanobacteria (Anabaena)*; (iv) *Bacteroidetes (Tenacibaculum);* (v) *Proteobacteria*

(*Acinetobacter, Agrobacterium tumefaciens, Alteromonas, Comomonas, Halomonas, Hyphomonas, Klebsiella pneumoniae, P. aeruginosa, Ralstonia, Stappia*, and *Variovorax paradoxus*) [50–56].

Four types of enzymes are known to degrade AHLs [57, 58], a phenomenon sometimes described as "quorum quenching" (QQ) [59]; these include AHL-lactonases and decarboxylases that attack the lactone ring (*Bacillus indicus, B. pumilus*, and *B.* sp. SS4 cause significant inhibition of QS-dependent activities in Gram-negative bacteria such as *P. aeruginosa* PAO1, *Serratia marcescens*, and *Vibrio*), AHL-acylases that cleave the acyl side chain (*B. pumilus* S8-07 degrades 3-oxo-C12-HSL into the corresponding lauric acid [60]), and deaminases that separate the lactone ring from the acyl side chain. Recently, lactonases and acylases were identified in *Erythrobacter, Labrenzia*, and *Bacterioplanes* found in Red Sea sediments; these both degrade AHLs of different acyl chain lengths, particularly the 3-oxo-C12-HSL, and inhibit the formation *P. aeruginosa* PAO1 biofilm [59].

Mycobacteroides abscessus subspecies, emerging pathogens, are capable of degrading both PQS and HHQ. *M. abscessus* subsp. *abscessus*, in coculture with *P. aeruginosa* PAO1, reduced PQS levels through a PQS dioxygenase (encoded by the *aqdC* gene), *M. abscessus* subsp. *massiliense*, a recombinant strain overexpressing the *aqdC* gene, reduces the level of the virulence factors pyocyanin, pyoverdine, and rhamnolipids, suggesting that AqdC is a QQ enzyme [61]. Its impact on biofilm formation would have been interesting to investigate as another dioxygenase, the 2-alkyl-3-hydroxy-4(1H)-quinolone 2,4-dioxygenase (HodC), was described to cleave PQS, attenuate the production of virulence factors but conversely increase the viable biomass, in both newly formed and established biofilms, by increasing iron availability [62].

5.1.2 Organic acids

The acetic and phenyl lactic acids, found in the supernatant of probiotic strains *Lactobacillus paracasei* subsp. *paracasei* CMGB isolated from newborn feces, were shown to inhibit, at nonbacteriostatic/bactericide levels, the expression of QS genes in *P. aeruginosa*, preventing adherence of bacteria to an inert substratum [63, 64]. Similarly, the lactic acid produced by a potential probiotic *Pediococcus acidilactici* M7 strain, also isolated from newborn feces, inhibits the production of *P. aeruginosa* short-chain AHLs, elastase, protease, pyocyanin, and biofilm as well as the swarming-swimming-twitching motilities [65].

5.2 From fungi

5.2.1 Antibiotics and mycotoxins

Penicillin produced by *Penicillium* spp. has been shown to be effective in controlling a bacterial infection. Recently, about 33 *Penicillium* spp. have been recognized as producers of QS inhibitors such as the small lactone mycotoxins patulin and penicillic acid. The use of patulin can significantly reduce lung infection caused by *P. aeruginosa* on a mouse model. Interestingly, a synergy has been shown *in vitro* between patulin and tobramycin toward *P. aeruginosa* PAO1 biofilms, whereas patulin alone does not affect the development of biofilm [66]. Although the anti-infective property of patulin has been demonstrated, its genotoxicity and potential carcinogenic properties [67] probably preclude clinical applications.

Erythromycin, a macrolide antibiotic isolated from *Saccharopolyspora erythraea*, has been recently demonstrated to reduce virulence factors in *P. aeruginosa* PAO1, including various motilities, biofilm formation, and production of rhamnolipids, total protease, elastase, and pyocyanin at nonmicrobicidal level (1.6 μg/mL) [68]. Comparably, the erythromycin derivate, azithromycin, shows a strong *P. aeruginosa* QS and biofilm inhibitory effect [69–71] with inhibition of alginate synthesis [69], a reduction of each type of bacteria movement [72] and a diminution of *gacA* gene expression [73]. At weak antibiotic concentration (2 μg/mL), a biofilm inhibition is observed, probably explained by a lower production of both AHL signal molecules, C4-HSL and 3-oxo-C12-HSL, and of virulence factors [74–76].

5.2.2 Alkylcyclopentanone

Recently, Kim et al. [77] indicated that the alkylcyclopentanone terrein, isolated from *Aspergillus terreus*, reduced virulence factors (elastase, pyocyanin, and rhamnolipids) and biofilm formation via antagonizing QS receptors without affecting *P. aeruginosa* cell growth. Beyond a negative impact on the production of QS signaling molecules and expression of QS-related genes, terrein also reduced c-di-GMP levels, an important secondary messenger for the switch from planktonic to biofilm lifestyle mode, by decreasing the activity of a diguanylate cyclase required for c-di-GMP biosynthesis [78].

5.3 From Plants

5.3.1 Derivatives of shikimic acid, phenols, and polyphenols

Many phenolic compounds and derivatives with anti-QS and antibiofilm activities have been isolated from plants [79]. Cinnamaldehyde [the dominant compound of certain essential oils, in particular *Cinnamomum camphora* (L.) J. Presl] and its derivatives modulate a wide range of anti-QS and antibiofilm activities of *P. aeruginosa* [80–82]. *Curcuma longa* L. produces curcumin, which inhibits the expression of virulence genes of *P. aeruginosa* PA01 [83].

Ellagic acid derivatives from *Terminalia chebula* Retz. downregulate *lasIR* and *rhlIR* genes expression and decrease AHLs production, leading to an attenuation of virulence factor production and to an enhanced sensitivity of biofilm facing a tobramycin treatment [84].

Flavonoids have been investigated for their roles as QS modulating compounds. From these, naringenin and taxifolin reduced the expression of several QS-controlled genes (i.e., *lasI, lasR, rhlI, rhlR, lasA, lasB, phzA1,* and *rhlA*) in *P. aeruginosa* PAO1. Similarly, the flavan-3-ol catechin, extracted from the bark of *Combretum albiflorum* (Tul.) Jongkind, reduces the production of QS-dependent virulence factors, such as pyocyanin, elastase, and the formation of biofilm by *P. aeruginosa* PAO1 [85]. Interestingly, baicalin, an active natural compound extracted from the traditional Chinese medicinal *Scutellaria baicalensis*, has been demonstrated to inhibit the formation of *P. aeruginosa* biofilms and enhance the bactericidal effects of antibiotics such as amikacin. Moreover, at sub-minimal inhibitory concentration (256 μg/mL), this flavonoid has been shown to reduce LasA protease, LasB elastase, pyocyanin, rhamnolipids, and exotoxin A production and to downregulate the three QS-regulatory genes, including *lasI, lasR, rhlI, rhlR, pqsR*, and *pqsA* [86]. Consistently, *in vivo* experiments indicated that baicalin treatment reduces *P. aeruginosa* pathogenicity in *Caenorhabditis elegans* and enhances the clearance of *P. aeruginosa* from the peritoneal implants of infected mice.

Furocoumarins from grapefruit can inhibit the QS signaling (AHLs and AI-2) of *V. harveyi* BB886 and BB170 strains as well as biofilm formation in pathogens such as *E. coli* O157:H7, *Salmonella typhimurium* and *P. aeruginosa* [87]. These purified furocoumarins (dihydroxybergamottin and bergamottin), tested at the concentration of 1 μg/mL, cause 94% inhibition of autoinducers (AHLs) without affecting bacterial viability. Biofilm inhibition was up to 58.3 and 72%, respectively, for *E. coli* O157:H7 but modest for *P. aeruginosa* (27.3 and 18.1%, respectively).

Malabaricone C, a diarylnonanoid isolated from the bark of *Myristica cinnamomea* King inhibited the QS-regulated pyocyanin production and biofilm formation in *P. aeruginosa* PAO1 [88].

A screening of various herbs revealed that a clove extract [*Syzygium aromaticum* (L.) Merr. Et Perry] inhibits QS-controlled gene expression (*las* and PQS systems) in *P. aeruginosa* with eugenol as major active constituent [89]. Recently, the effects of eugenol and its nanoemulsion on *P. aeruginosa* QS-mediated virulence factors and biofilm formation have been identified by Lou et al. [90] at a 0.2 mg/mL concentration. Similarly, the anthraquinone emodin from *Rheum palmatum* L., a traditional Chinese medicinal plant, was found to inhibit the *P. aeruginosa* biofilm formation at 20 μM, increasing the antibiotic activity of ampicillin [91]. Finally, the 6-gingerol, isolated from fresh ginger oil, reduces the production of several virulence factors, decreasing the mortality induced in mice by *P. aeruginosa*. A DNA microarray analysis revealed that the application of the 6-gingerol on biofilm-encapsulated cells down-regulates several QS-related genes, notably those involved in the production of rhamnolipids, elastase, pyocyanin, all of which are involved in biofilm formation [92].

5.3.2 Alkaloids

Recently, caffeine (a purine alkaloid) has been shown to inhibit AHLs production and swarming mobility in *P. aeruginosa* PAO1 without causing AHLs degradation [93].

5.3.3 Terpenoids and Triterpenoids

The pentacyclic triterpenoid ursolic acid was identified as an inhibitor of biofilm formation from *Diospyros dendo* Welw, the tree used for ebony from Gabon, Africa [94]. Tested at a dose of 10 μg/mL, ursolic acid reduces biofilm formation by 79% in *E. coli* and 57–95% in *V. harveyi* and *P. aeruginosa* PAO1. Similarly, oleanolic acid inhibits the *in vitro* biofilm formation by *S. aureus* and *P. aeruginosa* [95]. However, these triterpenoids showed no inhibitory effect on QS mechanisms contrarily to triterpenoid coumarate esters isolated from *Dalbergia trichocarpa*, a tropical legume from Madagascar. Indeed, oleanolic aldehyde coumarate at 200 μM inhibits the formation/maintenance of *P. aeruginosa* PAO1 biofilm and the expression of the *las* and *rhl* QS systems as well as *gacA* gene [96]. Consequently, the production of QS-controlled virulence factors, including, rhamnolipids, pyocyanin, elastase, and extracellular polysaccharides, as well as twitching and swarming motilities is reduced. Other African plants harbor terpenoids and triterpenoids with antivirulence properties. Indeed, cassipourol and β-sitosterol (both at 100 μM), isolated from *Platostoma rotundifolium* (Briq.) A. J. Paton, a Burundian medicinal plant, inhibit quorum sensing-regulated and -regulatory gene expression in *las* and *rhl* systems. These triterpenoids can still disrupt the formation of biofilms at concentrations down to 12.5 and 50 μM [97].

5.3.4 Isothiocyanates and organosulfur compounds

Isothiocyanates produced by many plants are also QS inhibitors in *P. aeruginosa* PAO1. For example, iberin, isolated from horseradish (*Armoracia rusticana* G. Gaertn et al.), specifically blocks the expression of QS-regulated genes in *P. aeruginosa* PAO1 at the concentration of 100 μM; its impact on biofilm formation has not been investigated [98]. Sulforaphane and erucin, two isothiocyanates isolated from broccoli, inhibit the *P. aeruginosa* PAO1 *las* and *rhl* system as well as biofilm formation at concentrations of 50 and 100 μM, respectively [99].

A further compound known to affect the QS-regulated genes in *P. aeruginosa*, including the rhamnolipids production, is ajoene, an allyl sulfide isolated from *Allium sativum* L. Ajoene, at the concentration of 100 μg/mL and combined with the antibiotic tobramycin, leads to killing of biofilm-encapsulated *P. aeruginosa*. In a mouse model of pulmonary infection, this synergy improves the clearance of *P. aeruginosa* from lungs [100]. The S-phenyl-L-cysteine sulfoxide and its derivatives, notably diphenyl disulfide, have shown a significant impact on the biofilm formation by *P. aeruginosa* [101]; the sulfoxide derivative seems to interfere with both *las* and *rhl* systems whereas the diphenyl sulfide only disturbs the *las* system.

5.4 From marine organisms

5.4.1 Furanones

A series of studies have indicated that marine organisms are a potential source of anti-QS [102–104]. The halogenated furanones produced by the red alga *Delisea pulchra* inhibit QS-induced activities in bacteria by competing with AHL signals related to their receptor site (LuxR) [104]. This protein-ligand binding is destabilized, causing rapid receptor recycling [102]. Inspired from natural compounds, the halogenated furanones C-30 and C-56 have been demonstrated to exhibit biofilm reduction and target the *las* and *rhl* systems in *P. aeruginosa* [105].

5.4.2 Terpenoids

Following a screening of 284 extracts from the marine sponge *Luffariella variabilis*, 36 extracts were revealed as inhibitors of *P. aeruginosa* QS, targeting the *las* system [103]; from these, the sesterterpenoids manoalide displays anti-biofilm activities. Note that this molecule does not generate bactericidal effects on *P. aeruginosa* [103], but presents an antibiotic activity against Gram-positive bacteria [106].

5.5 From animals and human

5.5.1 Enzymes

Type I porcine kidney acylase inactivates QS signals such as C6-HSL and 3-oxo-C12-HSL but not C4-HSL [50]. This type I acylase moderately reduces biofilm formation in *Aeromonas hydrophila*, *P. putida*, and probably *P. aeruginosa* [107]. This degradation is dependent on the length of the acyl chain, since only C6-HSL and 3-oxo-C12-HSL are degraded [108].

Mammalian cells release enzymes called paraoxonases 1 (extracted from human and murine sera) that have lactonase activity; degrading *P. aeruginosa* AHLs. They prevent, in an indirect way, QS and biofilm formation [109]. Similarly, human epithelial cells and particularly human respiratory epithelia have the capacity to inactivate a *P. aeruginosa* QS signal by inactivating AHLs (3-oxo-C12HSL) produced by *P. aeruginosa* [108, 110]. However, the enzyme or enzyme-like compound involved in acyl-homoserine lactone inactivation have not been identified and characterized yet. Recently, Losa et al. [111] demonstrated that polarized airway epithelial monolayers, in contrast to nonpolarized cells, are also able to degrade 3-oxo-C12-HSL using membrane-associated paraoxonase 2 that catalyzes the opening of the lactone ring.

5.5.2 Alkaloids

The *P. aeruginosa* pyocyanin production is inhibited by a molecule found and isolated from the ant *Solenopsis invicta*, the piperidine alkaloid Solenopsin A alkaloid. The biofilm formation is also reduced in a dose-dependent manner. This molecule probably disrupts the signals from the *rhl* system [112].

6. Concluding remarks

This review presents natural compounds reported to exhibit anti-QS and antibiofilm properties against *P. aeruginosa* (summarized in **Table 1**); these highlight the great potentiality of living organisms as reservoir of compounds susceptible to modulate virulence mechanisms without affecting bacterial viability. Overall, it appears that prokaryotes as well as animals and humans are sources for enzymes that degrade or antagonize AHLs, whereas plants harbor larger panels of anti-QS and antibiofilm compounds with very diverse chemical structures, including alkaloids, organosulfurs, phenolics, and terpenoids. Contrarily to animals and humans, plants are not able to deploy elaborate defense through humoral and cell-mediated immunity (antibodies and phagocytes) to struggle against bacterial invasions [113]. Plants immune defenses rely on the secretion of antibacterial compounds (bactericide and/or bacteriostatic agents [114]), including resistance modulating compounds [115] (e.g., inhibitors of efflux pumps [116]), and mostly on their ability to recognize molecules released from pathogens through plant cell surface receptors. This recognition triggers specific signaling cascades, activating series of defense responses, including the synthesis of antimicrobial lytic proteins, enzymes, phytoalexins, and other secondary metabolites. Some of these exert nonmicrobicidal antivirulence properties [117, 118]. Finally, marine organisms and fungi produce also bioactive secondary metabolites (halogenated furanones and antibiotics, respectively) and other original and promising compounds, such as terrein which was identified as the first dual inhibitor of QS and c-di-GMP signaling at 30 μM.

The increasing presence of antibiotic-resistant bacteria certainly pushes scientists to reorient the strategy of fight against bacterial infections to defer entry into a post-antibiotic era where major antibiotics would not be effective even for banal infections. Antivirulence approaches and antivirulence drugs are being increasingly considered as potential therapeutic alternatives and/or adjuvants to currently failing antibiotics. For example, oleanolic aldehyde coumarate and cassipourol, anti-QS compounds, exert interesting antibiofilm properties, restoring the effectiveness of the antibiotic tobramycin in the clearance of biofilm-encapsulated *P. aeruginosa* (**Figure 4**); also the association between biofilm formation and antimicrobial resistance has been highlighted in carbapenem-resistant *P. aeruginosa* [119]. Such nonmicrobicidal drugs inhibit virulence factors essential for establishing infection and pathogenesis through targeting nonessential metabolic pathways which should not lead to activation of bacterial evasion mechanisms. This approach should reduce the selective pressure and consequently could slow down the development of resistance. Compounds that target QS may be particularly interesting as they impact planktonic and biofilm lifestyles, by reducing at the same time the production of virulence factors and the generation of biofilms. This should lead to less severe infections at levels that can be cleared by the host's immune defense and with increased activity of antibiotics.

Despite these important prospects, however, the big breakthrough in antibacterial strategies is still out of reach. This is probably due to a very complex

	Origin	Compounds (class)	Target (QS)	Synergy with antibiotics
Prokaryotes	*Bacillus indicus, B. pumilus, B.* sp. [60]; *Erythrobacter, Labrenzia, Bacterioplanes* [59]	AHL-acylase (Enzyme)	AHL degradation	NC
		AHL-lactonase (Enzyme)		NC
	Lactobacillus paracasei subsp. Paracasei [64]; *Pediococcus acidilactici* M7 [65]	Acetic acid, lactic acid, phenyl lactic acid	AHL antagonist	NC
Fungi	*Penicillium* species [66]	Penicillic acid (Furanone)	LasR and RhlR	NC
		Patulin (Furopyranone)	LasR and RhlR‡	+[1]
	Saccharopolyspora erythraea [68]	Erythromycin (Macrolide)	*rhl system* and *GacA*	NC
	Aspergillus terreus [77]	Terrein (alkylcyclopentanone)	LasR and RhlR antagonist; c-di-GMP	NC
marine organisms	*Delisea pulchra* [102, 104]	halogenated furanones and derivative	AHL antagonist	+[1]
	Luffariella variabilis (Polejaeff, 1884) [103]	Manoalide (Sesterterpenoid)	*las* system	NC
Plants	*Platostoma rotundifolium* (Briq,) A, J, Paton [97]	Cassipourol (terpenoid), β-sitosterol (terpenoid)	*las* and *rhl* systems	+[1]
	Combretum albiflorum (Tul.) Jongkind [85]	Catechin (Flavonoid)	*las* and *rhl* systems	NC
	Dalbergia trichocarpa Baker. [96]	Oleanolic aldehyde Coumarate (Phenolic compound)	*las* and *rhl* systems	+[1]
	Allium sativum L. [100]	Ajoene (Organosulfur)	*las* and *rhl* systems	+[1]
	Armoracia rusticana G. Gaertn et al. [98]	Iberin (Isothiocyanate)	*las* and *rhl* systems	NC
	Terminalia chebula Retz. [84]	Ellagic acid derivatives (Phenolic compound)	*las* and *rhl* systems	+[1]
	Syzygium aromaticum (L.) Merr. Et Perry [89, 90]	Eugenol (Phenylpropanoid)	*las* and PQS systems	NC
	Curcuma longa L. [83]	Curcumin (Phenolic compound)	AHLs inhibition	NC
	Citrus paradisi Macfad. (Rio Red and Marsh White grapefruits) [87]	Bergamottin and dihydroxybergamottin (Furocoumarins)	AHLs inhibition	NC
	Rheum palmatum L. [91]	Emodin (Anthraquinone)	docking traR*	+[2]
	Scutellaria baicalensis Georgi. [86]	Baicalin (Flavonoid)	*las*, *rhl* and PQS systems	+[1]
	Zingiber officinale Rosc. [92]	6-gingerol (Phenolic compound)	docking lasR	NC

	Origin	Compounds (class)	Target (QS)	Synergy with antibiotics
Animals and Human	Porcine kidney [50, 107]	Type I acylase	AHL degradation	NC
	Human and murine sera [109, 110]	Paraoxonases 1 Enzyme (lactonase)	AHL degradation	NC
	Solenopsis invicta (insect; ant) [112]	Solenopsin A (Alkaloid)	*rhl* system	NC

+, yes; NC, not communicated.
†Patulin alone does not affect the development of biofilm.
**LuxR-type transcription factor of Agrobacterium tumefaciens.*
[1]Aminoglycosides.
[2]Ampicillin.

Table 1.
Natural compounds inhibiting P. aeruginosa *QS and biofilm formation.*

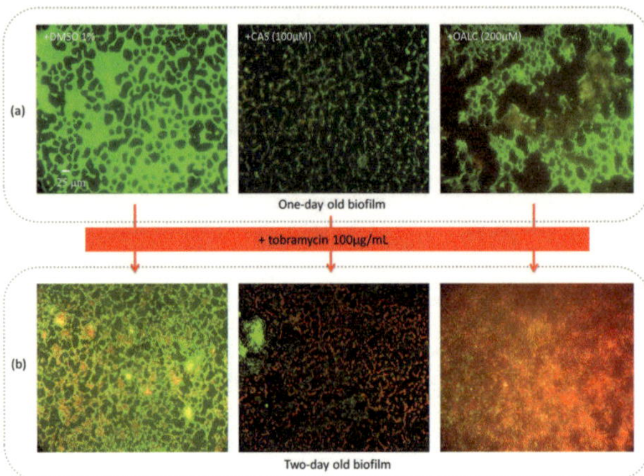

Figure 4.
P. aeruginosa *biofilm phenotypes and effectiveness of tobramycin treatment in presence of dimethyl sulfoxide (DMSO 1%) or, cassipourol (CAS: 100 μM) or oleanolic aldehyde coumarate (OALC: 200 μM). (a) After 1 day of incubation,* P. aeruginosa *fails to form structured confluent aggregate in presence of CAS or OALC as compared to DMSO treatment. (b) CAS and OALC considerably increase the susceptibility of* P. aeruginosa *to tobramycin (100 μg/mL), as shown by the increased proportion of dead cells compared with DMSO. Similar results are observed when tobramycin is added simultaneously with CAS or OALC to one-day old untreated biofilms. The bacterial viability was assessed by staining the cells with SYTO-9 (green areas zones—live living bacteria) and propidium iodide (red areas zones—dead bacteria) furnished in the LIVE/DEAD BacLight kit. Cells were visualized using a LeicaDMIRE2 inverted fluorescence microscope using equipped with a 40× objective lens and colored images were assembled using Adobe Photoshop.*

entanglement between different QS systems, to the ability of *Pseudomonas* to compensate deficient systems and to the intervention of key actors involved in biofilm formation, outside of QS circuitry [12]. Millenia of coevolution between plants and bacteria have led to complex defense strategies, with plants producing cocktails of bioactive compounds with multiple targets [114] and/or compounds such as terrein that impact dual inhibitory targets. In the current state of research, much remains to be done in understanding these mechanisms and the real impact of such combinations before arriving at a commercial use. Nevertheless, following a combined approach for "adjuvant antibiotherapy" and "combined antibiotherapy" will undeniably lead to a renewed concept of "complex drugs for complex diseases," a well-known presupposed in traditional medicines [120].

Acknowledgements

The authors would like to thank ARES (Académie de Recherche et d'Enseignement Supérieur, Belgium) for financial support throughout PRD projects.

Conflict of interest

The authors declare that there is no conflict of interests regarding the publication of this paper.

Author details

Julie Carette[1], Amandine Nachtergael[1], Pierre Duez[1], Mondher El Jaziri[2] and Tsiry Rasamiravaka[3]*

1 University of Mons, Mons, Belgium

2 Université Libre de Bruxelles, Brussels, Belgium

3 University of Antananarivo, Antananarivo, Madagascar

*Address all correspondence to: travaka@yahoo.fr

References

[1] Wu H-J, Wang AHJ, Jennings MP. Discovery of virulence factors of pathogenic bacteria. Current Opinion in Chemical Biology. 2008;**12**:93-101

[2] Ciofu O, Tolker-Nielsen T. Tolerance and resistance of Pseudomonas aeruginosa biofilms to antimicrobial agents-how P. aeruginosa can escape antibiotics. Frontiers in Microbiology. 2019;**10**:913

[3] Pashang R, Yusuf F, Zhao S, Deljoomanesh S, Gilbride KA. Widespread detection of antibiotic-resistant bacteria from natural aquatic environments in southern Ontario. Canadian Journal of Microbiology. 2018;**65**(4):322-331

[4] Torres-Barceló C, Hochberg ME. Evolutionary rationale for Phages as complements of antibiotics. Trends in Microbiology. 2016;**24**(4):249-256. DOI: 10.1016/j.tim.2015.12.011

[5] Valdivieso-Ugarte M, Gomez-Llorente C, Plaza-Díaz J, Gil Á. Antimicrobial, antioxidant, and Immunomodulatory properties of essential oils: A systematic review. Nutrients. 2019;**11**(11):2786. DOI: 10.3390/nu11112786

[6] Rai M, Paralikar P, Jogee P, Agarkar G, Ingle AP, Derita M, et al. Synergistic antimicrobial potential of essential oils in combination with nanoparticles: Emerging trends and future perspectives. International Journal of Pharmaceutics. 2017;**519** (1-2):67-78. DOI: 10.1016/j.ijpharm.2017.01.013

[7] Cotter PD, Ross RP, Hill C. Bacteriocins- a viable alternative to antibiotics? Nature Reviews Microbiology. 2013;**11**(2):95-105. DOI: 10.1038/nrmicro2937

[8] Nuti R, Goud NS, Saraswati AP, Alvala R, Alvala M. Antimicrobial peptides: A Promissing therapeutic strategy in tackling antimicrobial resistance. Current Medicinal Chemistry. 2017;**24**(38):4303-4314. DOI: 10.2174/0929867324666170815102441

[9] Rutherford ST, Bassler BL. Bacterial quorum sensing: Its role in virulence and possibilities for its control. Cold Spring Harb, Perspect, Med. 2012;**2**:a012427

[10] Turkina MV, Vikström E. Bacteria-host crosstalk: Sensing of the quorum in the context of Pseudomonas aeruginosa infections. Journal of Innate Immunity. 2019;**11**(3):263-279

[11] Bricha S, Ounine K, Oulkheir S, Haloui N, Attarassi B. Virulence factors and epidemiology related to Pseudomonas aeruginosa. Tunisian Journal of Infectious Diseases. 2009;**2**:7-14

[12] Rasamiravaka T, Labtani Q, Duez P, El Jaziri M. The formation of biofilms by Pseudomonas aeruginosa: A review of the natural and synthetic compounds interfering with control mechanisms. BioMed Research International 2015; 2015: a759348

[13] Filloux A, Vallet I. [biofilm: Set-up and organization of a bacterial community] article in french. Médecine Sciences. 2003;**19**:77-83. DOI: 10,1051/medsci/200319177

[14] Vu B, Chen M, Crawford RJ, Ivanova EP. Bacterial extracellular polysaccharides involved in biofilm formation. Molecules. 2009;**14**(7):2535-2554. DOI: 10.3390/molecules14072535

[15] Sauer K. The genomics and proteomics of biofilm formation. Genome Biology. 2003;**4**:219. DOI: 10.1186/gb-2003-4-6-219

[16] Häussler S, Parsek MR. Biofilms 2009: New perspectives at the heart of surface-associated microbial communities. Journal of Bacteriology. 2010;**192**(12):2941-2949. DOI: 10.1128/JB.00332-10

[17] Ghafoor A, Hay ID, Rehm BHA. Role of exopolysaccharides in Pseudomonas aeruginosa biofilm formation and architecture. Applied and Environmental Microbiology. 2011;**77**(15):5238-5246

[18] Evans LR, Production LA. Characterization of the slime polysaccharide of Pseudomonas aeruginosa. Journal of Bacteriology. 1973;**116**(2):915-924

[19] Hentzer M, Wu H, Andersen JB. Attenuation of Pseudomonas aeruginosa virulence by quorum sensing inhibitors. EMBO Journal. 2003;**22**(15):3803-3815

[20] Wozniak DJ, Wyckoff TJO, Starkey M, Keyser R, Azadi P, O'Toole GA, et al. Alginate is not a significant component of the extracellular polysaccharide matrix of PA14 and PAO1 Pseudomonas aeruginosa biofilms. Proceedings of the National Academy of Sciences of the United States of America. 2003;**100**(13):7907-7912

[21] Stapper AP, Narasimhan G, Ohman DE, Barakat J, Hentzer M, Molin S, et al. Alginate production affects Pseudomonas aeruginosa biofilm development and architecture, but is not essential for biofilm formation. Journal of Medical Microbiology. 2004;**53**(7):679-690

[22] Leid JG, Willson CJ, Shirtliff ME, Hassett DJ, Parsek MR, Jeffers AF. The exopolysaccharide alginate protects Pseudomonas aeruginosa biofilm Bacteria from IFN-γ-mediated macrophage killing. Journal of Immunology. 2005;**175**(11):7512-7518

[23] Friedman L, Kolter R. Genes involved in matrix formation in Pseudomonas aeruginosa PA14 biofilms. Molecular Microbiology. 2004;**51**(3):675-690

[24] Holloway BW, Krishnapillai V, Morgan AF. Chromosomal genetics of pseudomonas. Microbiological Reviews. 1975;**43**(1):73-102

[25] Yang L, Liu Y, Wu H. Combating biofilms. FEMS Immunology & Medical Microbiology. 2012;**65**(2):146-157

[26] Okshevsky M, Meyer RL. The role of extracellular DNA in the establishment, maintenance and perpetuation of bacterial biofilms. Critical Reviews in Microbiology. 2015;**41**(3):341-352

[27] Swartjes JJTM, Das T, Sharifi S. A functional DNase coating to prevent adhesion of bacteria and the formation of biofilm. Advanced Functional Materials. 2013;**23**(22):2843-2849

[28] Yang L, Barken KB, Skindersoe ME, Christensen AB, Givskov M, Tolker-Nielsen T. Effects of iron on DNA release and biofilm development by Pseudomonas aeruginosa. Microbiology. 2007;**153**(5):1318-1328

[29] Gloag ES, Turnbull L, Huang A. Self-organization of bacterial biofilms is facilitated by extracellular DNA. Proceedings of the National Academy of Sciences of the United States of America. 2013;**110**(28):11541-11546

[30] Mulcahy H, Charron-Mazenod L, Lewenza S. Extracellular DNA chelates cations and induces antibiotic resistance in Pseudomonas aeruginosa biofilms. PLoS Pathogens. 2008;**4**(11):e1000213

[31] Jimenez PN, Koch G, Thompson JA, Xavier KB, Cool RH, Quax WJ. The multiple signaling systems regulating

virulence in Pseudomonas aeruginosa. Microbiology and Molecular Biology Reviews. 2012;**76**(1):46-65

[32] Davies DG, Parsek MR, Pearson JP, Iglewski BH, Costerton JW, Greenberg EP. The involvement of cell-to-cell signals in the development of a bacterial biofilm. Science. 1998;**280**(5361):295-298

[33] Gilbert KB, Kim TH, Gupta R, Greenberg EP, Schuster M. Global position analysis of the Pseudomonas aeruginosa quorum-sensing transcription factor LasR. Molecular Microbiology. 2009;**73**(6):1072-1085

[34] Sakuragi Y, Kolter R. Quorum-sensing regulation of the biofilm matrix genes (pel) of Pseudomonas aeruginosa. Journal of Bacteriology. 2007;**189**(14):5383-5386

[35] Davey ME, Caiazza NC, O'Toole GA. Rhamnolipid surfactant production affects biofilm architecture in Pseudomonas aeruginosa PAO1. Journal of Bacteriology. 2003;**185**(3):1027-1036

[36] Pamp SJ, Tolker-Nielsen T. Multiple roles of biosurfactants in structural biofilm development by Pseudomonas aeruginosa. Journal of Bacteriology. 2007;**189**(6):2531-2539

[37] Dusane DH, Zinjarde SS, Venugopalan VP, Mclean RJC, Weber MM, Rahman PKSM. Quorum sensing: Implications on Rhamnolipid biosurfactant production. Biotechnology and Genetic Engineering Reviews. 2010;**27**:159-184

[38] O'Toole GA, Kolter R. Flagellar and twitching motility are necessary for Pseudomonas aeruginosa biofilm development. Molecular Microbiology. 1998;**30**(2):295-304

[39] Boles BR, Thoendel M, Singh PK. Rhamnolipids mediate detachment of Pseudomonas aeruginosa from biofilms. Molecular Microbiology. 2005;**57**(5):1210-1223

[40] Schooling SR, Charaf UK, Allison DG, Gilbert P. A role for rhamnolipid in biofilm dispersion. Biofilms. 2004;**1**:91-99

[41] Diggle SP, Stacey RE, Dodd C, Camara M, Williams P, Winzer K. The galactophilic lectin, LecA, contributes to biofilm development in Pseudomonas aeruginosa. Environmental Microbiology. 2006;**8**(6):1095-1104

[42] Tielker D, Hacker S, Loris R. Pseudomonas aeruginosa lectin LecB is located in the outer membrane and is involved in biofilm formation. Microbiology. 2005;**151**(5):1313-1323

[43] Fraser GM, Hughes C. Swarming motility. Current Opinion in Microbiology. 1999;**2**(6):630-635

[44] Daniels R, Vanderleyden J, Michiels J. Quorum sensing and swarming migration in bacteria. FEMS Microbiology Reviews. 2004;**28**(3):261-289

[45] Shrout JD, Chopp DL, Just CL, Hentzer M, Givskov M, Parsek MR. The impact of quorum sensing and swarming motility on Pseudomonas aeruginosa biofilm formation is nutritionally conditional. Molecular Microbiology. 2006;**62**(5):1264-1277

[46] Mattick JS. Type IV pili and twitching motility. Annual Review of Microbiology. 2002;**56**:289-314

[47] Patriquin GM, Banin E, Gilmour C, Tuchman R, Greenberg EP, Poole K. Influence of quorum sensing and iron on twitching motility and biofilm formation in Pseudomonas aeruginosa. Journal of Bacteriology. 2008;**190**(2):662-671

[48] Parkins MD, Ceri H, Storey DG. Pseudomonas aeruginosa GacA, a factor in multihost virulence, is also essential for biofilm formation. Molecular Microbiology. 2001;**40**(5):1215-1226

[49] Merighi M, Lee VT, Hyodo M, Hayakawa J, Lory S. The second messenger bis-(3′-5′)-cyclic-GMP and its PilZ domain-containing receptor Alg44 are required for alginate biosynthesis in Pseudomonas aeruginosa. Molecular Microbiology. 2007;**65**(4):876-895

[50] Dong YH, Zhang LH. Quorum sensing and quorum-quenching enzymes. The Journal of Microbiology. 2005;**43**(Spec):101-109

[51] Huma N, Pratap S, Jyoti K, Ashish B, Jayadev J, Tanmoy M, et al. Diversity and polymorphism in AHL-Lactonase gene (aiiA) of bacillus. Journal of Microbiology and Biotechnology. 2011;**21**(10):1001-1011

[52] Kalia VC, Purohit HJ. Quenching the quorum sensing system: Potential antibacterial drug targets. Critical Reviews in Microbiology. 2011;**37**(2):121-140

[53] Kang Y, Durfee T, Glasner JD, Qiu Y, Frisch D, Winterberg KM, et al. Systematic mutagenesis of the Escherichia coli genome. Journal of Bacteriology. 2004;**186**:4921-4930

[54] Park JK, Jung JY, Park YH. Cellulose production by Gluconacetobacter hansenii in a medium containing ethanol. Biotechnology Letters. 2003;**25**(24):2055-2059

[55] Romero D, Aguilar C, Losick R, Kolter R. Amyloid fibers provide structural integrity to Bacillus subtilis biofilms. Proceedings of the National Academy of Sciences of the United States of America. 2010;**107**:2230-2234

[56] Uroz S, Dessaux Y, Oger P. Quorum sensing and quorum quenching : The Yin and Yang of bacterial communication. Chembiochem. 2009;**10**(2):205-216

[57] Musthafa KS, Ravi AV, Annapoorani A, Packiavathy ISV, Pandian SK. Evaluation of anti-quorum-sensing activity of edible plants and fruits through inhibition of theN-acyl-homoserine lactone system in Chromobacterium violaceum and Pseudomonas aeruginosa. Chemotherapy. 2010;**56**:333-339

[58] Nithya C, Pandian SK. The in vitro antibiofilm activity of selected marine bacterial culture supernatants against vibrio spp. Archives of Microbiology. 2010;**192**:10

[59] Rehman ZU, Leiknes T. Quorum quenching bacteria isolated from Red Sea sediments reduces biofilm formation by Pseudomonas aeruginosa. Frontiers in Microbiology. 2018;**9**:1354

[60] Nithya C, Begum MF, Pandian SK. Marine bacterial isolates inhibit biofilm formation and disrupt mature biofilms of Pseudomonas aeruginosa PAO1. Applied Microbiology and Biotechnology. 2010;**88**(1):341-358

[61] Birmes FS, Säring R, Hauke MC, Ritzmann NH, Drees SL, Daniel J, et al. Interference with Pseudomonas aeruginosa quorum sensing and virulence by the mycobacterial pseudomonas quinolone signal Dioxygenase AqdC in combination with the N-Acylhomoserine lactone Lactonase QsdA. Infection and Immunity. 2018;**87**(10):e00278-e00219

[62] Tettmann B, Niewerth C, Kirschhöfer F, Neidig A, Dötsch A, Brenner-Weiss G, et al. Enzyme-mediated quenching of the pseudomonas quinolone signal (PQS) promotes biofilm formation of Pseudomonas aeruginosa by increasing iron

availability. Frontiers in Microbiology. 2016;**7**:1978

[63] Cotar A, Chifiriuc MC, Dinu S, Pelinescu D, Banu O. Quantitative real-time PCR study of the influence of probiotic culture soluble fraction on the expression of Pseudomonas aeruginosa quorum sensing genes. Roumanian Archives of Microbiology and Immunology. 2010;**69**:213-223

[64] Chifiriuc MC, Ditu ML, Banu O, Bleotu C, Dracea O. Subinhibitory concentrations of phenyl lactic acid interfere with the expression of virulence factors in Staphylococcus aureus and Pseudomonas aeruginosa clinical strains. Roumanian Archives of Microbiology and Immunology. 2009;**68**:27-33

[65] Kiymaci ME, Altanlar N, Gumustas M, Ozkan SA, Akin A. Quorum sensing signals and related virulence inhibition of Pseudomonas aeruginosa by a potential probiotic strain's organic acid. Microbial Pathogenesis. 2018;**121**:190-197

[66] Rasmussen TB, Skindersoe ME, Bjarnsholt T. Identity and effects of quorum-sensing inhibitors produced by Penicillium species. Microbiology. 2005;**151**(5):1325-1340

[67] Glasser N. Patulin: Mechanism of genotoxicity. Food and Chemical Toxicology. 2012;**50**(5):1796-1801. DOI: 10.1016/j.fct.2012.02.096

[68] Shusaku S, Yoko M, Katsumi F, Nobuhiko F. Effects of long- term, low-dose macrolide treatment on Pseudomonas aeruginosa PAO1 virulence factors In vitro. Archives in Clinical Microbiology. 2017;**8**(4):50. DOI: 10.21767/1989-8436.100050

[69] Skindersoe ME, Alhede M, Phipps R. Effects of antibiotics on quorum sensing in Pseudomonas aeruginosa. Antimicrobial Agents and Chemotherapy. 2008;**52**(10):3648-3663

[70] Tateda K, Comte R, Pechere J-C, Köhler T, Yamaguchi K, van Delden C. Azithromycin inhibits quorum sensing in Pseudomonas aeruginosa. Antimicrobial Agents and Chemotherapy. 2001;**45**(6):1930-1933

[71] Ichimiya T, Takeoka K, Hiramatsu K, Hirai K, Yamasaki T, Nasu M. The influence of azithromycin on the biofilm formation of Pseudomonas aeruginosa in vitro. Chemotherapy. 1996;**42**(3):186-191

[72] Bala A, Kumar R, Harjai K. Inhibition of quorum sensing in Pseudomonas aeruginosa by azithromycin and its effectiveness in urinary tract infections. Journal of Medical Microbiology. 2011;**60**(3):300-306

[73] Pérez-Martìnez I, Haas D. Azithromycin inhibits expression of the GacA-dependent small RNAs RsmY and RsmZ in Pseudomonas aeruginosa. Antimicrobial Agents and Chemotherapy. 2011;**55**(7):3399-3405

[74] Pechère J-C. Azithromycin reduces the production of virulence factors in Pseudomonas aeruginosa by inhibiting quorum sensing. Japanese Journal of Antibiotics. 2001;**54**:87-89

[75] Sofer D, Gilboa-Garber N, Belz A, Garber NC. 'Subinhibitory' erythromycin represses production of Pseudomonas aeruginosa lectins, autoinducer and virulence factors. Chemotherapy. 1999;**45**(5):335-341

[76] Favre-Bonté S, Köhler T, van Delden C. Biofilm formation by Pseudomonas aeruginosa: Role of the C4-HSL cell-to-cell signal and inhibition by azithromycin. Journal of Antimicrobial Chemotherapy. 2003;**52**(4):598-604

[77] Kim B, Park J-S, Choi HY, Yoon SS, Kim WG. Terrein is an inhibitor of quorum sensing and c-di-GMP in Pseudomonas aeruginosa: A connection between quorum sensing and c-di-GMP. Scientific Reports. 2018;**8**(1):8617

[78] Merighi M, Lee VT, Hyodo M, Hayakawa Y, Lory S. The second messenger bis-(3′-5′)-cyclic-GMP and its PilZ domain containing receptorAlg44 are required for alginate biosynthesis in Pseudomonas aeruginosa. Molecular Microbiology. 2007;**65**(4):876-895

[79] Silva LN, Zimmer KR, Macedo AJ, Trentin DS. Plant natural products targeting bacterial virulence factors. Chemical Reviews. 2016;**116**:9162-9236

[80] Brackman G, Defoirdt T, Miyamoto C, Bossier P, Van Calenbergh S, Nelis H, et al. Cinnamaldehyde and cinnamaldehyde derivatives reduce virulence in vibrio spp, by decreasing the DNA-binding activity of the quorum sensing response regulator LuxR. BMC Microbiology. 2008;**8**:149. DOI: 10.1186/1471-2180-8-149

[81] Niu C, Gilbert ES. Colorimetric method for identifying plant essential oil components that affect biofilm formation and structure. Applied and Environmental Microbiology. 2004;**70**(12):6951-6956

[82] Niu C, Afre S, Gilbert ES. Subinhibitory concentrations of cinnamaldehyde interfere with quorum sensing. Letters in Applied Microbiology. 2006;**43**(5):489-494

[83] Rudrappa T, Bais HP. Curcumin, a known phenolic from Curcuma longa, attenuates the virulence of Pseudomonas aeruginosa PAO1 in whole plant and animal pathogenicity models. Journal of Agricultural and Food Chemistry. 2008;**56**(6):1955-1962

[84] Sarabhai S, Sharma P, Capalash N. Ellagic acid derivatives from Terminalia chebula Retz, Downregulate the expression of quorum sensing genes to attenuate Pseudomonas aeruginosa PAO1 virulence. PLoS One. 2013;**8**(1):e53441

[85] Vandeputte OM, Kiendrebeogo M, Rajaonson S, Diallo B, Mol A, El Jaziri M, et al. Identification of catechin as one of the flavonoids from Combretum albiflorum bark extract that reduces the production of quorum-sensing-controlled virulence factors in Pseudomonas aeruginosa PAQ1. Applied and Environmental Microbiology. 2010;**76**(1):243-253

[86] Luo J, Dong B, Wang K, Cai S, Liu T, Cheng X, et al. Baicalin inhibits biofilm formation, attenuates the quorum sensing-controlled virulence and enhances Pseudomonas aeruginosa clearance in a mouse peritoneal implant infection model. PLoS One. 2017;**12**(4):e0176883

[87] Girennavar B, Cepeda ML, Soni KA. Grapefruit juice and its furocoumarins inhibits autoinducer signaling and biofilm formation in bacteria. International Journal of Food Microbiology. 2008;**125**(2):204-208

[88] Chong YM, Yin WF, Ho CY, Mustafa MR, Hadi AH, Awang K. Malabaricone C from Myristica cinnamomea exhibits anti-quorum sensing activity. Journal of Natural Products. 2011;**74**:2261-2264

[89] Zhou L, Zheng H, Tang Y, Yu W, Gong Q. Eugenol inhibits quorum sensing at sub-inhibitory concentrations. Biotechnology Letters. 2013;**35**(4):631-637

[90] Lou Z, Letsididi KS, Yu F, Pei Z, Wang H, Letsididi R. Inhibitive effect of eugenol and its nanoemulsion on quorum sensing–mediated virulence

factors and biofilm formation by pseudomonas aeruginosa. Journal of Food Protection. 2019;**82**(3):379-389

[91] Ding X, Yin B, Qian L. Screening for novel quorum sensing inhibitors to interfere with the formation of Pseudomonas aeruginosa biofilm. Journal of Medical Microbiology. 2011;**60**(12):1827-1834

[92] Kim H-S, Lee S-H, Byun Y, Park H-D. 6-Gingerol reduces Pseudomonas aeruginosa biofilm formation and virulence via quorum sensing inhibition. Scientific Reports. 2015;**5**:8656

[93] Maisarah Norizan SN, Chan KG, Yin W-F, Ping TS, Nafiah MA. The study of caffeine as novel quorum sensing inhibitor. The Open Conference Proceedings Journal. 2013;**4**:185. DOI: 10.2174/2210289201304010185

[94] Ren D, Zuo R, Gonzàlez-Barrios AF. Differential gene expression for investigation of Escherichia coli biofilm inhibition by plant extract ursolic acid. Applied and Environmental Microbiology. 2005;**71**(7):4022-4034

[95] Kiplimo JJ, Koorbanally NA, Chenia HY. Triterpenoids from Vernonia auriculifera Hiern exhibit antimicrobial activity. African Journal of Pharmacy and Pharmacology. 2011;**5**:1150-1156

[96] Rasamiravaka T, Vandeputte OM, Pottier L, Huet J, Rabemanantsoa C, Kiendrebeogo M, et al. Pseudomonas aeruginosa biofilm formation and persistence, along with the production of quorum sensing dependent virulence factors, are disrupted by a triterpenoid coumarate ester isolated from Dalbergia trichocarpa, a tropical "legume". PLoS One. 2015;**10**:e0132791

[97] Rasamiravaka T, Ngezahayo J, Pottier L, Oliveira Ribeiro S, Souard F, Hari L, et al. Terpenoids from Platostoma rotundifolium (Briq,) a, J, Paton Alter the expression of quorum sensing-related virulence factors and the formation of biofilm in Pseudomonas aeruginosa PAO1. International Journal of Molecular Sciences. 2017;**18**:1270

[98] Jakobsen TH, Bragason SK, Phipps RK, Christensen LD, van Gennip M, Alhede M, et al. Food as a source for quorum sensing inhibitors: Iberin from horseradish revealed as a quorum sensing inhibitor of Pseudomonas aeruginosa. Applied Environmental Microbiology. 2012;**78**(7):2410-2421

[99] Ganin H, Rayo J, Amara N, Levy N, Krief P, Meijler MM. Sulforaphane and erucin, natural isothiocyanates from broccoli, inhibit bacterial quorum sensing. Medicinal Chemistry Communications. 2013;**4**:175-179

[100] Jakobsen TH, van Gennip M, Phipps RK. Ajoene, a sulfur-rich molecule from garlic, inhibits genes controlled by quorum sensing. Antimicrobial Agents and Chemotherapy. 2012;**56**(5):2314-2325

[101] Cady NC, McKean KA, Behnke J. Inhibition of biofilm formation, quorum sensing and infection in Pseudomonas aeruginosa by natural products-inspired organosulfur compounds. PLoS One. 2012;7(6):e38492

[102] Manefield M, de Nys R, Naresh K, Roger R, Givskov M, Peter S, et al. Evidence that halogenated furanones from Delisea pulchra inhibit acylated homoserine lactone (AHL)-mediated gene expression by displacing the AHL signal from its receptor protein. Microbiology. 1999;**145**(2):283-291

[103] Skindersoe ME, Ettinger-Epstein P, Rasmussen TB, Bjarnsholt T, de Nys R, Givskov M. Quorum sensing antagonism from marine organisms. Marine Biotechnology. 2008;**10**(1):56-63

[104] Manefield M, Rasmussen TB, Henzter M, Andersen JB, Steinberg P, Kjelleberg S, et al. Halogenated furanones inhibit quorum sensing through accelerated LuxR turnover. Microbiology. 2002;**148**(4):1119-1127

[105] Hentzer M, Riedel K, Rasmussen TB. Inhibition of quorum sensing in Pseudomonas aeruginosa biofilm bacteria by a halogenated furanone compound. Microbiology. 2002;**148**(1):87-102

[106] Ebada SS, Lin W, Proksch P. Bioactive sesterterpenes and triterpenes from marine sponges: Occurrence and pharmacological significance. Marine Drugs. 2010;**8**(2):313-346

[107] Paul D, Kim YS, Ponnusamy K, Kweon JH. Application of quorum quenching to inhibit biofilm formation. Environmental Engineering Science. 2009;**26**(8):1319-1324

[108] Chun CK, Ozer EA, Welsh MJ, Zabner J, Greenberg EP. Inactivation of a Pseudomonas aeruginosa quorum-sensing signal by human airway epithelia. Proceedings of the National Academy of Sciences. 2004;**101**(10):3587-3590

[109] Ozer EA, Pezzulo A, Shih DM. Human and murine paraoxonase 1 are host modulators of Pseudomonas aeruginosa quorum-sensing. FEMS Microbiology Letters. 2005;**253**(1):29-37

[110] Stoltz DA, Ozer EA, Ng CJ, Yu JM, Reddy ST, Lusis AJ, et al. Paraoxonase-2 deficiency enhances Pseudomonas aeruginosa quorum sensing in murine tracheal epithelia. American Journal of Physiology-Lung Cellular and Molecular Physiology. 2007;**292**:852-860

[111] Losa D, Kohler T, Bacchetta M, Saab JB, Frieden M, van Delden C, et al. Airway epithelial cell integrity protects from cytotoxicity of Pseudomonas aeruginosa quorum-sensing signals. American Journal of Respiratory Cell and Molecular Biology. 2015;**53**(2):265-275

[112] Park J, Kaufmann GF, Bowen JP, Arbiser JM, Janda KD. Solenopsin a, a venom alkaloid from the fire ant Solenopsis invicta, inhibits quorum-sensing signaling in Pseudomonas aeruginosa. Journal of Infectious Diseases. 2008;**198**(8):1198-1201

[113] Villena J, Kitazawa H, Van Wees SC, Pieterse CM, Takahashi H. Receptors and signaling pathways for recognition of bacteria in livestock and crops: Prospects for beneficial microbes in healthy growth strategies. Frontiers in Immunology. 2018;**9**:2223. DOI: 10.3389/fimmu.2018.02223

[114] Ngezahayo J, Pottier L, Ribeiro SO, Delporte C, Fontaine V, Hari L, et al. Plastotoma rotundifolium aerial tissue extract has antibacterial activities. Industrial Crops and Products. 2016;**86**:301-310

[115] Okusa PN, Penge O, Devleeschouwer M, Duez P. Direct and indirect antimicrobial effects and antioxidant activity of Cordia gilletii De wild (Boraginaceae). Journal of Ethnopharmacology. 2007;**112**(3):476-481

[116] Okusa PN, Stévigny C, Névraumont M, Gelbcke M, Van Antwerpen P, Braekman J-C, et al. Ferulaldehyde and lupeol as direct and indirect antimicrobial compounds from Cordia gilletii (Boraginaceae) root barks. Natural Product Communications. 2014;**9**(5):619-622

[117] Nobori T, Mine A, Tsuda K. Molecular networks in plant–pathogen holobiont. FEBS Letters. 2018;**592**(12):1937-1953

[118] Ahmed SA, Rudden M, Smyth TJ, Dooley JS, Marchant R, Banat IM. Natural quorum sensing

inhibitors effectively downregulate gene expression of Pseudomonas aeruginosa virulence factors. Applied Microbiology and Biotechnology. 2019;**103**(8):3521-3535

[119] Cho HH, Kwon KC, Kim S, Park Y, Koo SH. Association between biofilm formation and antimicrobial resistance in carbapenem-resistant Pseudomonas aeruginosa. Annals of Clinical and Laboratory Science. 2018;**48**(3):363-368

[120] Xu Q, Bauer R, Hendry BM, Fan T-P, Zhao Z, Duez P, et al. The quest for modernisation of traditional Chinese medicine. BMC Complementary and Alternative Medicine. 2013;**13**:132-143

Chapter 4

Inhibition of Bacterial Biofilm Formation

Angela Di Somma, Antonio Moretta, Carolina Canè, Arianna Cirillo and Angela Duilio

Abstract

Biofilm is a complex matrix consisting of extracellular polysaccharides, DNA, and proteins that protect bacteria from a variety of physical, chemical, and biological stresses allowing them to survive in hostile environments. Biofilm formation requires three different stages: cell attachment to a solid substrate, adhesion, and growth. The inhibition of one of these steps by small molecules, such as antimicrobial peptides, or their action on specific targets will leave pathogens armless against classical antibiotics. Any drug impairing crucial processes for bacterial life will inevitably lead to the development of drug-resistant strains, whereas the inhibition of biofilm formation might prevent the onset of bacterial resistance. In this section, we will focus on proteins involved in biofilm formation as useful targets for the development of new drugs that can effectively and specifically impair biofilm formation with slight effects on cell survival, thus avoiding the generation of drug-resistant strains.

Keywords: bacterial biofilms, biofilm inhibition, antimicrobial peptides, protein target, mechanism of action

1. Introduction

Microorganisms have the extraordinary ability to live in almost all environments and to protect themselves from external agents through sophisticated survival mechanisms. Bacteria can be found in planktonic form or in specific conditions, as sessile aggregates on both biotic and abiotic surfaces originating complex structures known as biofilm.

Biofilms are an ensemble of microbial cells irreversibly associated with a surface and enclosed in an essentially self-produced matrix. The biofilm matrix consists of polysaccharides, proteins, and DNA and constitutes a stubborn source that protects bacteria from a variety of physical, chemical, and biological stresses. One of its characteristics is the capability to impair antimicrobial molecules to spread through the polymer matrix or the ability of the matrix material to inactivate antibacterial molecules. Today, the increase and spread of antibiotic resistance among microorganisms (bacteria, fungi, viruses, and parasites) represent one of the greatest emergencies for human health worldwide [1]. Based on these characteristics, biofilm plays crucial roles in humans and nonhuman infections and represents the most important adaptive mechanism closely related to pathogenicity.

An antibiofilm agent must display several specific characteristics to target the biofilm lifestyle. First, due to the temporal biofilm heterogeneity, it must show a rapid killing ability to face a changing entity and to target cells before their entry into the biofilm community; it must be able to act in different environmental niches and to target different growth rate cells. The cells located in the periphery of biofilm are directly in contact with nutrients and oxygen, while those placed deepest in the biofilm layers may undergo lack of nutrients, anoxia, and acidic conditions. In this way, a metabolic and spatial heterogeneity is generated including both rapidly and slowly growing cells. In particular, due to environmental conditions, inside the biofilm, it is possible to find the so-called persister, dormant, quiescent cells characterized by a low rate of cell division that are believed to play an essential role in the biofilm resistance to antibiotics [2]. Other important characteristics for a good antimicrobial candidate are the ability to interfere with the production of the extracellular matrix and the possibility to penetrate the biofilm architecture. This matrix consists for 90% of EPS, whose principal components are proteins, polysaccharides, lipids, and extracellular DNA, and it is involved into the biofilm architecture maintenance. An antibiofilm agent should also be able to interfere with bacterial cell communication machinery.

This chapter aims to investigate and clarify in detail the inhibition of biofilm formation by different approaches.

Other additional aspects to consider the identification of potential antimicrobial agents are the ability to recruit immune cells and/or modulate the host immune response and the synergy with other conventional and unconventional antimicrobial compounds [3, 4].

Biofilms are very dynamic and spatially heterogeneous structures originating gradients of oxygen, nutrients, and pH, and their formation occurs through three phases: adhesion, maturation, and dispersal phase as described earlier.

2. Small molecules capable to inhibit biofilm formation

The inhibition or prevention of biofilm formation has been a subject of study for a long time. The first important action against biofilm formation is to prevent bacterial adhesion to surfaces and host tissues to reduce infection [5]. Preventing bacterial adhesion is an attractive target [6] for hampering bacterial infection, and several different strategies have been proposed including hindering cellular receptors from recognizing adhesion surfaces or inhibiting the process of bacterial adhesion. Blocking the primary colonizers can prevent initial biofilm colonization and the subsequent infection produced by planktonic cells released from the biofilm itself.

The adhesion process consists of various distinct steps. In the first step, bacterial cell establishes reversible adhesion interactions on host surfaces [7], while in the second step, a stronger type of adhesion is carried out, which involves specific molecules that bind in a complementary manner [5]. In particular, in Gram-negative bacteria, adhesion is mediated by special proteins known as adhesins associated with cell surface structures such as fimbriae or pili [8, 9]. Initial adhesion is then followed by a complex colonization process that offers a number of advantages to bacteria, including increased protection against dislocation by hydrokinetic forces from fluid surfaces or better access to nutrients released by the host cells [10]. Finally, in these favorable conditions, the development of the elaborate biofilm structures can take place.

For a long time, the first strategies used to inhibit the adhesion process were focused on the use of adhesin analogues that bind to the receptor and competitively block bacterial adhesion [5]. However, this strategy resulted unpractical because adhesin proteins are not readily available, and they become toxic at the relatively high concentrations that had to be used. An attempt to overcome this problem consisted in the design and use of synthetic peptides mimicking the sequence of cell surface adhesins. For example, the small peptide p1025 inhibits *Streptococcus mutans* binding to dental surfaces [11]. Analogously, a fragment of the fimbrillin adhesin was found to inhibit the adhesion of *Porphyromonas gingivalis* to hydroxyapatite [12]. However, this approach showed several drawbacks as different adhesins usually mediate the adhesion process and the expression of carbohydrates or cell surface ligands may vary depending on environmental conditions, originating a large number of variables and making this approach more difficult and not very applicable.

A novel and interesting approach to inhibit bacterial adhesion consists in the use of cell coatings with antimicrobial peptides that alter the chemical properties of the surface [13, 14], thus interfering with bacterial adhesion and preventing surface binding. Although "passive," this method is rather attractive and may serve as a novel approach to address the biofilm problem on artificial medical devices. However, limited successes have been achieved so far due to attachment variability among different bacterial strains. Recently, many active polymeric coatings were designed to bind the surface and release a variety of antimicrobial molecules such as antibiotics, bacteriocins, and metal ions [15–18]. A significant reduction in biofilm formation of *Staphylococcus epidermidis* on hydrogel-coated and serum/hydrogel-coated silicone catheters was observed following the release of bacteriophagic factors from the polymer with and without supplemental divalent cations [19]. Similarly, treatment of piperacillin-tazobactam coated tympanostomy tubes reduces biofilm infection of ciprofloxacin-resistant *Pseudomonas aeruginosa* (CRPA) [20]. The negative aspect of this approach might be the continuous release of high concentration of antimicrobials in a short time by the active polymer often higher than the MIC values without a specific target. However, target release polymer can be foreseen as the new era of biofilm treatments in industrial food safety and packaging [21].

Recently, great attention was paid to a different approach addressed to killing planktonic cells for prevention and treatment of biofilms. The new catheter lock solution C/MB/P (citrate, methylene blue, and parabens) was able to act against planktonic and sessile bacteria within a biofilm preventing bacterial colonization of hemodialysis catheters [22]. Killing planktonic cells might represent a good approach, but this strategy cannot be carried out on long term because any drug targeting crucial processes for bacterial life will unavoidably lead to the development of resistant strains.

An effective and positive control of biofilm formation might be obtained by interfering with specific cellular process crucial for biofilm formation. Biofilm formation is often associated with the phenomenon of quorum sensing (QS), in which bacterial cells communicate with each other by small diffusible signal molecules [23]. Moreover, bacterial gene expression has to be synchronized to form biofilms, and to achieve this goal, the quorum-sensing (QS) mechanism is used by bacteria, producing and responding to a several intra and intercellular signals called autoinducers [24]. At low-cell densities, the autoinducer is present in the extracellular media in a small amount that is too dilute to be detected. When the cell density increases, the autoinducer concentration reaches a threshold, and the autoinducer-receptor complex (the regulatory protein) acts to induce or repress the expression of target genes. The QS controls some physiological processes such as secretion of virulence factors, biofilm formation, and antibiotic resistance in several bacterial species [25, 26]. Investigation and elucidation of the molecular mechanisms underlying the QS effects on biofilms including the production of virulence factors may help to control bacterial infection. More than 70 species of Gram-negative bacteria communicate and control their population density and mobility via N-acyl homoserine lactones (AHLs) mediated QS and represented one of the primary scaffolds studied for the design of potential biofilm inhibitors [27]. N-butanoylhomoserine lactone 1 (C4-AHL, for the rhl system) and 3-oxo-C12-AHL 5 (for the las system) are among the most important AHLs involved in QS (REF Small molecule control of bacterial biofilms). In *P. aeruginosa*, one of the most important bacteria involved in human infections, different antibiofilm molecules focused on AHL analogues were designed to develop new strategies to impair biofilm formation. The Blackwell et al. identified, designed, and synthesized several different AHLs capable to significantly reduce biofilm formation and virulence factor production in *P. aeruginosa* [28, 29].

A different approach consisted in the use of the synthetic halogenated furanone produced by secondary metabolism of the Australian macroalga *Delisea pulchra*, which is able to penetrate the biofilm matrix and to alter its architecture in flow chambers [30, 31]. Furthermore, T315, an integrin-linked kinase inhibitor previously identified as a potential therapeutic agent against chronic lymphocytic leukemia [32], was shown to selectively inhibit biofilm formation in both *Salmonella typhi* and *Salmonella Typhimurium* at early stages of biofilm development without affecting bacterial viability. T315 was also demonstrated to reduce biofilm formation in *Acinetobacter baumannii* but had no effect on *P. aeruginosa* suggesting a bacterial specificity [33].

3. Biofilm inhibition by antimicrobial peptides

Antimicrobial peptides (AMPs) are small molecules (10–100 amino acids) widespread in nature that play an essential role in the innate immunity. Recently, much attention has been paid to AMPs as they exert a broad spectrum of action, exploiting different activities as antibacterial, antifungal, antiparasites, anticancer, and antibiofilm factors [34]. This paragraph will focus on the ability of some antimicrobial peptides to inhibit biofilm formation.

The use of antimicrobial peptides to impair biofilm formation is attracting great interest, and many peptides have already been tested on different bacterial biofilms. In particular, the molecular mechanism of biofilm inhibition by AMPs is very much under investigation. The AMPs tested on biofilms so far derive from different natural sources, such as humans, mammals, bacteria, plants, and amphibians, but many synthetic peptides have also been studied. For example, it was demonstrated that the human cathelicidin LL-37 and indolicidin peptides could prevent biofilm formation of *P. aeruginosa* by downregulating the transcription of Las and Rhl, two quorum-sensing systems [35]. Moreover, AMPs could inhibit biofilm formation by increasing twitching motility in *P. aeruginosa* through the stimulation of the expression of genes needed for type IV pilli biosynthesis and function. Type IV pilli has the main function to increase bacteria movement on surfaces, which could facilitate cell removal [35]. The synthetic antimicrobial peptide meta-phenylene ethynylene (mPE), based on magainin, was active against biofilms of *Streptococcus mutans*, both as an intracellular antibiotic by binding to DNA and as a membrane-active molecule inhibiting lipopolysaccharides (LPSs), similar to magainin action [36].

In addition, the LL-37 peptide can also inhibit initial biofilm attachment. In *Pseudomonas aeruginosa*, this peptide downregulates the expression of genes associated with the assembly of flagella involved in the process of initial adherence [37]. Antiadhesion could be one of the major AMPs antibiofilm properties leading to their potential use as an effective pretreatment strategy. For example, the nisin peptide, which interferes with cell wall synthesis and is capable to form membrane pores, delays biofilm formation, but it does not inhibit the *Staphylococcus aureus* growth when it is immobilized in multiwalled carbon nanotubes [38].

AMPs can also cause biofilm matrix disruption. The human liver-derived hepcidin 20 peptide can reduce the mass of extracellular matrix and can alter the *S. epidermidis* biofilm architecture by targeting polysaccharide intercellular adhesin (PIA). Being endowed with nucleosidase activity, the fish-derived piscidin-3 peptide can degrade *P. aeruginosa* extracellular DNA by coordinating with Cu^{2+} through its N-terminus [39, 40].

Although several antimicrobial peptides have nowadays been studied for the inhibition of biofilm formation, a further aspect needs to be considered. Several biofilms have developed defense mechanisms to protect themselves from antimicrobial agents. The interaction with EPS is thought to be the principal reason of biofilm resistance to AMPs even if the exact mechanism is not well understood. Gram-negative bacteria, such as *P. aeruginosa*, can secrete alginate, an anionic extracellular polysaccharide consisting of uronic acid D-mannuronate and C-5 epimer-L guluronate. Alginate can interact with cationic AMPs and protect *P. aeruginosa* biofilm from the effect of the antimicrobial peptides [41]. Moreover, the peptide sensing system known as aps, first recognized in *S. epidermidis*, can protect Gram-positive bacteria from AMP action. This system upregulates the D-alanylation of teichoic acid and increases the expression of putative AMP efflux pumps. It was demonstrated that *Enterococcus faecalis* D-alanine deficient mutant is more resistant to AMPs than the wild type even if they produce less biofilm [42].

4. Biofilm inhibition by protein targets

Planktonic bacteria can adhere to different cells or tissues starting biofilm formation via production of a multitude of proteins, which act at different stages of biofilm formation. Some proteins contribute to biofilm accumulation, while others are involved into the mediation of primary attachment to surfaces [43, 44]. For this reason, thc formation and the development of bacterial biofilm can be associated with the production of specific proteins, which play essential roles in the bacterial biofilm formation and development. Strategies leading to the identification of these proteins are fundamental as they could represent interesting targets to inhibit biofilm formation, allowing the development of new antibiofilm agents and procedures [45]. In this paragraph, we will focus on some target proteins involved in the production of biofilms in different bacteria: the N-acetylneuraminate lyase (NanA) in *Escherichia coli*, the bifunctional enzyme N-acetyl-D-glucosamine-1-phosphate acetyltransferase (GlmU) in *Mycobacterium smegmatis*, and the surface protein G (SasG) in *S. aureus*.

The NanA protein of *E. coli* is an enzyme able to recognize the sialic acid, a molecule essential to a number of critical biological processes, such as cell recognition, adhesion, and immune system evasion. NanA catalyzes the transformation of sialic acid into pyruvate and N-acetyl-D-mannosammine [46, 47], favoring cell-cell adhesion. Therefore, NanA plays a fundamental role in the adhesion development of host cells a process of great importance in the formation of biofilm. This enzyme is then considered an important target for developing molecules able to reduce biofilm accumulation. Recently, a relationship between methylation stress in *E. coli* and the reduction of bacterial adhesion properties thus decreasing its ability to form biofilm was reported. This phenomenon was associated with a drastic reduction in the expression levels of the NanA protein, suggesting a possible role of NanA in biofilm formation and bacteria host interactions. Using a null NanA mutant and DANA, a substrate analog acting as competitive inhibitor, it was demonstrated that the downregulation of NanA or inhibition of its enzymatic activity affects biofilm formation and adhesion properties of *E. coli* [48, 49].

Another important protein target is GlmU, a bifunctional enzyme with acetyltransferase activity involved in the biosynthesis of Uridine diphosphate N-acetylglucosamine (UDP-GlcNAc), a key precursor of β-1,6-N-acetyl-D-glucosamine polysaccharide adhesin required for biofilm formation [50, 51]. GlmU is a possible factor involved in biofilm production in *M. smegmatis*, a nonpathogenic bacterium homologous to the pathogenic *M. tuberculosis*. The response of *M. smegmatis* to alkylating stress is different from *E. coli*, resulting in an increase in biofilm formation possibly due to a very strong defense mechanism. In this contest, GlmU has an important role in the process of biofilm production in *M. smegmatis*, being its expression highly upregulated when the bacterium needs to activate defense mechanisms [52]. Experiments with both conditional deletion and overexpressing glmU mutants demonstrated that the downregulation of GlmU decreased *M. smegmatis* capabilities to produce biofilm, whereas the overexpression of enzyme increased biofilm formation. These results were supported by inhibition of GlmU acetyltransferase activity with two different inhibitors, suggesting the involvement of this enzyme in the *M. smegmatis* defense mechanisms. Focusing on the inhibition of GlmU might then be an efficient method to disable the bacterium defense mechanism.

S. aureus is a common pathogen responsible for nosocomial and community infections being able to colonize the squamous epithelium of the anterior nares. One of the adhesins likely to be responsible for this ability is the *S. aureus* surface protein G (SasG), which promotes cellular aggregation leading to biofilm formation [53, 54]. SasG comprises an N-terminal A domain and repeated B domains with only the B domain required for the accumulation of biofilm. Expression of SasG does not increase the adherence of bacteria, and it is not involved in primary attachment but plays a role in the accumulation phase of biofilm formation [55]. For different aspects and playing different roles, NanA, GlmU, and SasG may all represent interesting targets to address the inhibition of biofilm production.

5. Conclusions

Currently, biofilm infections constitute a serious medical problem, and their treatment is far from being satisfactory. Biofilm formation inhibitors have several potential therapeutic applications as coatings in medical devices or in the prophylaxis of implanted surgery. In this respect, the identification of new strategies to counteract biofilm formation is a broad subject of study. The antibiofilm activity of many molecules such as proteins, peptides, and small organic molecules is currently under investigation. Each of these molecules is endowed with specific characteristics and can exert its ability to inhibit bacterial biofilm formation with different mechanisms. Antibiofilm agents are able to act both at the initial stages of biofilm formation, such as bacterial adhesion to the host surface, and on preformed biofilm, leading to the disruption of the EPS architecture. Many small organic molecules are able to interfere with the bacterial QS system, but their lack of activity in *in vivo* models and the high toxicity make these molecules of limited use in clinical applications.

As antimicrobial peptides show a broad spectrum of action, exploiting different activities including antibiofilm capabilities, these molecules might be considered as new promising factors to impair biofilm formation that exploit different mechanisms to hamper biofilms at different stages.

The administration of a single antibiotic is often not enough to eradicate bacterial invasions, and a high concentration of the antibiotic can be extremely toxic. A possible solution might be the coadministration of antibiotics with antibiofilm peptides that allow the use of low antibiotic concentrations. This strategy can be tuned to affect biofilms without killing bacteria, thus avoiding the emergence of drug-resistant populations through synergy with existing antibiotics.

Acknowledgements

This work was supported in part by MIUR grants ARS01_00597 Project "NAOCON" and PRIN 2017 "Identification and characterization of novel antitumoral/antimicrobial insect-derived peptides: a multidisciplinary, integrated approach from in silico to *in vivo*."

Conflict of interest

The authors declare no conflict of interest.

Author details

Angela Di Somma[1*], Antonio Moretta[2], Carolina Canè[3], Arianna Cirillo[3] and Angela Duilio[1]

1 Department of Chemical Sciences, Federico II University, Naples, Italy

2 Department of Science, University of Basilicata, Potenza, Italy

3 CEINGE Biotecnologie Avanzate, Naples, Italy

*Address all correspondence to: angela.disomma@unina.it

References

[1] Cepas V, López Y, Munoz E, Rolo D, Ardanuy C, Martí S, et al. Relationship between biofilm formation and antimicrobial resistance in gram-negative bacteria. Microbial Drug Resistance. 2019;**25**(1):72-79

[2] Lebeaux D, Ghigo JM, Beloin C. Biofilm-related infections: Bridging the gap between clinical management and fundamental aspects of recalcitrance toward antibiotics. Microbiology and Molecular Biology Reviews. 2014;**78**(3):510-543

[3] Anderson JM, Patel JD. Biomaterial-dependent characteristics of the foreign body response and *S. epidermidis* biofilm interactions. In: Biomaterials Associated Infection. New York, NY: Springer; 2013. pp. 119-149

[4] Beloin C, Renard S, Ghigo JM, Lebeaux D. Novel approaches to combat bacterial biofilms. Current Opinion in Pharmacology. 2014;**18**:61-68

[5] Ofek I, Hasty DL, Sharon N. Anti-adhesion therapy of bacterial diseases: Prospects and problems. FEMS Immunology and Medical Microbiology. 2003;**38**(3):181-191

[6] Bavington C, Page C. Stopping bacterial adhesion: A novel approach to treating infections. Respiration. 2005;**72**(4):335-344

[7] Hasty DL, Ofek I, Courtney HS, Doyle RJ. Multiple adhesins of streptococci. Infection and Immunity. 1992;**60**(6):2147

[8] Krogfelt KA. Bacterial adhesion: Genetics, biogenesis, and role in pathogenesis of fimbrial adhesins of *Escherichia coli*. Reviews of Infectious Diseases. 1991;**13**(4):721-735

[9] Kline KA, Fälker S, Dahlberg S, Normark S, Henriques-Normark B. Bacterial adhesins in host-microbe interactions. Cell Host & Microbe. 2009;**5**(6):580-592

[10] Zafriri D, Oron Y, Eisenstein BI, Ofek I. Growth advantage and enhanced toxicity of *Escherichia coli* adherent to tissue culture cells due to restricted diffusion of products secreted by the cells. The Journal of Clinical Investigation. 1987;**79**(4):1210-1216

[11] Kelly CG, Younson JS, Hikmat BY, Todryk SM, Czisch M, Haris PI, et al. A synthetic peptide adhesion epitope as a novel antimicrobial agent. Nature Biotechnology. 1999;**17**(1):42

[12] Lee JY, Sojar HT, Bedi GS, Genco RJ. Synthetic peptides analogous to the fimbrillin sequence inhibit adherence of *Porphyromonas gingivalis*. Infection and Immunity. 1992;**60**(4):1662-1670

[13] Hetrick EM, Schoenfisch MH. Reducing implant-related infections: Active release strategies. Chemical Society Reviews. 2006;**35**(9):780-789

[14] Tamilvanan S, Venkateshan N, Ludwig A. The potential of lipid-and polymer-based drug delivery carriers for eradicating biofilm consortia on device-related nosocomial infections. Journal of Controlled Release. 2008;**128**(1):2-22

[15] La AS, Ercolini D, Marinello F, Mauriello G. Characterization of bacteriocin-coated antimicrobial polyethylene films by atomic force microscopy. Journal of Food Science. 2008;**73**(4):T48-T54

[16] Ruggeri V, Francolini I, Donelli G, Piozzi A. Synthesis, characterization, and *in vitro* activity of antibiotic releasing polyurethanes to prevent bacterial resistance. Journal of biomedical materials research part a: An official journal of the Society for Biomaterials, the Japanese Society for

Biomaterials, and the Australian Society for Biomaterials and the Korean society for. Biomaterials. 2007;**81**(2):287-298

[17] Xu Q, Czernuszka JT. Controlled release of amoxicillin from hydroxyapatite-coated poly (lactic-co-glycolic acid) microspheres. Journal of Controlled Release. 2008;**127**(2):146-153

[18] Taylor EN, Webster TJ. The use of superparamagnetic nanoparticles for prosthetic biofilm prevention. International Journal of Nanomedicine. 2009;**4**:145

[19] Curtin JJ, Donlan RM. Using bacteriophages to reduce formation of catheter-associated biofilms by *Staphylococcus epidermidis*. Antimicrobial Agents and Chemotherapy. 2006;**50**(4):1268-1275

[20] Jang CH, Park H, Cho YB, Choi CH, Park IY. The use of piperacillin–tazobactam coated tympanostomy tubes against ciprofloxacin-resistant *Pseudomonas* biofilm formation: An *in vitro* study. International Journal of Pediatric Otorhinolaryngology. 2009;**73**(2):295-299

[21] Balasubramanian A, Rosenberg LE, Yam KIT, Chikindas ML. Antimicrobial packaging: Potential vs. reality—A review. The Journal of Applied Packaging Research. 2009;**3**(4):193-221

[22] Steczko J, Ash SR, Nivens DE, Brewer L, Winger RK. Microbial inactivation properties of a new antimicrobial/antithrombotic catheter lock solution (citrate/methylene blue/parabens). Nephrology, Dialysis, Transplantation. 2009;**24**(6):1937-1945

[23] Li YH, Tian X. Quorum sensing and bacterial social interactions in biofilms. Sensors. 2012;**12**(3):2519-2538

[24] Camilli A, Bassler BL. Bacterial small-molecule signaling pathways. Science. 2006;**311**(5764):1113-1116

[25] Miller MB, Bassler BL. Quorum sensing in bacteria. Annual Review of Microbiology. 2001;**55**(1):165-199

[26] Irie Y, Parsek MR. Quorum sensing and microbial biofilms. In: Bacterial Biofilms. Berlin, Heidelberg: Springer; 2008. pp. 67-84

[27] Worthington RJ, Richards JJ, Melander C. Small molecule control of bacterial biofilms. Organic & Biomolecular Chemistry. 2012;**10**(37):7457-7474

[28] Geske GD, Wezeman RJ, Siegel AP, Blackwell HE. Small molecule inhibitors of bacterial quorum sensing and biofilm formation. Journal of the American Chemical Society. 2005;**127**(37):12762-12763

[29] Geske GD, O'Neill JC, Miller DM, Wezeman RJ, Mattmann ME, Lin Q, et al. Comparative analyses of N-acylated homoserine lactones reveal unique structural features that dictate their ability to activate or inhibit quorum sensing. Chembiochem. 2008;**9**(3):389-400

[30] Hentzer M, Riedel K, Rasmussen TB, Heydorn A, Andersen JB, Parsek MR, et al. Inhibition of quorum sensing in *Pseudomonas aeruginosa* biofilm bacteria by a halogenated furanone compound. Microbiology. 2002;**148**(1):87-102

[31] Roy R, Tiwari M, Donelli G, Tiwari V. Strategies for combating bacterial biofilms: A focus on anti-biofilm agents and their mechanisms of action. Virulence. 2018;**9**(1):522-554

[32] Lee SL, Hsu EC, Chou CC, Chuang HC, Bai LY, Kulp SK, et al. Identification and characterization of a novel integrin-linked kinase inhibitor. Journal of Medicinal Chemistry. 2011;**54**(18):6364-6374

[33] Moshiri J, Kaur D, Hambira CM, Sandala JL, Koopman JA, Fuchs JR,

et al. Identification of a small molecule anti-biofilm agent against *Salmonella enterica*. Frontiers in Microbiology. 2018;**9**:2804

[34] Malik E, Dennison S, Harris F, Phoenix D. pH dependent antimicrobial peptides and proteins, their mechanisms of action and potential as therapeutic agents. Pharmaceuticals. 2016;**9**(4):67

[35] Overhage J, Campisano A, Bains M, Torfs EC, Rehm BH, Hancock RE. Human host defense peptide LL-37 prevents bacterial biofilm formation. Infection and Immunity. 2008;**76**(9):4176-4182

[36] Beckloff N, Laube D, Castro T, Furgang D, Park S, Perlin D, et al. Activity of an antimicrobial peptide mimetic against planktonic and biofilm cultures of oral pathogens. Antimicrobial Agents and Chemotherapy. 2007;**51**(11):4125-4132

[37] Dean SN, Bishop BM, Van Hoek ML. Susceptibility of *Pseudomonas aeruginosa* biofilm to alpha-helical peptides: D-enantiomer of LL-37. Frontiers in Microbiology. 2011;**2**:128

[38] Qi X, Poernomo G, Wang K, Chen Y, Chan-Park MB, Xu R, et al. Covalent immobilization of nisin on multi-walled carbon nanotubes: Superior antimicrobial and anti-biofilm properties. Nanoscale. 2011;**3**(4):1874-1880

[39] Brancatisano FL, Maisetta G, Di Luca M, Esin S, Bottai D, Bizzarri R, et al. Inhibitory effect of the human liver-derived antimicrobial peptide hepcidin 20 on biofilms of polysaccharide intercellular adhesin (PIA)-positive and PIA-negative strains of *Staphylococcus epidermidis*. Biofouling. 2014;**30**(4):435-446

[40] Libardo MDJ, Bahar AA, Ma B, Fu R, McCormick LE, Zhao J, et al. Nuclease activity gives an edge to host-defense peptide piscidin 3 over piscidin 1, rendering it more effective against persisters and biofilms. The FEBS Journal. 2017;**284**(21):3662-3683

[41] Chan C, Burrows LL, Deber CM. Helix induction in antimicrobial peptides by alginate in biofilms. Journal of Biological Chemistry. 2004;**279**(37):38749-38754

[42] Fabretti F, Theilacker C, Baldassarri L, Kaczynski Z, Kropec A, Holst O, et al. Alanine esters of enterococcal lipoteichoic acid play a role in biofilm formation and resistance to antimicrobial peptides. Infection and Immunity. 2006;**74**(7):4164-4171

[43] Pizarro-Cerdá J, Cossart P. Bacterial adhesion and entry into host cells. Cell. 2006;**124**(4):715-727

[44] Speziale P, Pietrocola G, Foster TJ, Geoghegan JA. Protein-based biofilm matrices in staphylococci. Frontiers in Cellular and Infection Microbiology. 2014;**4**:171

[45] Sintim HO, Smith JA, Wang J, Nakayama S, Yan L. Paradigm shift in discovering next-generation anti-infective agents: Targeting quorum sensing, c-di-GMP signaling and biofilm formation in bacteria with small molecules. Future Medicinal Chemistry. 2010;**2**(6):1005-1035

[46] Daniels AD, Campeotto I, van der Kamp MW, Bolt AH, Trinh CH, Phillips SE, et al. Reaction mechanism of N-acetylneuraminic acid lyase revealed by a combination of crystallography, QM/MM simulation, and mutagenesis. ACS Chemical Biology. 2014;**9**(4):1025-1032

[47] Uchida Y, Tsukada Y, Sugimori T. Purification and properties of N-acetylneuraminate lyase from *Escherichia coli*. The Journal of Biochemistry. 1984;**96**(2):507-522

[48] Di Pasquale P, Caterino M, Di Somma A, Squillace M, Rossi E, Landini P, et al. Exposure of *E. coli* to DNA-methylating agents impairs biofilm formation and invasion of eukaryotic cells via down regulation of the N-acetylneuraminate lyase NanA. Frontiers in Microbiology. 2016;7:147

[49] Volkert MR, Landini P. Transcriptional responses to DNA damage. Current Opinion in Microbiology. 2001;**4**(2):178-185

[50] Burton E, Gawande PV, Yakandawala N, LoVetri K, Zhanel GG, Romeo T, et al. Antibiofilm activity of GlmU enzyme inhibitors against catheter-associated uropathogens. Antimicrobial Agents and Chemotherapy. 2006;**50**(5):1835-1840

[51] Itoh Y, Wang X, Hinnebusch BJ, Preston JF, Romeo T. Depolymerization of β-1, 6-N-acetyl-d-glucosamine disrupts the integrity of diverse bacterial biofilms. Journal of Bacteriology. 2005;**187**(1):382-387

[52] Di Somma A, Caterino M, Soni V, Agarwal M, di Pasquale P, Zanetti S, et al. The bifunctional protein GlmU is a key factor in biofilm formation induced by alkylating stress in *Mycobacterium smegmatis*. Research in Microbiology. 2019;**170**:171-181

[53] Corrigan RM, Rigby D, Handley P, Foster TJ. The role of *Staphylococcus aureus* surface protein SasG in adherence and biofilm formation. Microbiology. 2007;**153**(8):2435-2446

[54] Roche FM, Meehan M, Foster TJ. The *Staphylococcus aureus* surface protein SasG and its homologues promote bacterial adherence to human desquamated nasal epithelial cells. Microbiology. 2003;**149**(10):2759-2767

[55] Geoghegan JA, Corrigan RM, Gruszka DT, Speziale P, O'Gara JP, Potts JR, et al. Role of surface protein SasG in biofilm formation by *Staphylococcus aureus*. Journal of Bacteriology. 2010;**192**(21):5663-5673

Chapter 5

Approaches for Modeling Anaerobic Granule-Based Reactors

Jixiang Yang

Abstract

Anaerobic granule sludge is a self-forming biofilm. This biofilm can be developed without the presence of bio-carriers. Anaerobic granule sludge-based technologies are dominant technologies in the field of anaerobic wastewater treatment. Although they are successful technologies, many efforts are still needed for a better understanding of the granules and granule-based reactors because reactor failure can occur. Here, reactor modeling is highly helpful in understanding the performance of anaerobic bioreactors. A model that can accurately model bioprocesses in a sludge bed reactor and predict concentrations of effluent components is valuable. This is because the model can provide insights into the reactor and be useful in reactor control. Current models of granules are models on bioprocesses in a single granule sludge or on the hydrodynamics and biokinetics in a sludge bed reactor. Here, we review advances in reactor model and its applications as well as limitations and further improvements in the models.

Keywords: anacrobic, granule, model, wastewater, sludge, modeling

1. Introduction

The phenomenon of anaerobic sludge granulation was first observed in the 1990s. Extensive experimental works have been implemented since then. The culture conditions for forming the anaerobic granular sludge are well understood [1]. A high upflow velocity (usually >1 m/h) is usually required. The diameters of anaerobic granules can be up to 0.15–4 mm, which results in high free sedimentation velocities, that is, 15–50 m/h. The high sedimentation velocities can make a large amount of highly active granular sludge retained in a bioreactor in a highly efficient way. By 2007, the market share for anaerobic granule sludge-based technologies in the field of anaerobic wastewater treatment was 89%. Anaerobic granular sludge-based technologies have been extensively applied to treat wastewater from different industries, including agriculture, food, beverage, alcohol distillery, pulp, and papermaking.

Bioreactors involved in wastewater treatment are complex systems, and many nonlinear biokinetics occur in the bioreactors. A model that can successfully model bioprocesses in the bioreactors is effective in understanding the bioreactors and their manipulation. Versus aerobic wastewater treatment, modeling an anaerobic wastewater treatment is much more difficult. This chapter summarizes different model strategies for a granular sludge bed reactor. These strategies are beneficial for further model development and applications.

2. Bioparticle model

The distribution of microorganisms in an anaerobic granule has big impacts on modeling the bioactivity of this granule. Different microbial structures for granules are identified. A layered and a cluster granular sludge structures are observed [2]. Here, three layers are proposed. The outermost layer includes acidogens and hydrogen-consuming organisms. In the middle layer, hydrogen-producing organisms as well as hydrogen-consuming organisms both exist, while *Methanosaeta* locate in the core layer. In this clustered structure, *Methanosaeta* clusters and zones with syntrophic eubacteria and hydrogenotrophic methanogens scatter in the granule.

A granular sludge bed consists of numeral sludge granules. Modeling substrate degradation in a single sludge granule has other applications. Indeed, understanding bioreactions in a single granule can explain the operation of an entire bioreactor. Two strategies are used to model substrate degradation in a single granule. Modeling strategies are both termed bioparticle models in this study. The bioparticle models are discussed below.

2.1 Diffusion-reaction model

A diffusion-reaction model couples mass transfer and substrate degradation kinetics in a single granule. Some assumptions need to be made to establish a diffusion-reaction model. The shape of real granules in reactors is irregular and nonuniform. In addition, the biogas that results from bioprocesses contributes to the formation of pores in the inner space of a granule. Water and biomass are different materials and constitute a granule. Therefore, substrate diffusion in the inner space of a granule is different at different locations. Nevertheless, some assumptions are adopted for building a typical diffusion-reaction model to simplify the difficulty in modeling and ensure model accuracy. The assumptions are listed here: (1) the granules are spherical and uniform; (2) only radial diffusion transport is considered and is described by Fick's law; (3) the diffusion coefficient is constant; and (4) there are no active biomass gradients in the granules at time zero [3].

A representative granule is assumed in a diffusion-reaction model [3, 4]. A typical diffusion-reaction model is characterized by the following equations:

$$D_i\left(\frac{d^2 S_i(r)}{dr^2} + \frac{2}{r}\frac{dS_i(r)}{dr}\right) + r_i = 0. \tag{1}$$

with two boundary conditions:

$$\begin{aligned} \frac{dS_i}{dr} &= 0, \quad at \quad r = 0 \\ S_i &= S_{i,sur}, \quad at \quad r = R \end{aligned} \tag{2}$$

where S_i is the substrate concentration of component i in the granule, $S_{i,sur}$ is the substrate concentration of component *i* in the granule surface, r_i is the volumetric substrate conversion rate in the granule, and D_i is the diffusion coefficient of substrate *I*; *r* is the distance from the granule center.

The diffusion-reaction model was successfully applied in an anaerobic ammonium oxidation (ANAMMOX) granule [3]. However, the above diffusion-reaction model must be revised accordingly, while other sludge granules are modeled. The ANAMMOX reaction is a simple and single reaction that involves simple substrates. If a complex substrate is involved in a diffusion-reaction model, then a hydrolysis process as well as other downstream processes are involved, and it is hard to calibrate the kinetic parameters for each process. In addition, assumption (4) for a diffusion-reaction model may not be true for complex substrates such as carbohydrates. If a carbohydrate is used as the substrate, then a multilayer model could be a better alternative. In the multilayer model, a granule is divided into three layers: H_2, producing acetogens; H_2, consuming organisms (*Methanothrix*); and acidogens. Furthermore, the boundary conditions (Eq. (2)) should be revised accordingly, while substrate concentrations at the core of a representative granule are not zero or there is a pore at the core of the representative granule.

2.2 Individual-based model

In the other model, substrate degradation can be coupled with the dynamic growth of a sludge granule. In the dynamic growth process, the sludge granule consists of many bacteria, and the granular surface growth and detachment are involved. The model is called an individual-based model (IBM) because the model is based on each single individual bacterium.

The IBM significantly differs from the above diffusion-reaction model. The size and shape of a single granule are not constant in the IBM. Bacteria grow and can be sheared off in the model, which mimics the natural growth of a single granule. The model has clear and active biomass gradients because the growth of different bacterial species interacts with substrate degradation. The IBM can be one-dimensional, two-dimensional, or three-dimensional.

Figure 1 shows the model strategy of the IBM model. The IBM model was applied to model the biofilm development in a reverse osmosis module. This data verified the validity of the IBM model [5, 6]. In principal, any kind of microorganisms can be applied in this model strategy.

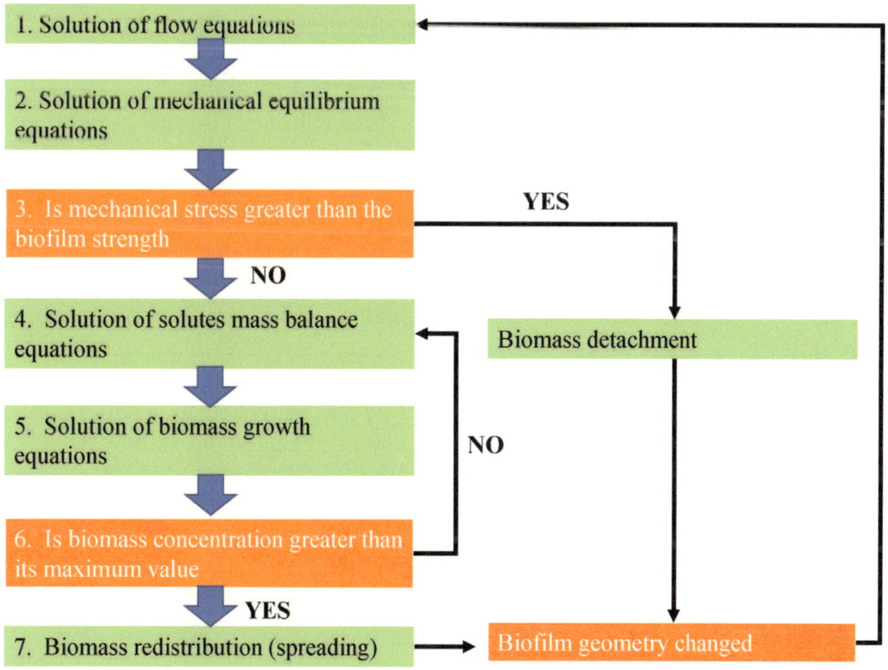

Figure 1.
Algorithm steps for the biofilm model including substrate convection, substrate diffusion, substrate reaction, biomass growth, and biofilm detachment.

The implementation of an IBM model requires a big computational workload because a modeling domain must be divided into numerous micro grids. Hence, the implementation of an IBM model at a reactor scale would require huge computational workload and appear to be impossible.

3. Reactor model

3.1 Integration of hydrodynamic and biokinetics

3.1.1 Applied hydrodynamic models

The modeling of wastewater treatment at the reactor scale usually requires a hydrodynamic model. The hydrodynamic model tries to explain water flow in a bioreactor. There are two major strategies for constructing a hydrodynamic model. On one hand, a reactor can be treated as a connection of continuous stirred reactors (CSTRs) and/or plug-flow (PF) reactors. This is termed the reactor compartmentalization (RC) strategy. On the other hand, computational fluid dynamics (CFD) can be applied instead of reactor compartmentalization. This is termed the CFD strategy.

3.1.1.1 RC strategy

Many different flow schemes have been applied to model hydrodynamics in granular sludge bed reactor. In each of these models, CSTRs are widely applied to model a sludge bed and a blanket, while a PF reactor is usually applied to model a settler in a reactor. The flow schemes do not have to fit the real physical flow conditions. A flow scheme is considered acceptable if the resulting tracer concentrations fit the tracer concentrations measured at the outlet of reactors [7].

Figure 2 shows that four major flow schemes have been applied to model the hydrodynamics in granular-based reactors [8, 9, 11]. A sludge bed can be modeled by using a combination of a CSTR and a dead volume. The sludge blanket can be modeled via the other CSTR. A bypass flow always starts from the inlet of a sludge bed but ends at different compartments in different flow schemes. The settler can be modeled as a plug-flow reactor. Other flow schemes are also applied but with less applications. An upflow anaerobic sludge bed (UASB) reactor was treated as the connection of several CSTRs, and there was a good agreement between experimental and simulated results. This shows that this variation is acceptable [12].

A CSTR is often applied when a blanket zone is modeled because biomass transport and rising bubbles are two important factors that lead to turbulence in this compartment. In a settler, a degree of mixing can be expected due to movement of rising gas bubbles. Therefore, this zone is modeled as a dispersed plug-flow reactor. The choice of a plug-flow reactor or a CSTR should depend on the flow conditions in the sludge bed. Although many researchers use a CSTR to represent a sludge bed, a plug-flow reactor can be an alternative [13]. Similarly, a sludge bed in a reactor is often modeled as a CSTR when a high recirculation rate is applied [4, 7].

3.1.1.2 CFD strategy

If the characteristics of each sludge granule can be obtained by applying basic equations such as Navier-Stokes equations, then the exact hydrodynamic

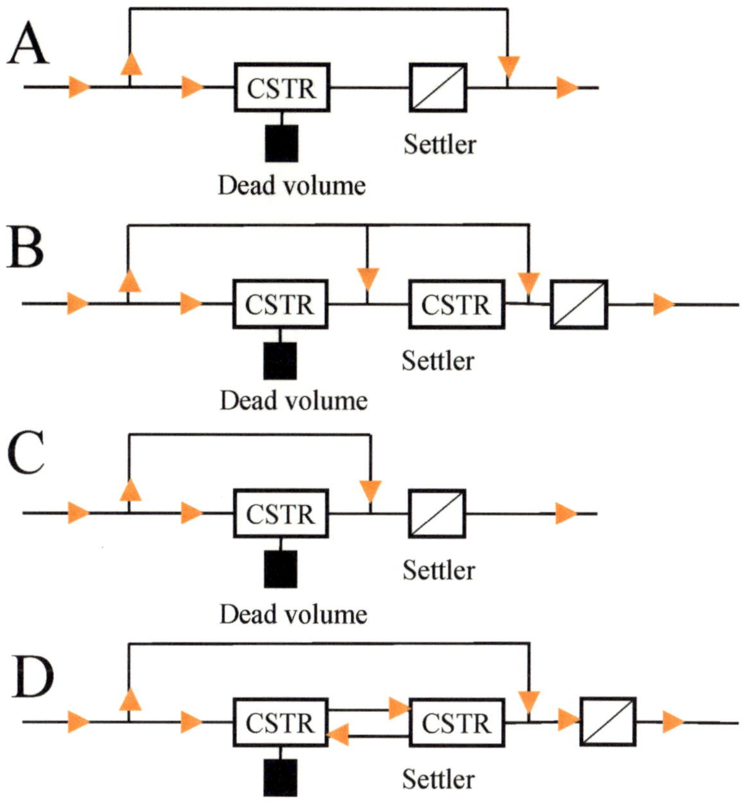

Figure 2.
Flow schemes for UASB reactors. (A) [8], (B) [9, 10], (C) [11], and (D) arrows show flow directions.

modeling of a granular sludge bed can be obtained. However, it is impossible to obtain details of each sludge particle; therefore, an exact model cannot be obtained and is not even necessary.

The hydrodynamic modeling of a granular sludge bed by CFD requires certain techniques. On one hand, a sludge bed can be a liquid that is different from water. The sludge bed and the blanket over the sludge bed can then be modeled separately but simultaneously using different approaches [14]. The sludge concentration has a big impact on sludge hydraulic characteristics such as viscosity [15], and the concentration of the sludge decreases along the reactor height even at lab scales [16]. However, Wang et al. overlooked this difference. In contrast, the sludge bed can be treated as a porous bed. Influent penetrates through the porous bed and flows into a blanket over the porous bed [16]. The permeability and porosity of the sludge bed can be measured and then included in the second method that treats the sludge bed as a porous bed.

The CFD strategy has a few advantages over the RC strategy. First, an optimum flow scheme must be selected from many different flow schemes with the RC strategy; however, there is no need to manually divide a real reactor into virtual reactors while applying a CFD strategy. An RC strategy is difficult to establish with a very complex reactor structure, that is, internal circulation reactor. Second, the CFD strategy can provide many more details about water flow. Complex hydraulic calculations were applied to obtain head loss in a granular bed anaerobic baffled reactors [17]. Such reactor details can be simply extracted from a CFD model, which is more efficient than a complicated manual calculation. Third, the RC strategy cannot provide flow details for each part of a reactor; however, CFD can be applied to obtain the details [18–20].

3.1.2 Applied biokinetics

The biokinetics regarding wastewater treatment are nicely represented by a series of mathematical equations. Either the RC strategy or the CFD strategy is ready to be combined with the equations to model bioprocesses in a bioreactor.

3.1.2.1 Reactor modeling with anaerobic digestion model 1 (ADM1)

The combination of CFD and a simple bioprocess was used to model an expanded granular sludge bed (EGSB) reactor [21]. The biokinetics usually interact with each other. ADM1 is the widely applied model for modeling bioprocesses in anaerobic wastewater treatment. When calibrated, ADM1 can be integrated with hydrodynamic models to obtain an integrated model for reactor modeling. An integrated model can be obtained when a granular sludge bed reactor is treated as a connection of virtual CSTRs and PF reactors and by applying the ADM1 model to each of these virtual reactors. An integrated model was used to model an UASB reactor treating traditional Chinese medicine wastewater [22]. In the integrated model, values for nonsensitive parameters were adopted from public reports, while sensitive parameters were calibrated. Similarly, sensitive parameters were calibrated while modeling a UASB reactor treating wastewater from a molasses-based ethanol distillery [23]. In these two studies, the original form of ADM1 was maintained. However, the ADM1 can be extended to be more practicable. The ADM1 can be extended by extending the number of microbial species [7, 24] or by including new soluble fermentable substrates [25].

3.1.2.2 Reactor modeling with a bioparticle model

Strategies for reactor modeling based on a bioparticle model are reported. However, this reactor modeling strategy is case-specific, and relevant road maps of each strategy are not clearly stated [3, 4, 12]. By cross-checking these models, a general model strategy is summarized below:

i. First, a representative granular size is assumed and applied to all granules in the model.

ii. Second, a RC strategy is applied to divides a real reactor into a single or a series of virtual reactors, that is, CSTRs and/or PF reactors.

iii. Then, the number of representative granules can be obtained in each virtual reactor in the model by measuring the total sludge mass in a real reactor and calculating the mass of the representative granule.

iv. Fourth, the substrate degradation rates in each virtual reactor are obtained by adding substrate degradation rates of all representative granules in each virtual reactor.

v. Finally, the substrate degradation rates in each virtual reactor can be added together to obtain a reactor model that models the operation of a real granular sludge bed.

The bioparticle model applied here is a diffusion-reaction model rather than an IBM because the implementation of an IBM will encounter a huge computational workload. In addition, this strategy can be enriched by including other sub-models such as sludge concentration distribution along reactor height and sludge bed expansion at different upflow velocities [4]. Furthermore, the size of a representative granule is manually but carefully selected while applying a bioparticle model. The weakness of reactor modeling with a bioparticle model is that a reactor model cannot be obtained when a bioparticle model cannot be obtained. A bioparticle model has not been convincingly established for complex substrates. Therefore, reactor modeling with a bioparticle model for complex substrate is still difficult.

Types	Reactors and wastewater	No. of layers	Inputs	Outputs	Ref
BP	UASB, domestic	3	TSS, VSS, COD, alkalinity and VFA concentration, T, pH in the influent	COD of effluent	[30]
BP	UASB, domestic	3	BOD, COD, NH_4–N, TKN	Effluent BOD and COD	[29]
BP	EGSB, denitrifying sulfide removal	4	NO_3^-, NO_2^-, S^{2-}, pH, HRT	Nitrate, nitrite, sulfide acetate	[31]
BP	UASB, denitrifying sulfide removal	3	Influent sulfide, nitrate concentration, S/N mole ratio, pH, and HRT	Sulfide, nitrate removal percentage, sulfate and nitrogen production percentage	[32]
BP	UASB, pharmaceutical	3	Influent COD, HRT, pH, COD loading rate	Effluent COD	[33]
BP	UASB, cotton textile	3	HRT, influent COD, pH, T, alkalinity, VFA, dilution rate, organic load, TSS	COD removal	[34]
NARX	UASB, bagasse wash	4	Influent, flow rate, inlet and outlet COD	Biogas production rate	[35]
ANFIS	EGSB, corn processing	5	Influent COD, Q, TKN, effluent VFA and bicarbonate	Effluent COD	[27]
BP	UASB, molasses	3	OLR, VFA of effluent, influent–effluent alkalinity and pH, T	Biogas production	[36]
AMIMO	UASB, molasses	—	OLR, TCOD removal rate, influent alkalinity and pH, effluent pH	Biogas and methane production rates	[37]
Not clear	UASB, molasses	3	OLR, influent and effluent pH, T, alkalinity effluent COD and VFA concentrations	Biogas and methane production rates	[38]

ANFIS, adaptive neuro-fuzzy inference system; OLR, volumetric organic loading rate; TCOD, volumetric total chemical oxygen demand; AMIMO, multiple inputs and multiple outputs; TSS, total suspended solids; VSS, volatile suspended solids; COD, chemical oxygen demand; VFA, volatile fatty acid; T, temperature; BOD, biological oxygen demand; TKN, total Kjeldahl nitrogen; HRT, hydraulic retention time; Q, reactor flow rate; and OLR, organic loading rate.

Table 1.
Overview of neural networks applying to sludge bed reactor modeling.

3.2 Neural network modeling

Although a anaerobic digestion model can be applied, model calibration is difficult and laborious, while errors between model results and measured results still cannot be ignored [26, 27]. The nonlinear regression method shows an empirical relationship between effluent chemical oxygen demand (COD) and operation parameters such as organic load and hydraulic retention time (HRT) [28]. This empirical relation can be treated as a weak artificial neural network (ANN). ANN can effectively model non-linear systems such as bioreactors for wastewater treatment. ANN is a very powerful computational technique for modeling complex nonlinear relationships [29].

There are a few types of ANNs. **Table 1** shows that the most popular type for granular sludge bed modeling is back propagation (BP). The main difference between various types of ANNs is the network structure and the method for determining the weights and functions for inputs and neurons [29]. **Figure 3** shows the schematic structure of a BP model. In the BP model structure, there is an input layer that is applied for inputting measured data for model training (calibration). An output layer is also required for model results. The selected reactor operation parameters for the input layer and output layer are case-specific. Generally, influent COD and effluent COD are usually applied in an input layer and an output layer to model a wastewater treatment reactor, respectively; pH is not always included. **Table 1** shows the selected parameters for the two layers. In addition, a few layers or a hidden layer is applied to bridge the input layer and output layer. The number of hidden layers in an ANN model is usually determined automatically by a trial and error method, while a single hidden layer network is commonly sufficient for most of the problems [29]. Therefore, three layers are generally applied (**Table 1**). In the structure of an ANN, each layer consists of a few neurons that are shown as circles in **Figure 3**. The connections between neurons in each two nearby layers are usually determined while training the system [29]. The modeled results in the output layer are different from measured data, and the weights are recalculated until the model results that fit the measured results.

Table 1 shows that BP has been effectively applied to model granular sludge-based reactors. Although these reactors treat different kinds of wastewater, the model results can accurately fit the measured results. However, while reactors are treating the same kind of wastewater, model inputs and output can vary significantly [29, 30, 32]. As a result, different ANN models can be applied to UASB reactors to treat the same kind of wastewater [36–38].

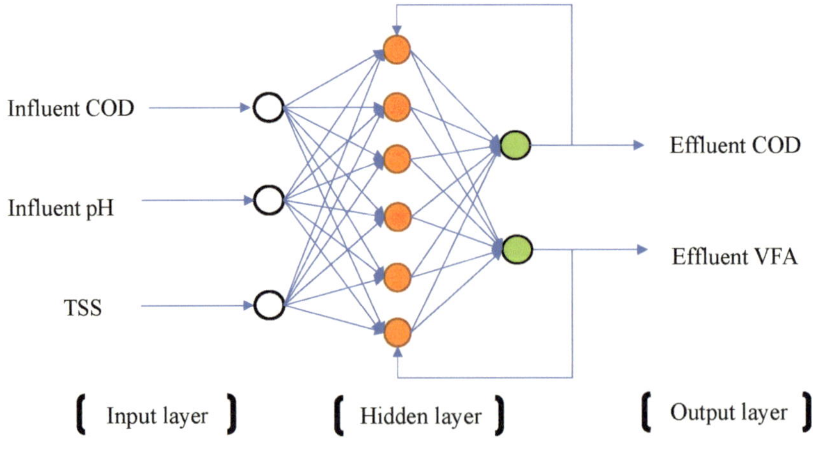

Figure 3.
Schematic view of the structure of BP model.

4. Challenges and discussions

A bioparticle model can explain biokinetics in a sludge granule. Although a bioparticle model can theoretically be applied for reactor modeling, ANN and models integrating reactor hydrodynamics and ADM1 are much more suitable.

A model was established for modeling ANAMMOX process in a single granule based on a few assumptions. However, neither the layered structure nor the cluster structure of a single granule can model complex substrate degradation in a single granule. The IBM model successfully modeled VFA degradation in a granule and shows a clustered structure. The IBM model is based on accurate relevant anaerobic kinetics and can hopefully model complex substrate degradation in a single granule.

The influent COD should be nicely characterized when applying the ADM1 for reactor modeling. The ADM1 requires a detailed characterization of influent organic matter. Nevertheless, such a detailed characterization is generally very difficult [25, 26].

Furthermore, there are 86 parameters in the ADM1. While nonsensitive parameter values can be adopted from the literature, sensitive parameters—which vary significantly—must be calibrated, which is extremely time-consuming and laborious [27]. In addition, the mass of microbial species in bioreactors are not measurable, which challenges the implementation of ADM1 [30].

Calibrating ANN models is easier than ADM1. When the measured variables begin to show differences in the response of ANN models, the model can be retrained using the newer data employed for cross-checking [27]. Numerous applications of ANNs have been successfully utilized in wastewater treatment modeling [38–40]. This is because of the reliable and robust characteristics of ANNs in capturing the nonlinear relationships between variables (multi-input and output) in complex systems.

The other benefit of applying an ANN model for reactor modeling is that an ANN model does not need well-established biokinetics. Currently, the production of extracellular polymeric substances (EPS) is not well understood—a modified ADM1 still cannot effectively model the production of EPS [41]. Therefore, the ANN can be hopefully applied to model the EPS production in an anaerobic sludge bed to provide better effluent quality modeling. Nevertheless, although an ANN model is convenient and reliable in reactor modeling, an applied ANN model cannot explain reactor operation failure because it always treats a bioreactor as a black box.

ANNs are better and more convenient tools for reactor modeling than the integration of hydrodynamics and ADM1. Nevertheless, a calibrated ADM1 can provide more details regarding reactor operation. A calibrated ADM1 can nicely control reactor operation. An algorithm could be developed for ADM1 calibration considering the difficulty in manual ADM1 calibration and efficient calibration of ANN. These have been successfully achieved [42]. This makes applying ADM1 much easier because parameter calibration is not as difficult as it used to be.

5. Conclusions

A bioparticle model is beneficial for providing insights into reactions in the inner space of a granule. Anaerobic ammonia oxidation processes are a simple process and have been modeled in an ANAMMOX granule. However, when complex substrates are involved, a model including relevant bioprocesses in a single granule has not been available. This calls for further research in this field.

The RC and CFD strategies can both be applied to obtain a reactor hydraulic model that can be further integrated with a kinetic model for modeling effluent quality. The RC strategy manually divides a sludge bed reactor into several virtual reactors. The division does not have to fit the real flow conditions in the reactor. Alternatively, the CFD strategy can provide more details for reactor understanding and manipulation while being integrated with a kinetic model.

Parameter calibration for ADM1 is required before being integrated with a hydraulic model—this is a difficult task. Alternatively, most applied BP neural networks can accurately model concentrations of components in effluent, although the involved reactor is still a black box because the BP neural network completely ignores all bioprocesses in the reactor. An algorithm could be programed for ADM1 calibration by applying the high calibrating capacity of the ANN.

Acknowledgements

This chapter was supported by the Youth Innovation Promotion Association (NO.2019375) and Key Research Project from Chongqing City (cstc2018jszx-zdyfxmX0013).

Appendices and nomenclature

ADM1	Reactor modeling with anaerobic digestion model 1
AMIMO	Multiple inputs and multiple outputs
ANAMMOX	An anaerobic ammonium oxidation
ANFIS	Adaptive neuro-fuzzy inference system
ANN	Artificial neural network
BOD	Biological oxygen demand
BP	Back propagation
CFD	Computational fluid dynamics
COD	Chemical oxygen demand
CSTRs	Continuous stirred reactors
Di	The diffusion coefficient of substrate I
EPS	Extracellular polymeric substances
HRT	Hydraulic retention time
IBM	Individual-based model
OLR	Volumetric organic loading rate
OLR	Organic loading rate
PF	Plug-flow reactors
Q	Reactor flow rate
r	The distance from the granule center
RC	Reactor compartmentalization
r_i	The volumetric substrate conversion rate in the granule
S_i	The substrate concentration of component i in the granule
$S_{i,sur}$	The substrate concentration of component i in the granule surface
T	Temperature
TCOD	Volumetric total chemical oxygen demand
TKN	Total Kjeldahl nitrogen
TSS	Total suspended solids
VFA	Volatile fatty acid
VSS	Volatile suspended solids

Author details

Jixiang Yang[1,2]

1 School of Urban Construction and Environmental Engineering, Chongqing University, Chongqing, China

2 Chongqing Institute of Green and Intelligent Technology, Chinese Academy of Sciences, Chongqing, China

*Address all correspondence to: jixiang504@hotmail.com

References

[1] Hulshoff Pol LW, de Castro Lopes SI, Lettinga G, Lens PNL. Anaerobic sludge granulation. Water Research. 2004;**38**:1376-1389. DOI: 10.1016/j.watres.2003.12.002

[2] Gonzalez-Gil G, Lens PNL, Van Aelst A, Van As H, Versprille AI, Lettinga G. Cluster structure of anaerobic aggregates of an expanded granular sludge bed reactor. Applied and Environmental Microbiology. 2001;**67**:3683-3692. DOI: 10.1128/aem.67.8.3683-3692.2001

[3] Ni B-J, Chen Y-P, Liu S-Y, Fang F, Xie W-M, Yu H-Q. Modeling a granule-based anaerobic ammonium oxidizing (ANAMMOX) process. Biotechnology and Bioengineering. 2009;**103**:490-499. DOI: 10.1002/bit.22279

[4] Fuentes M, Scenna NJ, Aguirre PA. A coupling model for EGSB bioreactors: Hydrodynamics and anaerobic digestion processes. Chemical Engineering and Processing. 2011;**50**:316-324. DOI: 10.1016/j.cep.2011.01.005

[5] Picioreanu C, Vrouwenvelder JS, van Loosdrecht MCM. Three-dimensional modeling of biofouling and fluid dynamics in feed spacer channels of membrane devices. Journal of Membrane Science. 2009;**345**:340-354. DOI: 10.1016/j.memsci.2009.09.024

[6] Vrouwenvelder JS, Picioreanu C, Kruithof JC, van Loosdrecht MCM. Biofouling in spiral wound membrane systems: Three-dimensional CFD model based evaluation of experimental data. Journal of Membrane Science. 2010;**346**:71-85. DOI: 10.1016/j.memsci.2009.09.025

[7] Rajinikanth R, Ramirez I, Steyer JP, Mehrotra I, Kumar P, Escudie R, et al. Experimental and modeling investigations of a hybrid upflow anaerobic sludge-filter bed (UASFB) reactor. Water Science and Technology. 2008;**58**:109-117. DOI: 10.2166/wst.2008.342

[8] Wu M, Hickey R. Dynamic model for UASB reactor including reactor hydraulics, reaction, and diffusion. Journal of Environmental Engineering. 1997;**123**:244-252. DOI: 10.1061/(ASCE)0733-9372(1997)123:3(244)

[9] Bolle WL, Vanbreugel J, Vaneybergen GC, Kossen NWF, Zoetemeyer RJ. Modelling the liquid flow in up-flow anaerobic sludge blanket reactors. Biotechnology and Bioengineering. 1986;**28**:1615-1620. DOI: 10.1002/bit.260281105

[10] Pontes RFF, Pinto JM. Analysis of integrated kinetic and flow models for anaerobic digesters. Chemical Engineering Journal. 2006;**122**:65-80. DOI: 10.1016/j.cej.2006.02.018

[11] Singhal A, Gomes J, Praveen VV, Ramachandran KB. Axial dispersion model for upflow anaerobic sludge blanket reactors. Biotechnology Progress. 1998;**14**:645-648. DOI: 10.1021/bp980042f

[12] Rodriguez-Gomez R, Renman G, Moreno L, Liu L. A model to describe the performance of the UASB reactor. Biodegradation. 2014;**25**:239-251. DOI: 10.1007/s10532-013-9656-z

[13] Narayanan CM, Narayan V. Multiparameter models for performance analysis of UASB reactors. Journal of Chemical Technology and Biotechnology. 2008;**83**:1170-1176. DOI: 10.1002/jctb.1959

[14] Wang X, Ding J, Ren N-Q, Liu B-F, Guo W-Q. CFD simulation of an expanded granular sludge bed (EGSB) reactor for biohydrogen production.

International Journal of Hydrogen Energy. 2009;**34**:9686-9695. DOI: 10.1016/j.ijhydene.2009.10.027

[15] Yang J, Vedantam S, Spanjers H, Nopens I, van Lier JB. Analysis of mass transfer characteristics in a tubular membrane using CFD modeling. Water Research. 2012;**46**:4705-4712. DOI: 10.1016/j.watres.2012.06.028

[16] Yang J, Yang Y, Ji X, Chen Y, Guo J, Fang F. Three-dimensional modeling of hydrodynamics and biokinetics in EGSB reactor. Journal of Chemistry. 2015;**501**:1-7. DOI: 10.1155/2015/635281

[17] Shanmugam AS, Akunna JC. Modelling head losses in granular bed anaerobic baffled reactors at high flows during start-up. Water Research. 2010;**44**:5474-5480. DOI: 10.1016/j.watres.2010.06.062

[18] Wu R-M, Lee T-H, Yang W-J. Study of flow in a blanket clarifier using computational fluid dynamics. Journal of Environmental Engineering. 2008;**134**:443-455. DOI: 10.1061/(asce)07733-9372(2008)14.6443

[19] Qi W-K, Hojo T, Li Y-Y. Hydraulic characteristics simulation of an innovative self-agitation anaerobic baffled reactor (SA-ABR). Bioresource Technology. 2013;**136**:94-101. DOI: 10.1016/j.biortech.2013.02.033

[20] Wang H, Ding J, Liu X, Ren N. The impact of water distribution system on the internal flow field of EGSB by using CFD simulation. International Conference Machinery, Electronics and Control Simulation. 2014;**614**:596-604. DOI: 10.4028/www.scientific.net/AMM.614.596

[21] Wang X, Ding J, Guo WQ, Ren NQ. A hydrodynamics-reaction kinetics coupled model for evaluating bioreactors derived from CFD simulation. Bioresource Technology. 2010;**101**:9749-9757. DOI: 10.1016/j.biortech.2010.07.115

[22] Chen Z, Hu D, Zhang Z, Ren N, Zhu H. Modeling of two-phase anaerobic process treating traditional Chinese medicine wastewater with the IWA anaerobic digestion model no. 1. Bioresource Technology. 2009;**100**:4623-4631. DOI: 10.1016/j.biortech.2009.04.066

[23] Taruyanon K, Tejasen S. Modelling of two-stage anaerobic treating wastewater from a molasses-based ethanol distillery with the IWA anaerobic digestion model no. 1. Engineering Journal. 2010;**14**:25-36. DOI: 10.4186/ej.2010.14.1.25

[24] Ramirez I, Steyer JP. Modeling microbial diversity in anaerobic digestion. Water Science and Technology. 2008;**57**:265-270. DOI: 10.2166/wst.2008.055

[25] Garcia-Gen S, Lema JM, Rodriguez J. Generalised modelling approach for anaerobic co-digestion of fermentable substrates. Bioresource Technology. 2013;**147**:525-533. DOI: 10.1016/j.biortech.2013.08.063

[26] Dereli RK, Ersahin ME, Ozgun H, Ozturk I, Aydin AF. Applicability of anaerobic digestion model no 1 (ADM1) for a specific industrial wastewater opium alkaloid effluents. Chemical Engineering Journal. 2010;**165**:89-94. DOI: 10.1016/j.cej.2010.08.069

[27] Yetilmezsoy K, Ozgun H, Dereli RK, Ersahin ME, Ozturk I. Adaptive neuro-fuzzy inference-based modeling of a full-scale expanded granular sludge bed reactor treating corn processing wastewater. Journal of Intelligent Fuzzy Systems. 2015;**28**:1601-1616. DOI: 10.3233/ifs-141445

[28] Yetilmezsoy K, Sakar S. Development of empirical models for performance evaluation of UASB reactors treating poultry manure wastewater under different operational conditions. Journal of Hazardous

Materials. 2008;**153**:532-543. DOI: 10.1016/j.jhazmat.2007.08.087

[29] Singh KP, Basant N, Malik A, Jain G. Modeling the performance of "up-flow anaerobic sludge blanket" reactor based wastewater treatment plant using linear and nonlinear approaches—A case study. Analytica Chimica Acta. 2010;**658**:1-11. DOI: 10.1016/j.aca.2009.11.001

[30] Mendes C, da Silva Magalhes R, Esquerre K, Queiroz LM. Artificial neural network modeling for predicting organic matter in a full-scale up-flow anaerobic sludge blanket (UASB) reactor. Environmental Modeling and Assessment. 2015;**20**(6):625-635. DOI: 10.1007/s10666-015-9450-x

[31] Wang A, Liu C, Han H, Ren N, Lee D-J. Modeling denitrifying sulfide removal process using artificial neural networks. Journal of Hazardous Materials. 2009;**168**:1274-1279. DOI: 10.1016/j.jhazmat.2009.03.006

[32] Cai J, Zheng P, Qaisar M, Luo T. Prediction and quantifying parameter importance in simultaneous anaerobic sulfide and nitrate removal process using artificial neural network. Environmental Science and Pollution Research. 2015;**22**:8272-8279. DOI: 10.1007/s11356-014-3976-3

[33] Chen Z, Wang H, Chen Z, Ren N, Wang A, Shi Y, et al. Performance and model of a full-scale up-flow anaerobic sludge blanket (UASB) to treat the pharmaceutical wastewater containing 6-APA and amoxicillin. Journal of Hazardous Materials. 2011;**185**:905-913. DOI: 10.1016/j.jhazmat.2010.09.106

[34] Yetilmezsoy K, Sapci-Zengin Z. Stochastic modeling applications for the prediction of COD removal efficiency of UASB reactors treating diluted real cotton textile wastewater. Stochastic Environmental Research and Risk Assessment. 2009;**23**:13-26. DOI: 10.1007/s00477-007-0191-5

[35] Jain V, Sambi S, Kumar S, Kumar B. Modeling of a UASB reactor by NARX networks for biogas production. Chemical Product and Process Modeling. 2015;**10**:113-121. DOI: 10.1515/cppm-2014-0035

[36] Kanat G, Saral A. Estimation of biogas production rate in a thermophilic UASB reactor using artificial neural networks. Environmental Modeling and Assessment. 2009;**14**:607-614. DOI: 10.1007/s10666-008-9150-x

[37] Turkdogan-Aydinol FI, Yetilmezsoy K. A fuzzy-logic-based model to predict biogas and methane production rates in a pilot-scale mesophilic UASB reactor treating molasses wastewater. Journal of Hazardous Materials. 2010;**182**:460-471. DOI: 10.1016/j.jhazmat.2010.06.054

[38] Yetilmezsoy K, Turkdogan FI, Temizel I, Gunay A. Development of ANN-based models to predict biogas and methane productions in anaerobic treatment of molasses wastewater. International Journal of Green Energy. 2013;**10**:885-907. DOI: 10.1080/15435075.2012.727116

[39] Hamed MM, Khalafallah MG, Hassanien EA. Prediction of wastewater treatment plant performance using artificial neural networks. Environmental Modelling & Software. 2004;**19**:919-928. DOI: 10.1016/j.envsoft.2003.10.005

[40] Raduly B, Gernaey KV, Capodaglio AG, Mikkelsen PS, Henze M. Artificial neural networks for rapid WWTP performance evaluation: Methodology and case study. Environmental Modelling & Software. 2007;**22**:1208-1216. DOI: 10.1016/j.envsoft.2006.07.003

[41] Aquino SF, Stuckey DC. Integrated model of the production of soluble microbial products (SMP) and extracellular polymeric substances (EPS) in anaerobic chemostats during transient conditions. Biochemical Engineering Journal. 2008;**38**:138-146. DOI: 10.1016/j.bej.2007.06.010

[42] Yang J, Lu L, Ouyang W, Gou Y, Chen Y, Ma H, et al. Estimation of kinetic parameters of an anaerobic digestion model using particle swarm optimization. Biochemical Engineering Journal. 2017;**120**:25-32. DOI: 10.1016/j.bej.2016.12.022

Chapter 6

Innovative Strategies for the Control of Biofilm Formation in Clinical Settings

Aqsa Shahid, Maria Rasool, Naheed Akhter, Bilal Aslam, Ali Hassan, Sadia Sana, Muhammad Hidayat Rasool and Mohsin Khurshid

Abstract

Biofilm formation in clinical settings is an increasingly important issue particularly due to the emergence of multidrug-resistant strains, as it resulted in increased mortality, which poses a considerable financial burden on healthcare systems. The bacterial biofilms are quite resistant to the routine antimicrobial-based therapies; therefore, the novel strategies are desired in addition to the conventional antibiotics for the effective control of infections caused by biofilm-forming microbes. So far, the approaches being proposed to control the biofilm formation in clinical practice settings include the use of biofilm inhibitors and the use of modified biomaterials for the development of medical devices to thwart the formation of biofilms. In this chapter, we have focused on the latest developments in the anti-biofilm strategies through the interruption of the quorum-sensing system, which is crucial for biofilm formation and have summarized the various classes of antibacterial compounds for the control of biofilm formation. This agrees with the recent approaches suggested by the National Institute of Health (NIH) that advocates the use of combinational therapies based on the conventional methods and complementary treatment to explore the potential utility and safety concerns of the natural products. The studies regarding these emerging strategies could possibly lead to the establishment of better therapeutic alternates compared to conventional treatments.

Keywords: biofilms, infections, catheters, antimicrobials, quorum sensing, implants

1. Introduction

Group of microbial cells that are surface-attached and embedded within the extracellular matrix (self-produced), and are strikingly resistant to antimicrobials are called biofilms [1]. Biofilms can adhere to almost different types of surfaces like body tissue and plant, plastics, metals, implant objects as well as medical devices [2]. Formation of biofilm on implants and medical equipment and implants, for example, vascular grafts, prosthetic joints, heart valves, catheters, intrauterine devices, pace-makers, and contact lenses can cause infection. Central line-associated bloodstream infection (CLABSI) can occur due to use of intravascular catheters, furthermore,

CLABSI can cause an increased rate of mortality and morbidity, and every year in the USA almost 250,000 cases of bloodstream infections are reported [3].

When cells adhere and attach to surfaces biofilm formation begins. Several factors can promote the attachment of microorganisms to biomaterials including increased shear forces, bacterial motility, and electrostatic as well as hydrodynamic interactions between the surface and microbial cells [4]. It has been observed that adherence of biomaterials to bacteria via biomaterial-surface interactions and cell-surface is facilitated by numerous factors, such as protein autolysin, surface, and adhesion proteins and capsular polysaccharides, etc. For example, '*Staphylococcus* species' show cell-surface proteins that are vital for adherence of '*Staphylococcus epidermidis*' to polystyrene which is named as staphylococcal surface protein-1 and -2 (SSP-1 and SSP-2). After attachment to the extracellular surfaces, microbial cells will start aggregate, multiply, and eventually differentiate into the biofilm network [5]. Such microbial cells can then be separated from mature biofilms, can cause chronic infections and can spread to other organs also [6].

Another worrying characteristic of infections associated with biofilm formation is increased biofilm cell tolerance to biocides. As biofilms provide an excellent niche for exchange of plasmid, so increased resistance to the drug can affect genes containing plasmids which results in multidrug resistance (MDR) phenotypes. Enhanced drug resistance mechanisms include incomplete or slow infiltration of antimicrobials within the extracellular matrix, the formation of dormant cells during the non-dividing phase, reduced cell's growth rate within the biofilm, hence ultimately decreasing total targets for antimicrobial molecules [7]. Furthermore, it is difficult to treat biofilm formation with the traditional antimicrobial approach and the therapy is further inhibited by increased resistance to the antibiotic because under antibiotic selective pressure microbial cells develop resistance. For instance, it has been observed that almost above 70% of hospital isolate of '*Staphylococcus epidermidis*' show resistance to methicillin and surprisingly there are many strategies to prevent infections associated with biofilm formation other than antibiotic treatment [8]. In this chapter, we have focused on anti-biofilm approaches and some promising efforts for controlling these biofilm-based infections.

2. The process of biofilm development

The production and maturation of biofilm are complex, subsequent and dynamic processes, depending upon several factors i.e. cellular metabolism, intrinsic properties of the cells, genetic control, the substratum, and the medium signaling molecules. Biofilm formation is introduced with a conditioning film of inorganic or organic material on the cell surface; furthermore, this layer modifies the surface feature of substratum which ultimately favors microbes for colonization on the cell surface. The formation of biofilm consist of several different steps: (i) initially the reversible attachment of microbial cells with biotic or abiotic surfaces through weak forces for example van der Waals forces, (ii) irreversible attachment to the cell surface with the help of different attachment structure i.e. lipopolysaccharides, flagella, adhesive proteins or fimbriae by hydrophobic or hydrophilic interactions, (iii) and then eventually biofilm architecture development due to the production and proliferation of extracellular polysaccharide (EPS) matrix which is self-produced and is made up of proteins, extracellular deoxyribonucleic acid (DNA) and polysaccharides [9] (iv) in the next step mature biofilm is formed which has water channels that are responsible for distribution of nutrients as well as signaling molecules within the biofilm [10], (v) and then due to extrinsic or intrinsic factors separation of biofilm cells occurs individually or in clumps and finally colonization of other niches and dispersal of the cells [11].

3. Inhibition of initial attachment

3.1 Altering physical properties of biomaterials

Biofilm development starts with a reversible weak adhesion of microbial cells to the exterior surface of medical equipment, however, if they are not removed from the exterior of devices, they adhered permanently through their adhesion structures i.e. fimbriae, pili and thereby forming biofilm matrix [12]. Surface charge and hydrophobicity of implant constituents play a significant role in controlling the ability of microbes to anchor to cell surfaces. Therefore, alteration in the hydrophobicity and surface charge of polymeric constituents are proved as efficient for controlling biofilm formation by using numerous antimicrobial agents and backbone compounds [13]. Poly N-vinylpyrrolidone and Hydrophilic polymers i.e. hyaluronic acid [14] on silicone shunt and polyurethane catheters have been widely used to decrease the adherence of '*Staphylococcus epidermidis*'. Furthermore, several hydrogel membranes have been introduced particularly for ureteral stents that decrease bacterial adherence because of their hydrophilic characteristics. It has been observed that due to very low wettability superhydrophobic coatings play a significant role to reduce the biofilm matrix formation and adhesion of bacteria [15]. Later, it has been suggested that *S. aureus* and *Pseudomonas aeruginosa* poorly attached on superhydrophobic fluorinated silica coating as well as on titanium coatings. However, it was demonstrated that *Escherichia coli* and *Staphylococcus aureus* poorly adhered on other superhydrophobic surfaces i.e. (AACVD) aerosol assisted chemical vapor deposition-coated [16]. In some cases, it was observed that hairpin coating affects colonization and adhesion of bacteria because it forms vascular catheter negatively charged, so contribute to reducing the catheter-related infections, inhibiting microbial colonization and thrombosis [17]. It has been described that the surface roughness can modulate hydrophobicity, which ultimately influences the bacterial adhesion [18].

3.2 Altering the chemical properties of biomaterials

There are several chemical approaches used to alter the exterior of biomedical equipment to inhibit the biofilm formation comprising ion coatings, biocides and also antibiotics [19]. Catheters that are impregnated with antibiotics, for example, rifampin and minocycline have been revealed to reduce the occurrence of biofilm-based infections by *Staphylococcus aureus*. Furthermore, catheters are coated with several antibiotics that play a significant role in biofilm production during urinary tract infections (UTI) like norfloxacin, nitrofurazone, and gentamicin [20]. Several chemical molecules are identified through screening of chemical libraries, these molecules are used as potential drugs to control infection and biofilm development. Furthermore, such molecules do not provoke antimicrobial action, and hence reduces the development of resistance due to no selective pressure against biofilm matrix formation. In *Staphylococcus aureus* and *Streptococcus pyogenes* a series of small chemical molecules have an inhibitory effect on the expression of different important virulent factors during infection and biofilm formation [21]. Several aryl rhodamines showed inhibitory effect on early stages of biofilm development in *Enterococcus faecalis, S. epidermidis,* and *S. aureus*. Moreover, it was reported that a mucolytic mediator N-acetylcysteine has inhibited the formation of exopolysaccharides in the biofilm layer in case of *S. epidermidis* [22]. In another microorganism *Vibrio cholerae*, small substances suppressed the initiation of cyclic di-GMP that acts as the second messenger to control switch in-between the aquatic and sessile way of living of microbes [23].

It has been observed that numerous antibacterial peptides also inhibit biofilm formation in several microbes. For instance, it is considered that peptide 1018 has inhibitory effects in different microbes such as in *Acinetobacter baumannii, Burkholderia cenocepacia, Klebsiella pneumoniae, P. aeruginosa, E. coli, Salmonella typhimurium* and *S. aureus* [24]. Furthermore, class of peptide antibiotics called lantibiotics i.e. gallidermin, epidermin, subtilin, and nisin has been reported and control the biofilm production in *S. aureus, S. epidermidis* and also in *Lactococcus lactis*.

Chelators hindering the role of metal ions in the production of biofilm are considered as biofilm inhibitors, for example, silver salts, metallic silver and also silver nanoparticles are commonly employed as antibacterial agents in clinical implants against *P. aeruginosa, Salmonella typhimurium, Klebsiella species, E. coli,* and *S. aureus* [25]. It is observed that antibiotics i.e. amoxicillin, clindamycin, vancomycin, penicillin G and erythromycin show increased antimicrobial activity against *Staphylococcus aureus* in the presence of nanoparticles [26]. Treatment with silver substances prevents DNA replication, expression of cellular as well as ribosomal proteins, and also respiration process that leads to death of the cell [27]. In addition, It is also suggested that silver-coated implants inhibit *Staphylococcus aureus* biofilm production without aggregating silver inside the host tissue [28].

4. Quorum quenching

In the majority of Gram-negative and Gram-positive bacteria, an essential cellular communicating system is presently called as Quorum sensing, which regulates a variety of genes in accordance with the density of signaling molecules furthermore, signaling molecules are called autoinducers [29]. On the bases of signaling molecules QS is classified into three i.e. autoinducing peptide (AIP-based) for Gram-positive bacteria, N-acyl homoserine lactones (AHLs-based) for Gram-negative bacteria and autoinducer-2 (AI-2-based) for both Gram-negative and Gram-positive bacteria [30]. When the biofilm is formed, after the initial attachment, cells secrete QS molecules that alter the expression of the microbial gene, thus changing planktonic form into a sessile form. Furthermore, QS plays a significant role in biofilm development, so It has been observed that QS inhibition i.e. quorum quenching (QQ) would be a striking approach to control biofilm formation [31]. QS system is thought to be a target for developing new antimicrobial agents, moreover, QS system plays a crucial role in regulating pathogenetic factors and also virulence factors production in several pathogens [32]. The most important benefit of preventing biofilm formation by QQ is that this approach decreases the risk of multidrug resistance (MDR) and thus creating this approach noticeable to prevent biofilm-based infections in clinical settings. The different approaches for the inhibition or removal of biofilms are summarized in **Table 1** and **Figure 1**.

Bacteria	Compound	Mechanism	Antibiofilm activity	References
P. aeruginosa	N-Acyl homoserine lactones	Transcriptional regulators (LuxR and LasR)	Decreased the production of QS signals and virulence factors	[49]
	Patriniae	Biofilm related genes	Reduced the production of exopolysaccharide	[50]
	Hordenine	Quorum sensing related genes	Blocked QS-controlled phenotypes like biofilm formation	[51]
	Quercetin	Transcriptional regulators of quorum sensing related genes	Inhibition of biofilm formation	[52, 53]
	'*Piper betle*' Leaves (Ethanolic Extract)	Pyocyanin	Inhibited Pyocyanin production and reduced twitching ability	[53]
	Parthenolide	Extracellular polymeric substance and transcriptional regulators of quorum sensing related genes	Inhibition of the expression of QS related genes expression and downregulation of extracellular polymeric substance production	[54]
E. coli O157:H7	Ginkgolic acids (GAs)	Curli gene expression, prophage genes	Biofilm formation was inhibited on the polystyrene, glass and nylon membrane	[55]
	Phloretin	Toxin genes, autoinducer-2 importer genes curli genes, prophage genes	Decreased biofilm formation and production of fimbria	[55]
	Cinnamaldehyde	LuxR-DNA-binding	Affected the biofilm formation and virulence	[56]
S. mutans	'*Zingiber officinale*' (Methanolic fraction)	F-ATPase activity, virulence genes, surface protein antigen (SpaP)	Affected the cell-surface hydrophobicity index, Inhibited surface protein antigen (SpaP)	[57]
	Leaf extract of '*Bergenia crassifolia*' (L.)	Exopolysaccharides (EPSs), glucosyltransferases (Gtfs)	Decreased adherence properties of bacterial cells	[58]
	Quercetin	pH	Disrupted the pH in biofilm	[59]
Staphylococcus aureus and *Staphylococcus epidermidis*	'*Rhodomyrtus tomentosa*' (Ethanol extract)	Not mentioned	Inhibition of biofilm formation and disruption of mature biofilm	[60]
S. aureus strains	Phloretin	Efflux protein genes	Anti-biofilm formation at low	[61]

Bacteria	Compound	Mechanism	Antibiofilm activity	References
S. aureus and *C. albicans*	'*Hymenocallis littoralis*' leaf extract	Adhesin proteins	Antimicrobial and anti-biofilm activity	[62]
Streptococcus pneumoniae	Quercetin	SrtA gene	The blockage of SrtA gene function, impairment of biofilm formation	[63]
Enterococcus faecalis	Quercetin	Protein translation and folding pathways	Blocked the protein translation and folding pathways	[64]

Table 1.
Anti-biofilm compounds for various clinically important bacteria.

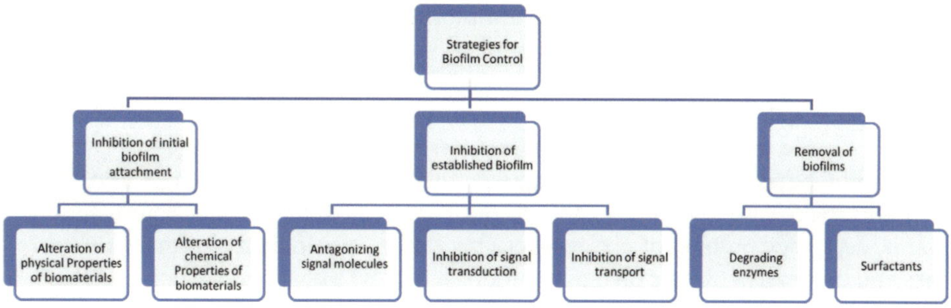

Figure 1.
An overview of the different anti-biofilm strategies.

5. Removal of biofilms

Another anti-biofilm approach is the dissociation of the biofilm matrix which accounts for around 90% of biofilm dry mass. This dissociation will ultimately expose the sessile bacteria to the antibiotics as well as host immune defense. The enzymes majorly employed for biofilm matrix-degradation can be divided into three categories Proteases, nucleases and polysaccharide degrading enzymes [33].

Moreover, the surfactants also possess the antibiofilm activities as the cetyltrimethylammonium bromide (CTAB), sodium dodecyl sulfate (SDS) and Tween 20 have been found to promote the detachment or dispersal of biofilms. Surfactin; a biosurfactant produced by the *Bacillus subtilis* was shown to inhibit the biofilm dispersal in *E. coli*, *Proteus mirabilis,* and *S. typhimurium* [33].

6. Bacteriophages as antibiofilm agents

Bacteriophages are considered as the largest creature in the biosphere, because of antibiotic resistance development, bacteriophages play an important role in the destruction of microbes. Use of bacteriophages is now considered as an alternative strategy to antibiotics, particularly for disruption or biofilm inhibition. Bacteriophages are beneficial than chemical agents and antibiotics. The isolation of bacteriophage is simple and fast, furthermore, its production is also cheap, and these are very distinct against a host or either host range, therefore, do not disrupt the normal flora. Bacteriophages are ecologically friendly, so with the persistence of host bacteria, they can replicate at the target site and have no adverse effects.

Bacteriophages also considered as potent antibiofilm mediators, e.g., phage T4 can cause infection and replicates within *E. coli* biofilms and by destroying microbial cells it can disturb the biofilm matrix. Doolittle and colleagues reported a study and demonstrated the interaction of phages with biofilms. The interaction among biofilm and phage is a dynamic as well as a sequential process. Phage adsorption with the target bacterial receptors is the significant phase in phage infection. The EPS matrix suggests a potent challenge for bacteriophages as EPS must be enough penetrated so that bacteriophages can attach with and reach to the particular host receptors. Furthermore, the EPS matrix also helps in the protection of bacteria in the biofilm. Moreover, by diffusion or through phage derived enzymes, for example, polysaccharide depolymerase can easily penetrate the EPS layer because these enzymes have the ability to destroy the structure of biofilm so that these phages can readily anchor to outer membrane receptors, lipopolysaccharides, or other proteins that are essential for replication process [34]. It is surely suggested that these phages induced depolymerizes can easily disrupt biofilms. Now genetically engineering for phages have been introduced that explicit biofilm degrading enzymes during infection. The scientist has engineered a gene namely "dispersion" (dspB) into an *E. coli* specific T7 phage to yield an engineered enzymatic phage, which shows more efficacy for the removal of biofilms as compared to non-cloned phages.

Despite the several benefits of phage use, there are some disadvantages also, for example, the release of a considerable amount of bacterial membrane-bound endotoxins, decreased number of phages encoding toxins, insufficient pharmacokinetic data and conversion of lytic phages to prophages is also a big concern. Some of the above-mentioned problems have been well determined through different processes like designing a recombinant phage from *Pseudomonas aeruginosa* filamentous phage to minimize the mortality rate in experimental animals and release of membrane-bound endotoxins to report the endotoxin release issue is major advances to overcome the above-mentioned concerns [35]. It has been observed that bacteriophages and antibiotics have a big potential to control biofilms such as phage PhilBB-PF 7A plays role in the removal of *Pseudomonas fluorescens* biomass and has shown almost 63–91% activity.

Different studies show some of the strongest inhibitions, for example, the existence of biofilm EPS matrix hindering the control of biofilm via antibacterial agents and higher antibiotic resistance can be controlled through phage use. Furthermore, there are many limitations of phage use such as microbial resistance to phages, virulence genes that are phage-encoded can incorporate inside the host bacterial genome and the narrow host range. Phage efficacy can also be reduced by the immune system, and phage preparations that are improperly obtained can also contain endotoxin. To control these obstacles engineered phages or phage mixtures can be an effective alternative. Moreover, after proper selection and several studies phages has become one of the most useful anti-biofilm agents.

7. Natural anti-biofilm strategies

7.1 Plant extracts

Many extracts of plants and their derivatives were widely studied to eliminate the '*Propionibacterium acne*' biofilm [36]. It has been reported that out of 119 plant extracts, five showed strong antibiofilm activity i.e. *Rhodiola crenulata, Dolichos lablab*, *Malus pumila, Epimedium brevicornum,* and *Polygonum cuspidatum.* These scientists also suggested that extracts of *P. cuspidatum* and *E. brevicornum* and their active derivatives i.e. resveratrol and icartin show a potential antibiofilm activity even when used at lowest MIC. Bark extracts of *Melia dubia* were evaluated with 30 mg/mL concentration [37]. Furthermore, these extracts exhibit potential suppression of hydrophobicity, swarming motility, hemolysis, and biofilm production in *E. coli*. Other colleagues also reported similar results about *Capparis spinosa* (caper bush) extract, this extract shows inhibitory effect on the EPS production and biofilm production in *Serratia marcescens, Pseudomonas aeruginosa, Proteus mirabilis,* and *Escherichia coli* at 2 mg/mL concentration [38]. In addition, well-known biofilm formation of 3 microbes was dispersed. A medically important plant '*Lagerstroemia speciosa*' usually present in Southeast Asia, fruit extract from this plant is capable of inhibiting biofilm formation by '*P. aeruginosa*' PAO1 at 10 mg/mL concentration [39].

Other two plant extracts Dandasa (*Juglans regia* Tree Bark) and green tea (*Camellia sinensis*) show a potential antibiofilm activity individually. Recently, researchers observed that both Green tea and Dandasa exhibit potential antibiofilm activity of *Streptococcus mutans* at 12.5 and 6.2 mg/mL concentration, respectively, and on *E. coli* at 3.1 and 12.5 mg/mL concentration, respectively.

Allium sativum extract i.e. fresh garlic extract (FGE) has a potential inhibitory effect against biofilm formation, it has been observed that FGE decreased '*P. aeruginosa*' biofilm formation [40]. In-vitro screening of antibiofilm activity of '*Staphylococcus epidermidis*' of different 45 aqueous extracts from twenty-four Caatinga (Brazilian xeric shrubland) medicinal species was published. Extremely favorable extracts were taken from *Chamaecrista desvauxii* fruits, *Pityrocarpa moniliformis* leaves, *Bauhinia acuruana* fruits and *B. acuruana* branches, which show decreased the formation of biofilm even when they were tested at the lowest concentration. In addition, it was also suggested that *Senna macranthera* and *Commiphora leptophloes* fruit extracts decreased biofilms by 66.7% and 67.3% respectively. *Mycobacterium smegmatis* which plays a significant role in biofilm development was observed by using many quantitative and qualitative techniques. Other scientists examined different plants i.e. *Vaccinium oxycoccos*, *Hippophae rhamnoides, Azadirachta indica* and *Juglans regia* and spices to look for useful antibiofilm natural substitutes. When the efficiency of plant extracts as an antibiofilm agent was checked it showed that the extract of *Azadirachta indica* usually named as "Neem" was surprisingly helpful at removing and lowering *M. smegmatis* biofilms [41].

Another plant extract 'casbane diterpene' isolated from "*Croton nepetaefolius*" extract, is used to suppress the biofilm production of five Gram-negative bacterial species (*Klebsiella oxytoca, Klebsiella pneumoniae, Escherichia coli, Pseudomonas fluorescens* and *Pseudomonas aeruginosa*), two Gram-positive bacterial species (*S. epidermidis* and *S. aureus*), and three yeast species (*Candida glabrata, Candida tropicalis* and *Candida albicans*) [42]. Furthermore, another study demonstrated that *Candida* biofilm formation was remarkably decreased by *Boesenbergia pandurata* also known as "finger root oil" almost by 63–98% when MIC levels were used from 4 to 32 μL/mL [43]. Later studies showed that different plant extracts were isolated against *Enterohemorrhagic E. coli* (EHEC) O157:H7 biofilm. Furthermore, this study displayed that out of 498 plant extracts, almost 16 of them showed an inhibitory effect on biofilm formation of EHEC above 85% with no-growth of planktonic cells [44]. Certainly, these results specify that these different plant extracts show maximum inhibitory effect on biofilm formation of several microbes. Hence, it is suggested that further efforts are required to study the potential of these plant extracts as antibiofilm agents in detail.

7.2 'Honey'

A natural product extracted by 'honey' bee from floral nectar is called as 'honey' however, 'honey' is generally common and is usually used for its remarkable activity in wound-healing, anti-inflammatory, and antibacterial activity and used as an antioxidant. It has antimicrobial activities against 60 species of fungi and bacteria. 'Honey' was reported as a useful agent to control the biofilm formation. Furthermore, it was described that 'honey' is effective in the prevention of *Enterococcus spp.* biofilm production and can also use as a therapeutic agent against many *Enterococcal* infections that are biofilm-related. It can also decrease the biofilm production of EHEC O157:H7. Recent studies show that very low quantity of 'honey' can significantly decrease the formation of biofilm, the virulence of *E. coli* O157:H7 and Quorum sensing. So, a very low 'honey' concentration can decrease the formation of biofilm by preventing the virulence genes transfer in microbes and the expression of biofilm-associated curling QS, without inhibiting the cell growth. Due to its antimicrobial properties, high concentration of 'honey' can also prevent biofilm formation as well as adhesion of bacteria. Despite its antibacterial activity, it is also observed that 'honey' inhibits biofilm formation by antibacterial peptide which is bee defensin 1 that prevents microbial viability as well as biofilm formation indirectly [41].

7.3 Essential oils

Naturally plant-derived volatile substances are called as essential oils (EOs). Because of their antibacterial and preservative properties, these are effective and favorable natural products for the food industry. These essential oils are commonly used against a wide diversity of microorganisms since ancient time. These oils exhibit antimicrobial impact on the cell wall of microbes, leading to the destruction of microbes. Furthermore, it is suggested that these oils are very effective in inactivating many microbes without producing antimicrobial resistance [45]. Because of little mammalian toxicity, rapid degradation in the environment and availability of many essential oils make them beneficial antibiofilm agent [46].

Cumin oil scientifically named *Cuminum cyminum,* a derivative of an aromatic, therapeutic plant of "Apiaceae" family, has various medicinal properties and in the digestive system, it acts as an astringent. It has been widely used for acute gastric diseases as a carminative and eupeptic, and as an analgesic. It is also widely used to flavor foods, for example, added in food for fragrance. Cumin seeds have been used since ancient time. The efficiency of cumin seed against biofilm development on *Klebsiella pneumoniae* strains was observed, which showed that cumin seeds has decreased biofilm activity with improved ciprofloxacin efficiency [47].

Cinnamon oil is derived from the inner bark of the "*Cinnamomum zeylanicum*' as well as "*Cinnamomum cassia*" and is mostly used in the food industry due to its specific fragrance [48]. It is suggested that this oil is efficient against biofilm cultures *Lactobacillus plantarum, S. mutans*, and *S. epidermidis*. *Oregano* also is known as *Origanum vulgare* has inhibitory activity on biofilm production in case of *E. coli* and *Staphylococci*. A study revealed that Oregano essential oil exerts antimicrobial action on *E. coli, S. haemolyticus, S. sciuri, S. aureus,* and *S. lugdunensis* and could prevent biofilm formation. Moreover, it also able to detach active biofilm even at very low MIC. Inhibitory activity of *"Brazil nut oil"* named as *Bertholletia excelsa* (a vegetable oil) on commercially available dentifrice to prevent dental biofilm was also assessed. Scientists showed that by adding this vegetable oil to commercially available dentifrice, dental biofilm formation can be inhibited. Furthermore, this oil helps in preventing and controlling periodontal diseases [41].

The antimicrobial activity of "tea tree" essential oils scientifically named *Melaleuca alternifolia*, synergistically with ciprofloxacin was also evaluated against '*P. aeruginosa*' biofilms. The consequences showed that the combined effect of TTO with ciprofloxacin has decreased biofilm biomass significantly by more than 70% and lowered the number of cells at the lowest (1.25 μg/mL) ciprofloxacin concentration. The efficacy of essential oils from *cinnamon (Cinnamomum verum),* namely *thymol*, and *oregano* at sub-lethal concentrations on biofilm formation of 3 biofilm-forming bacterial strains i.e. *Stenotrophomonas*, *Acinetobacter* and *Sphingomonas* were assessed. Researchers showed that at the MIC, two out of three strains revealed resistance on microbial biofilm formation. Furthermore, among the three tested oils, "*thyme oil*" was considered as more efficient and showed inhibitory effect even on sub-lethal concentrations of 0.001% (w/v) [41].

8. Conclusion

Since biofilms are abundant in nature, the importance of biofilms in hospitals especially regarding their role in infections is often undervalued. Future studies should attempt to comprehend the biological forces controlling the colonization to develop innovative strategies for controlling biofilm biomass within a clinical context. Additionally, comprehensive research is required to recognize the potential of various synthetic and natural quorum sensing inhibitors (QSIs) for their applicability for humans. As these QSIs do not encourage the antibiotic resistance, therefore they can surely be the future therapeutic agents for the management of biofilm-based bacterial infections in clinical settings.

Conflict of interest

The authors declare no conflict of interest.

Author details

Aqsa Shahid[1], Maria Rasool[1,2], Naheed Akhter[1], Bilal Aslam[2], Ali Hassan[1], Sadia Sana[1], Muhammad Hidayat Rasool[2] and Mohsin Khurshid[2*]

1 College of Allied Health Professionals, Directorate of Medical Sciences, Government College University Faisalabad, Pakistan

2 Department of Microbiology, Government College University Faisalabad, Pakistan

*Address all correspondence to: mohsin.mic@gmail.com

References

[1] Costerton JW, Irvin RT, Cheng KJ. The bacterial glycocalyx in nature and disease. Annual Review of Microbiology. 1981;**35**:299-324

[2] Donlan RM. Biofilms: Microbial life on surfaces. Emerging Infectious Diseases. 2002;**8**:881-890

[3] Mauffrey C, Herbert B, Young H, Wilson ML, Hake M, Stahel PF. The role of biofilm on orthopaedic implants: The "holy grail" of post-traumatic infection management? European Journal of Trauma and Emergency Surgery. 2016;**42**:411-416

[4] Percival SL, Suleman L, Vuotto C, Donelli G. Healthcare-associated infections, medical devices and biofilms: Risk, tolerance and control. Journal of Medical Microbiology. 2015;**64**:323-334

[5] Stoodley P, Sauer K, Davies DG, Costerton JW. Biofilms as complex differentiated communities. Annual Review of Microbiology. 2002;**56**:187-209

[6] Fey PD, Olson ME. Current concepts in biofilm formation of 'Staphylococcus epidermidis'. Future Microbiology. 2010;**5**:917-933

[7] Lewis K. Persister cells, dormancy and infectious disease. Nature Reviews. Microbiology. 2007;**5**:48-56

[8] Diekema DJ, Pfaller MA, Schmitz FJ, et al. Survey of infections due to 'Staphylococcus species': Frequency of occurrence and antimicrobial susceptibility of isolates collected in the United States, Canada, Latin America, Europe, and the Western Pacific region for the SENTRY antimicrobial surveillance program, 1997-1999. Clinical Infectious Diseases. 2001;**32**(Suppl 2):S114-S132

[9] Flemming HC, Neu TR, Wozniak DJ. The EPS matrix: The "house of biofilm cells". Journal of Bacteriology. 2007;**189**:7945-7947

[10] Dufour D, Leung V, Lévesque CM. Bacterial biofilm: Structure, function, and antimicrobial resistance. Endodontic Topics. 2010;**22**:2-16

[11] Srey S, Jahid IK, Ha S-D. Biofilm formation in food industries: A food safety concern. Food Control. 2013;**31**:572-585

[12] Marlow VL, Porter M, Hobley L, et al. Phosphorylated DegU manipulates cell fate differentiation in the *Bacillus subtilis* biofilm. Journal of Bacteriology. 2014;**196**:16-27

[13] Jansen B, Kohnen W. Prevention of biofilm formation by polymer modification. Journal of Industrial Microbiology. 1995;**15**:391-396

[14] Cassinelli C, Morra M, Pavesio A, Renier D. Evaluation of interfacial properties of hyaluronan coated poly(methylmethacrylate) intraocular lenses. Journal of Biomaterials Science. Polymer Edition. 2000;**11**:961-977

[15] Crick CR, Ismail S, Pratten J, Parkin IP. An investigation into bacterial attachment to an elastomeric superhydrophobic surface prepared via aerosol assisted deposition. Thin Solid Films. 2011;**519**:3722-3727

[16] Privett BJ, Youn J, Hong SA, et al. Antibacterial fluorinated silica colloid superhydrophobic surfaces. Langmuir. 2011;**27**:9597-9601

[17] Abdelkefi A, Achour W, Ben Othman T, et al. Use of heparin-coated central venous lines to prevent catheter-related bloodstream infection.

The Journal of Supportive Oncology. 2007;**5**:273-278

[18] Meiron TS, Saguy IS. Adhesion modeling on rough low linear density polyethylene. Journal of Food Science. 2007;**72**:E485-E491

[19] Dror N, Mandel M, Hazan Z, Lavie G. Advances in microbial biofilm prevention on indwelling medical devices with emphasis on usage of acoustic energy. Sensors. 2009;**9**:2538-2554

[20] Schumm K, Lam TB. Types of urethral catheters for management of short-term voiding problems in hospitalised adults. Cochrane Database of Systematic Reviews. 2008;**2**:Cd004013

[21] Ma Y, Xu Y, Yestrepsky BD, et al. Novel inhibitors of *Staphylococcus aureus* virulence gene expression and biofilm formation. PLoS One. 2012;**7**:e47255

[22] Perez-Giraldo C, Rodriguez-Benito A, Moran FJ, Hurtado C, Blanco MT, Gomez Garcia AC. Influence of N-acetylcysteine on the formation of biofilm by '*Staphylococcus epidermidis*'. The Journal of Antimicrobial Chemotherapy. 1997;**39**:643-646

[23] Romling U, Balsalobre C. Biofilm infections, their resilience to therapy and innovative treatment strategies. Journal of Internal Medicine. 2012;**272**:541-561

[24] de la Fuente-Nunez C, Reffuveille F, Haney EF, Straus SK, Hancock RE. Broad-spectrum anti-biofilm peptide that targets a cellular stress response. PLoS Pathogens. 2014;**10**:e1004152

[25] Besinis A, Hadi SD, Le HR, Tredwin C, Handy RD. Antibacterial activity and biofilm inhibition by surface modified titanium alloy medical implants following application of silver, titanium dioxide and hydroxyapatite nanocoatings. Nanotoxicology. 2017;**11**:327-338

[26] Shahverdi AR, Fakhimi A, Shahverdi HR, Minaian S. Synthesis and effect of silver nanoparticles on the antibacterial activity of different antibiotics against *Staphylococcus aureus* and *Escherichia coli*. Nanomedicine. 2007;**3**:168-171

[27] Klasen HJ. A historical review of the use of silver in the treatment of burns. II. Renewed interest for silver. Burns. 2000;**26**:131-138

[28] Secinti KD, Ozalp H, Attar A, Sargon MF. Nanoparticle silver ion coatings inhibit biofilm formation on titanium implants. Journal of Clinical Neuroscience. 2011;**18**:391-395

[29] Frias J, Olle E, Alsina M. Periodontal pathogens produce quorum sensing signal molecules. Infection and Immunity. 2001;**69**:3431-3434

[30] Waters CM, Bassler BL. Quorum sensing: Cell-to-cell communication in bacteria. Annual Review of Cell and Developmental Biology. 2005;**21**:319-346

[31] Brackman G, Coenye T. Quorum sensing inhibitors as anti-biofilm agents. Current Pharmaceutical Design. 2015;**21**:5-11

[32] Hentzer M, Wu H, Andersen JB, et al. Attenuation of *Pseudomonas aeruginosa* virulence by quorum sensing inhibitors. The EMBO Journal. 2003;**22**:3803-3815

[33] Subhadra B, Kim DH, Woo K, Surendran S, Choi CH. Control of biofilm formation in healthcare: Recent advances exploiting quorum-sensing interference strategies and multidrug efflux pump inhibitors. Materials. 2018;**11**:E1676

[34] Hughes KA, Sutherland IW, Clark J, Jones MV. Bacteriophage and associated polysaccharide depolymerases--novel tools for study of bacterial biofilms.

Journal of Applied Microbiology. 1998;**85**:583-590

[35] Hagens S, Habel A, von Ahsen U, von Gabain A, Blasi U. Therapy of experimental pseudomonas infections with a nonreplicating genetically modified phage. Antimicrobial Agents and Chemotherapy. 2004;**48**:3817-3822

[36] Coenye T, Brackman G, Rigole P, et al. Eradication of *Propionibacterium acnes* biofilms by plant extracts and putative identification of icariin, resveratrol and salidroside as active compounds. Phytomedicine. 2012;**19**:409-412

[37] Ravichandiran V, Shanmugam K, Anupama K, Thomas S, Princy A. Structure-based virtual screening for plant-derived SdiA-selective ligands as potential antivirulent agents against uropathogenic *Escherichia coli*. European Journal of Medicinal Chemistry. 2012;**48**:200-205

[38] Issac Abraham SV, Palani A, Ramaswamy BR, Shunmugiah KP, Arumugam VR. Antiquorum sensing and antibiofilm potential of *Capparis spinosa*. Archives of Medical Research. 2011;**42**:658-668

[39] Singh BN, Singh HB, Singh A, Singh BR, Mishra A, Nautiyal CS. *Lagerstroemia speciosa* fruit extract modulates quorum sensing-controlled virulence factor production and biofilm formation in *Pseudomonas aeruginosa*. Microbiology. 2012;**158**:529-538

[40] Harjai K, Kumar R, Singh S. Garlic blocks quorum sensing and attenuates the virulence of *Pseudomonas aeruginosa*. FEMS Immunology and Medical Microbiology. 2010;**58**:161-168

[41] Sadekuzzaman M, Yang S, Mizan M, Ha S. Current and recent advanced strategies for combating biofilms. Comprehensive Reviews in Food Science and Food Safety. 2015;**14**:491-509

[42] Carneiro VA, Santos HS, Arruda FV, et al. Casbane diterpene as a promising natural antimicrobial agent against biofilm-associated infections. Molecules. 2010;**16**:190-201

[43] Taweechaisupapong S, Singhara S, Lertsatitthanakorn P, Khunkitti W. Antimicrobial effects of *Boesenbergia pandurata* and *Piper sarmentosum* leaf extracts on planktonic cells and biofilm of oral pathogens. Pakistan Journal of Pharmaceutical Sciences. 2010;**23**:224-231

[44] Lee JH, Cho HS, Joo SW, et al. Diverse plant extracts and trans-resveratrol inhibit biofilm formation and swarming of *Escherichia coli* O157:H7. Biofouling. 2013;**29**:1189-1203

[45] Ali SM, Khan AA, Ahmed I, et al. Antimicrobial activities of Eugenol and Cinnamaldehyde against the human gastric pathogen *Helicobacter pylori*. Annals of Clinical Microbiology and Antimicrobials. 2005;**4**:20

[46] Isman MB. Plant essential oils for pest and disease management. Crop Protection. 2000;**19**:603-608

[47] Derakhshan S, Sattari M, Bigdeli M. Effect of cumin (*Cuminum cyminum*) seed essential oil on biofilm formation and plasmid integrity of *Klebsiella pneumoniae*. Pharmacognosy Magazine. 2010;**6**:57-61

[48] Chang ST, Chen PF, Chang SC. Antibacterial activity of leaf essential oils and their constituents from *Cinnamomum osmophloeum*. Journal of Ethnopharmacology. 2001;**77**:123-127

[49] Bjarnsholt T, Jensen PO, Rasmussen TB, et al. Garlic blocks quorum sensing and promotes rapid clearing of pulmonary *Pseudomonas aeruginosa infections*. Microbiology. 2005;**151**:3873-3880

[50] Fu B, Wu Q, Dang M, Bai D, Guo Q, Shen L. Inhibition of *Pseudomonas*

aeruginosa biofilm formation by traditional Chinese medicinal herb *Herba patriniae*. BioMed Research International. 2017;**2017**:9584703

[51] Zhou JW, Luo HZ, Jiang H, Jian TK, Chen ZQ, Jia AQ. Hordenine: A novel quorum sensing inhibitor and Antibiofilm agent against *Pseudomonas aeruginosa*. Journal of Agricultural and Food Chemistry. 2018;**66**:1620-1628

[52] Krishnan T, Yin WF, Chan KG. Inhibition of quorum sensing-controlled virulence factor production in *Pseudomonas aeruginosa* PAO1 by Ayurveda spice clove (*Syzygium aromaticum*) bud extract. Sensors. 2012;**12**:4016-4030

[53] Ouyang J, Sun F, Feng W, et al. Quercetin is an effective inhibitor of quorum sensing, biofilm formation and virulence factors in *Pseudomonas aeruginosa*. Journal of Applied Microbiology. 2016;**120**:966-974

[54] Kalia M, Yadav VK, Singh PK, Sharma D, Narvi SS, Agarwal V. Exploring the impact of parthenolide as anti-quorum sensing and anti-biofilm agent against *Pseudomonas aeruginosa*. Life Sciences. 2018;**199**:96-103

[55] Lee JH, Kim YG, Ryu SY, Cho MH, Lee J. Ginkgolic acids and *Ginkgo biloba* extract inhibit *Escherichia coli* O157:H7 and *Staphylococcus aureus* biofilm formation. International Journal of Food Microbiology. 2014;**174**:47-55

[56] Niu C, Gilbert ES. Colorimetric method for identifying plant essential oil components that affect biofilm formation and structure. Applied and Environmental Microbiology. 2004;**70**:6951-6956

[57] Hasan S, Danishuddin M, Khan AU. Inhibitory effect of *Zingiber officinale* towards *Streptococcus mutans* virulence and caries development: in vitro and in vivo studies. BMC Microbiology. 2015;**15**:1

[58] Liu Y, Xu Y, Song Q, et al. Anti-biofilm activities from *Bergenia crassifolia* leaves against *Streptococcus mutans*. Frontiers in Microbiology. 2017;**8**:1738

[59] Zeng Y, Nikitkova A, Abdelsalam H, Li J, Xiao J. Activity of quercetin and kaemferol against *Streptococcus mutans* biofilm. Archives of Oral Biology. 2019;**98**:9-16

[60] Saising J, Ongsakul M, Voravuthikunchai SP. *Rhodomyrtus tomentosa* (Aiton) Hassk. ethanol extract and rhodomyrtone: A potential strategy for the treatment of biofilm-forming *Staphylococci*. Journal of Medical Microbiology. 2011;**60**:1793-1800

[61] Lopes LAA, Dos Santos Rodrigues JB, Magnani M, de Souza EL, de Siqueira-Junior JP. Inhibitory effects of flavonoids on biofilm formation by *Staphylococcus aureus* that overexpresses efflux protein genes. Microbial Pathogenesis. 2017;**107**:193-197

[62] Nadaf NH, Parulekar RS, Patil RS, et al. Biofilm inhibition mechanism from extract of *Hymenocallis littoralis* leaves. Journal of Ethnopharmacology. 2018;**222**:121-132

[63] Wang J, Song M, Pan J, et al. Quercetin impairs *Streptococcus pneumoniae* biofilm formation by inhibiting sortase a activity. Journal of Cellular and Molecular Medicine. 2018;**22**:6228-6237

[64] Qayyum S, Sharma D, Bisht D, Khan AU. Identification of factors involved in *Enterococcus faecalis* biofilm under quercetin stress. Microbial Pathogenesis. 2019;**126**:205-211

Chapter 7

Formation, Antibiotic Resistance, and Control Strategies of *Staphylococcus epidermidis* Biofilm

Wei Chen, Ting-Ting Xie and Hong Zeng

Abstract

Staphylococcus epidermidis, member of the group of coagulase-negative staphylococci, belongs to an opportunistic pathogen. It is reported that the major pathogenicity of *S. epidermidis* is attributed to its biofilm formed on the surface of infected tissues, which enhances bacterial resistance to antibiotics. Thus, how to inhibit biofilm formation and screening biofilm inhibitors will have great value in reducing bacterial drug-resistance, which is beneficial to prevent and treat biofilm-associated infections. In this chapter, we present the current knowledge on formation, antibiotic resistance, and control strategies of *S. epidermidis* biofilm. First, biofilm formation in *S. epidermidis*, including factors involved in different phases in the process of biofilm, is analyzed. Second, the mechanisms of antibiotic resistance in *S. epidermidis* biofilms, such as poor antibiotic penetration, slow growth, and formation of persister cells, are introduced. Finally, control strategies to *S. epidermidis* biofilm formation are provided.

Keywords: *Staphylococcus epidermidis*, biofilm, antibiotic resistance, biofilm inhibition

1. Introduction

Staphylococcus epidermidis is a commensal inhabitant of human and animal skin that rarely causes disease in healthy persons and animals. In recent years, however, *S. epidermidis* has been the most prevalent species isolated from device-associated infections [1]. The ability of biofilm formation by *S. epidermidis* is an important reason that investigators pay more attention to this emerging pathogen in recent years. It is reported that the major pathogenicity of *S. epidermidis* is attributed to its biofilm formed on the surface of infected tissues, which enhances bacterial resistance to antibiotics [2]. Biofilm formation by *S. epidermidis* involves two major steps. After finishing initial attachment, bacteria accumulate and form a multilayered architecture [3]. Bacteria develop biofilm by producing high-viscosity extracellular matrices including polysaccharides (EPS), proteins, and DNA (eDNA).

There is an increasing amount of biofilm research aimed at exploring how bacteria control their biofilm formation and to discover nontoxic compounds that can attenuate biofilm formation without allowing bacteria to develop drug resistance [4]. Special plants and Actinomycetes are both rich sources of bioactive substances, notably antibiotics, enzymes, enzyme inhibitors, and pharmacologically active agents [4, 5]. Moreover, some Actinomycete species were reported to produce inhibitors against biofilm formation by *Staphylococcus aureus*, *Escherichia coli*, and *Pseudomonas aeruginosa* [6–9].

With this background, we aim to present the current knowledge on biofilm formation of *S. epidermidis* and review the control strategies to biofilm.

2. Biofilm formation in *S. epidermidis*

S. epidermidis infections are regarded as prototypic biofilm infections. The process of biofilm formation by *S. epidermidis* is periodically dynamic. Also, surface adhesion between planktonic bacterial cells is a key for biofilm formation. Once several cells succeed in adhering on a surface, named initial attachment of cells, surface motility and binary division result in an aggregation of attached cells. These primary cell aggregates produce exopolymers, including exopolysaccharides and extracellular proteins, which form extracellular matrix. Some of those factors may also originate from lysed cells, such as extracellular DNA (eDNA) [10]. Subsequently, there is development of a multicellular, multilayered biofilm architecture. In the later phase of biofilm formation, biofilm cells and clusters can detach. This detachment process is of key importance for the dissemination of biofilm-associated infection [10].

2.1 Factors involved in primary attachment in *S. epidermidis* biofilm formation

Nonspecific adhesions between bacterial cells, which are mainly attributed to the composition of compounds on the surface of bacterial cells and their hydrophobicities, play an important role in biofilm formation. Additionally, autolysin (AtlE) and teichoic acids have influences on biofilm formation [11, 12]. It is reported that lots of autolysin enhanced the cell surface hydrophobicity and increased the biofilm formation. Also, teichoic acids correlated with increased cell surface hydrophobicity, so they contributed to biofilm formation [11, 12].

In vivo primary attachment occurs to host tissue or host matrix proteins. *S. epidermidis* produces a variety of surface proteins binding host proteins in a specific manner. Bacterial surface proteins with such capacities have been termed microbial components recognizing adhesive matrix molecules (MCRAMM) [13]. The C-terminus of such bacterial surface proteins consists of an LPxTG (Leu-Pro-x-Thr-Gly) motif containing Gram-positive cell wall anchor, which covalently links to the cell wall [1]. According to genomic analyses, *S. epidermidis* has at least 14 MCRAMMs with an LPxTG motif. Many of those belong to the serine-aspartate (SD)-repeat-containing protein family (called Sdr). The SD-repeat region spans the cell wall and extends the ligand-binding region from the surface of the bacteria [14]. Adequate SD repeats within proteins are essential for outstanding from bacterial cell surface, which are covalently anchored to the peptidoglycan of Gram-positive bacteria.

The SD repeat family protein Sdr G in *S. epidermidis,* which is very similar to a fibrinogen-binding protein (Fbe), is necessary and sufficient for binding to fibrinogen-coated material. SdrG knock-out mutant showed less adhesion on fibrinogen-coated surfaces. It is reported that in vivo anti-SdrG antibody decreased the numbers of *S. epidermidis* cells adherent to biomaterials [14]. One of Sdr proteins, SdrF, mediates binding to type I collagen via one or both α1 chains, named collagen-binding protein [15].

Some of surface proteins on bacterial cell wall are adherent to host cells via non-covalent interaction, such as hydrophobic bonds and Van der Waals' force, which of process are involved into the polymers on bacterial cell surface, e.g., teichoic acids. Teichoic acids are main components consisting of the cell wall of Gram-positive. They bind to peptidoglycan of cell wall and influence the activity of autolysin (AtlE). AtlE, encoded by the atlE gene, is a bifunctional autolysin: one is able to mediate bacterial adhesion, and the other is to promote bacterial cell autolysis, which releases DNA out of cells, named extracellular DNA (eDNA) [16].

2.2 Factors responsible for cellular aggregation in *S. epidermidis* biofilm formation

Following the primary attachment of cells to a surface, bacterial cells occur to accumulate with the help of a variety of associated-accumulation factors, such as polysaccharide intercellular adhesin (PIA), accumulation-associated protein (Aap), and so on.

In the process of biofilm formation by *S. epidermidis*, PIA plays an important role in cell aggregation. Studies with *S. epidermidis* mutant revealed that the accumulation-defective mutants were unable to form a biofilm as they were unable to display intercellular aggregation or to produce PIA [17]. Further characterization of this *S. epidermidis* mutant showed that a deletion of icaR gene was found to upregulate PIA expression, providing evidence that this gene negatively regulates the PIA expression [17]. However, it is reported that there is no ica operon in some of clinical *S. epidermidis* strains, which have capacity of biofilm formation, named ica or PIA-independent type. In these strains, the accumulation-associated protein (Aap) is a major factor contributing to exopolysaccharide-independent biofilms of *S. epidermidis* [1]. Aap protein promotes cell-cell adhesion via a Zn^{2+}-dependent mechanism [18]. It is reported that 90% of isolated *S. epidermidis* strains contain aap gene, which is implicated in both PIA-dependent and PIA-independent biofilm formations of *S. epidermidis* [18]. *S. epidermidis* ATCC 35984 is a ica^{+} strain and a biofilm former, whose biofilm formation mainly depends on PIA consisting of reducing polysaccharides in which dihydroxyl groups are unsubstituted. However, exopolysaccharides in ica^{-} *S. epidermidis* mainly consist of nonreducing polysaccharides [19].

2.3 Biofilm formation and maturation

Cellular aggregation constantly occurs and subsequently forms biofilm. Disruptive molecules create channels in the biofilm, which are essential for nutrient accessibility in deeper biofilm layers and give the biofilm its characteristic structure, often described as mushroom-like shapes [10]. The characteristic structure of mature biofilms with mushroom-like shapes and channels is dependent on the production of phenolsoluble modulins (PSMs) in *S. epidermidis*.

Of primary importance for dissemination of biofilm-associated infection, cells or cell aggregates may detach from a mature biofilm to reach the next infection sites. This may occur by mechanical forces under flow, such as present in a blood vessel, in a process often called sloughing [10]. Additionally, the bacteria can trigger detachment by PSM production. These surfactant-like molecules work by decreasing noncovalent adhesion between bacterial cells.

3. Mechanisms of antibiotic resistance in *S. epidermidis* biofilms

Several in vitro studies have demonstrated that bacteria within biofilms are more resistant against antibiotic treatment as compared to planktonic cultures of the same strains [20].

S. epidermidis and other bacterial species produce an extracellular matrix called glycocalyx or slime, which is a highly hydrated complex composed of teichoic acids, proteins, and exopolysaccharides. In biofilms, poor antibiotic penetration, nutrient limitation and slow growth, and formation of persister cells are hypothesized to be responsible for drug resistance.

3.1 Antibiotic penetration of biofilms

Biofilms are typically characterized by dense, highly hydrated clusters of bacterial cells enclosed in a self-produced polymeric matrix that is primarily composed of exopolysaccharides such as polysaccharide intercellular adhesin (PIA) in staphylococci and adherent to a surface. This matrix, also termed slime or extracellular polymeric substance (EPS), impairs the access of antimicrobial agents to the bacterial cells [21]. Additionally, either a reaction of EPS with or its adsorption to the components of the biofilm matrix can delay penetration of the antibiotics through the biofilm matrix. The effective diffusion coefficients of solutes in biofilms average about 40% of the respective diffusion coefficient in pure water [20]. *S. epidermidis* slime has been found to remarkably decrease the activity of the glycopeptides vancomycin and teicoplanin. The efficacy of cloxacillin, amoxicillin/clavulanic acid, imipenem, cefpirome, erythromycin, roxithromycin, clindamycin, fusidic acid, trimethoprim/sulfamethoxazole, doxycycline, gentamicin, tobramycin, netilmicin, amikacin, isepamicin, ofloxacin, ciprofloxacin, and daptomycin is also moderately affected by the exopolysaccharide matrix of *S. epidermidis*. Other studies have suggested that *S. epidermidis* glycocalyx reduces susceptibility to pefloxacin and moderately affects the activity of daptomycin, linezolid, and quinupristin/dalfopristin [22, 23]. The role of biofilm matrix in retarding the penetration of antibiotics is thereby contributed to the drug resistance of *S. epidermidis* biofilms.

3.2 Slow cell growth in biofilms

Slow cell growth of the bacterial has been found in mature biofilms [17]. This phenomenon is responsible for the decreased susceptibility of bacteria in biofilms to antibiotics requiring growing organisms for their bactericidal effects. For example, penicillins and cephalosporins prefer to killing the growing bacterial cells, and the rate of killing cells is proportional to the growth rate [17]. It is well known that most antimicrobial agents act on certain types of macromolecular synthesis to exert antimicrobial activities, such as the synthesis of enzymes, proteins, and nucleic acids (DNA or RNA). Thus, these antibiotics have little effects on bacteria with stagnant macromolecular synthesis, which leads to bacterial drug resistance.

Nutrition restriction is one of reasons that are responsible for slow cell growth. The mechanism of nutrition restriction is closely related to the osmotic restriction. Due to the existence of biofilm osmotic restriction, nutrients are not easy to pass through biofilm, which leads to the lack of nutrition in biofilm and slows down the growth rate of inner layer bacteria. This slow growth state of inner layer bacteria also forms a protective mechanism, which reduces the susceptibility of bacteria to antibiotics [24].

When the biofilm cells are exposed to antibiotics, the bacteria on the surface of the biofilm are killed by the drug, and the cells in the middle and deep layers of the biofilm are not affected. After the antibiotic treatment stops, the remaining bacteria will use dead bacteria as nutrients to reproduce rapidly, which can only take a few hours to reproduce [25, 26].

3.3 Formation of persister cells

Delayed penetration of the antibiotics through the biofilm matrix and slow rate of bacterial reproduction in biofilm cannot explain entirely the resistance of biofilms to one important class of antibiotics, namely the fluoroquinolones. This class of antimicrobial agents equilibrates across bacterial biofilms and exerts bactericidal effect on nondividing cells [17]. Although a dose-dependent bactericidal action was observed in *P. aeruginosa* biofilms by the fluoroquinolones ofloxacin and ciprofloxacin, a further increase in the antibiotic concentration or a prolonged drug action period did not improve killing rates after an initial 3- to 4-log drop bacterial counts. This result suggested that a small portion of "persister" cells occurs after administration of fluoroquinolones [17, 27, 28]. The most significant difference between persisters and mutant resistant strains is that the drug resistance of persisters is only a phenotypic variation without gene mutation, so this phenotype is not genetic. These strains were collected, recultured, and detected the drug resistance. It was intriguing that the drug resistance disappeared, and the minimum inhibitory concentrations (MICs) were the same level as those of parent strains. Meanwhile, the resistant strains caused by mutation showed a stable genetic drug resistance, and MICs were higher than those of parent strains [28].

Persister cells in biofilms are considered to the key in the extraordinary survival properties of biofilms. The dynamic features of biofilm formation and shedding of cells from one biofilm to form a new biofilm may also explain the chronic nature of biofilm infections and the need for extending antimicrobial agent treatment to disturb the dynamics of biofilm formation [17].

4. Control strategies to *S. epidermidis* biofilm formation

Because the expression of toxins and other virulence factors is less in *S. epidermidis*, the biofilm forming capacity is its major virulence factor. Biofilm growth is characterized by high resistance to antimicrobial agents and host immune responses, making biofilm eradication tremendously difficult. The increasing prevalence of multidrug-resistant *S. epidermidis* strains additionally hampers antimicrobial therapy. Therefore, targeting factors expressed at different phases in biofilm formation might offer new tools to combat *S. epidermidis* infections.

4.1 Inhibition of initial attachment

The first step of biofilm formation is bacterial adherence to the host cell surface. Direct binding to host cell surface is mediated by electrostatic and hydrophobic interactions and van der Waals forces and affected by physicochemical variables [29].

Found in our research, after investigating the antibiofilm activities of spent media from 185 Actinomycete strains using two *S. epidermidis* strains (ATCC 35984 and a clinical strain 5-121-2) as target bacteria, three strains of tested Actinomycete (TRM 46200, TRM 41337, and TRM 46814) showed a significant inhibition against *S. epidermidis* biofilm formation without affecting the growth of planktonic cells. Effect of Actinomycete supernatants on cell surface hydrophobicity (CSH) of *S. epidermidis* was measured by Microbial Adhesion to Hydrocarbon (MATH) assay. The adhesion of staphylococci to n-hexadecane was used to measure the hydrophobicity of *S. epidermidis*. All the crude proteins from spent media showed a reduction in the CSH against *S. epidermidis* ATCC 35984 and 5-121-2, which explain at least in part the inhibitory effect of Actinomycete supernatants on biofilm reduction [19].

Moreover, apart from physico-chemical determinants, it was demonstrated that the major autolysine AtlE is involved in attachment to polystyrene surfaces. Therefore, AtlE may be indirectly involved in cell adhesion via releasing DNA. Treatment of *S. epidermidis* cells with DNaseI was found to inhibit biofilm formation at an early time point, suggesting that release of DNA also contributes to the attachment of *S. epidermidis* to artificial surfaces [30]. In our research, we performed the degradation of the crude proteins from spent media against *S. epidermidis* DNA. The crude protein from spent media of TRM 46200 showed a significant DNA-degradation activity. Importantly, the crude protein from spent medium of TRM 41337 possessed the highest DNA-degradation activity as that of the positive control, 10 μg/ml of DNaseI [19].

S. epidermidis foreign-associated infections occurring early are thought to involve direct interactions of the bacterial surface with host extracellular matrix (ECM). Specific binding to surface ECM proteins involves cell wall-associated adhesins known as MSCRAMMs (microbial surface components recognizing adhesive matrix molecules) [31, 32]. The recent studies have shown that antibodies against cell surface components of *S. epidermidis* can affect the rate of biofilm formation or adherence of these bacteria to medical devices in vitro. Using polyclonal antibodies against a fibrinogen-binding protein from *S. epidermidis* (Fbe) could block adherence of *S. epidermidis* to fibrinogen-coated catheters in vitro [33, 34]. Consequently, all these surface-located components are good candidates for vaccine development aiming at the inhibition of the initial attachment step of biofilm formation.

4.2 Inhibition of bacterial accumulate

After adherence to the host cell surface, biofilms develop through intercellular aggregation. The major factor involved in intercellular adhesion is polysaccharide intercellular adhesin (PIA). The de-acetylation of PIA is not only essential for biofilm formation but also crucial for *S. epidermidis* virulence [29]. Hence, PIA was one of the first targets evaluated in view of biofilm-inhibiting *S. epidermidis* vaccine development. Pier and coworkers have significantly contributed to the evaluation of PIA as vaccine target. Following the evidence, high-molecular-weight PIA could elicit an antibody response accompanied by opsonophagocytic killing of the PIA-dependent biofilm-forming *S. epidermidis* M187 and three *S. aureus* strains. The PIA-specific antibodies can prevent biofilm formation or retard already initiated biofilm development [35].

PIA biosynthesis depends on the expression of the icaADBC operon, which is controlled by a complex regulatory network. Gomes et al. studied the effect of rifampicin+gentamicin and rifampicin+clindamycin combinations on the expression of icaA and rsbU genes, responsible for poly-N-acetylglucosamine/polysaccharide intercellular adhesin (PNAG/PIA) production. The results demonstrated that this combinatorial therapy can cause a lower genetic expression of the two specific genes tested and consequently can reduce biofilm formation recidivism [36, 37].

Nevertheless, *S. epidermidis* strains lacking icaADBC but still producing biofilm were isolated, indicating the existence of an ica-independent mechanism of cell accumulation. A proteinaceous intercellular adhesin involved in cell accumulation during biofilm formation was discovered. The accumulation-associated protein (Aap) can functionally substitute PIA as an intercellular adhesin, and there is good evidence that additional proteinaceous intercellular adhesins must exist. They showed that monoclonal antibodies against Aap can significantly reduce the accumulation but not initiation phase of *S. epidermidis* biofilm formation in vitro [38].

Biofilm formation is a result of bacterial interactions and group behavior. Quorum sensing (QS) is one of the regulatory mechanisms suggested to be involved in coordinating biofilm formation. The QS system is a cell-to-cell communication system used by many bacteria to assess the cell density. Quorum sensing inhibitors (QSI) could be a novel way to fight biofilm-associated infections. The study has identified furanones and thiophenones as inhibitors of quorum sensing and biofilm formation. In this study, the effect of both the furanone and the thiophenone could be abolished by the synthetic Autoinducer-2 (AI-2) molecule (S)-4,5-dihydroxy-2,3-pentanedione (DPD), indicating that furanone and thiophenone affect biofilm formation through interference with bacterial communication [39].

4.3 Promotion of biofilm detachment

For the biofilm that has been formed on the surface of the host, if the biofilm can be separated by antibacterial oranti-biofilm substances, the bacteria in the biofilm can be released, and the planktonic bacteria are more easily to be killed if the biofilm is exposed to antibiotics. Biofilms are composed primarily of microbial cells and extracellular polymeric substance (EPS). EPS may account for 50–90% of the total organic carbon of biofilms and can be considered the primary matrix material of the biofilm. The components of EPS include polysaccharides, nucleic acids, lipids, and proteins [36].

We initially determined the dependent type of biofilm formation by *S. epidermidis* ATCC 35984 and 5-121-2. The biofilm formation by *S. epidermidis* ATCC 35984 mainly depends on EPS consisting of reducing polysaccharides in which dihydroxyl groups are unsubstituted. Thus, sodium-meta-periodate, which specifically destroys sugars containing unsubstituted dihydroxyl groups, significantly decreased biofilm formation in *S. epidermidis* ATCC 35984. However, not only EPS but also proteins, eDNA, are responsible for the biofilm formation of *S. epidermidis* 5-121-2. Moreover, EPS in *S. epidermidis* 5-121-2, which mainly consists of nonreducing polysaccharides, is distinct with those in *S. epidermidis* ATCC 35984. Thus, three enzymes specific to nonreducing glycosides, amylase, β-glucanase, and β-glucosidase, worked effectively in the degradation of EPS, resulting in biofilm reduction in *S. epidermidis* 5-121-2 [19].

Since extracellular polysaccharides are the main compounds in biofilm matrices, namely in *S. epidermidis*, antimicrobial substances able to disrupt or inhibit EPS are of major interest. N-acetylcysteine (NAC) is an amino acid with strong antioxidant, antimucolytic, and antibacterial properties. As observed by researchers, NAC decreased biofilm formation and reduced the formation of extracellular polysaccharide matrix while promoting the disruption of mature biofilm. NAC has demonstrated not only to reduce adhesion but also to detach bacterial cells adhered to surfaces and to inhibit bacterial growth in vitro. The possible action of NAC in the biofilm matrix can result in the release of cells either individually or in cell clusters, becoming the biofilm and loose cells more exposed and susceptible to the host immune system and to other antimicrobial agents [40]. Kaplan et al. found an enzyme called dispersin B, which can promote biofilm detachment from *Actinobacillus actinomycetemconitans*, which rapidly and effectively removes biofilms formed by *S. epidermidis* on the host surface. Dispersin B is a β-1,6-N-Acetylglucosaminidase that causes *S. epidermidis* to detach from the biofilm matrix by degrading PIA [41].

Our results showed that EPS in *S. epidermidis* ATCC 35984 and 5-121-2 was degraded by crude proteins from three Actinomycete strains (TRM 41337, TRM 46200, and TRM 46814) supernatants. Specifically, for the strain ATCC 35984 when treated with crude proteins from spent medium of the strain TRM 41337, arabinose (Ara) was absent in the monosaccharide composition compared with the control. Furthermore, the proportion of mannose (Man) was decreased, while the proportions of glucosamine (GluN), galactosamine (GalN), and galactose (Gal) were increased. When treated with crude proteins from spent medium of the strain TRM 46814, three new monosaccharides, rhamnose (Rha), glucuronic acid (GluA), and galacturonic acid (GalA), appeared. Additionally, the proportions of Man and glucose (Glu) decreased obviously. For the strain 5-121-2, when treated with crude proteins from spent media of TRM 41337 and TRM 46814, a new monosaccharide, Rha, was present [19].

Acknowledgements

This work was supported by the Program for New Century Excellent Talents in University (grant number NCET-11-1071) of China, a NSFC (National Science Foundation of China) Grant 31260026, and a Fund for PhD in Xinjiang Production and Construction Corps (grant number 2009JC07) to W. Chen, a NSFC-Xinjiang joint Grant U1703236 to L. L. Zhang, and a Microbial Resources Utilization Innovation Team in Key Field of Xinjiang Production and Construction Crops (grant number 2017CB014) to C. X. Wan.

Conflict of interest

No conflict of interest declared.

Author details

Wei Chen[1*], Ting-Ting Xie[2] and Hong Zeng[2]

1 Key Laboratory of Tarim Animal Husbandy and Science Technology of Xinjiang Production and Construction Corps, College of Animal Sciences, Tarim University, Alar, China

2 Key Laboratory of Protection and Utilization of Biological Resources in Tarim Basin of Xinjiang Production and Construction Corps, College of Life Sciences, Tarim University, Alar, China

*Address all correspondence to: 379497687@qq.com

References

[1] Khodaparast L, Khodaparast L, Shahrooei M, Stijlemans B, Merckx R, Baatsen P, et al. The possible role of *Staphylococcus epidermidis* LPxTG surface protein SesC in biofilm formation. PLoS One. 2016;**11**:e0146704. DOI: 10.1371/journal.pone.0146704

[2] Vadyvaloo V, Otto M. Molecular genetics of *Staphylococcus epidermidis* biofilms on indwelling medical devices. The International Journal of Artificial Organs. 2005;**28**:1069-1078

[3] Rohde H, Burdelski C, Bartscht K, Hussain M, Buck F, Horstkotte MA, et al. Induction of *Staphylococcus epidermidis* biofilm formation via proteolytic processing of the accumulation associated protein by staphylococcal and host proteases. Molecular Microbiology. 2005;**55**:1883-1895. DOI: 10.1111/j.1365-2958.2005.04515.x

[4] Kim YG, Lee JH, Kim CJ, Lee JC, Ju YJ, Cho MH, et al. Antibiofilm activity of *Streptomyces* sp. BFI 230 and *Kribbella* sp. BFI 1562 against *Pseudomonas aeruginosa*. Applied Microbiology and Biotechnology. 2012;**96**:1607-1617. DOI: 10.1007/s00253-012-4225-7

[5] Zhang XX, Wu YP, Nan ZB. Antifungal activity of petroleum ether extracts from *Achnatherum inebrians* infected with *Neotyphodium gansuense*. Science in China Series C. 2014;**57**:1234-1235. DOI: 10.1007/s11427-014-4660-z

[6] Lee JH, Kim YG, Kim CJ, Lee JC, Cho MH, Lee J. Indole-3-acetaldehyde from *Rhodococcus* sp. BFI 332 inhibits *Escherichia coli* O157:H7 biofilm formation. Applied Microbiology and Biotechnology. 2012;**96**:1071-1078. DOI: 10.1007/s00253-012-3881-y

[7] Naik DN, Wahidullah S, Meena RM. Attenuation of *Pseudomonas aeruginosa* virulence by marine invertebrate-derived *Streptomyces* sp. Letters in Applied Microbiology. 2013;**56**:197-207. DOI: 10.1111/lam.12034

[8] Oja T, San Martin Galindo P, Taguchi T, Manner S, Vuorela PM, Ichinose K, et al. Effect antibiofilm polyketides against *Staphylococcus aureus* from the pyranonaphthoquinone biosynthetic pathways of *Streptomyces* species. Antimicrobial Agents and Chemotherapy. 2015;**59**:6046-6052. DOI: 10.1128/AAC.00991-15

[9] Suzuki N, Ohtaguro N, Yoshida Y. A compound inhibits biofilm formation of *Staphylococcus aureus* from *Streptomyces*. Biological & Pharmaceutical Bulletin. 2015;**38**:889-892. DOI: 10.1248/bpb.b15-00053

[10] Otto M. *Staphylococcus epidermidis* pathogenesis. Methods in Molecular Biology. 2014;**1106**:17-31. DOI: 10.1007/978-1-62703-736-5_2

[11] Mark ER, Paul DF, Christine H, Friedrich G. Characterization of the importance of *Staphylococcus epidermidis* autolysin and polysaccharide intercellular adhesin in the pathogenesis of intravascular catheter-associated infection in a rat model. The Journal of Infectious Diseases. 2001;**183**:1038-1042. DOI: 10.1086/319279

[12] Holland LM, Conlon B, O'Gara JP. Mutation of tagO reveals an essential role for wall teichoic acids in *Staphylococcus epidermidis* biofilm development. Microbiology. 2011;**157**:408-418. DOI: 10.1099/mic.0.042234-0

[13] Ko YP, Kang MS, Liang XW, Caná LR, Liu Q, Murray BE, et al. Collagen-binding MSCRAMMS of gram-positive bacteria inhibit complement activation via the classical pathway. Immunobiology.

2012;**217**:1151. DOI: 10.1016/j.imbio.2012.08.065

[14] Mccrea KW, Hartford O, Davis S, Eidhin DN, Lina G, Speziale P, et al. The serine-aspartate repeat (Sdr) protein family in *Staphylococcus epidermidis*. Microbiology. 2000;**146**:1535-1546

[15] Arrecubieta C, Lee MH, Macey A, Foster TJ, Lowy FD. SdrF, a *Staphylococcus epidermidis* surface protein, binds type I collagen. The Journal of Biological Chemistry. 2007;**282**:18767-18776. DOI: 10.1074/jbc.M610940200

[16] Sivadon V, Rottman M, Quincampoix JC, Prunier E, Le Moal M, De Mazancourt P, et al. Partial *atlE* sequencing of *Staphylococcus epidermidis* strains from prosthetic joint infections. Journal of Clinical Microbiology. 2009;**47**:2321-2324. DOI: 10.1128/JCM.01971-08

[17] Melchior MB, Vaarkamp H, Fink-Gremmels J. Biofilms: A role in recurrent mastitis infections? Veterinary Journal. 2006;**171**:398-407. DOI: 10.1016/j.tvjl.2005.01.006

[18] Conlon BP, Geoghegan JA, Waters EM, Mccarthy H, Rowe SE, Davies JR, et al. Role for the A domain of unprocessed accumulation-associated protein (Aap) in the attachment phase of the *Staphylococcus epidermidis* biofilm phenotype. Journal of Bacteriology. 2014;**196**:4268-4275. DOI: 10.1128/JB.01946-14

[19] Xie TT, Zeng H, Ren XP, Wang N, Chen ZJ, Zhang Y, et al. Antibiofilm activity of three *Actinomycete* strains against *Staphylococcus epidermidis*. Letters in Applied Microbiology. 2019;**68**:73-80. DOI: 10.1111/lam.13087

[20] Stewart PS. Mechanisms of antibiotic resistance in bacterial biofilms. International Journal of Medical Microbiology. 2002;**292**:107-113. DOI: 10.1078/1438-4221-00196

[21] Stewart PS. Theoretical aspects of antibiotic diffusion into microbial biofilms. Antimicrobial Agents and Chemotherapy. 1996;**40**:2517-2522. DOI: 10.1128/AAC.40.11.2517

[22] Singh R, Ray P, Das A, Sharma M. Penetration of antibiotics through *Staphylococcus aureus* and *Staphylococcus epidermidis* biofilms. The Journal of Antimicrobial Chemotherapy. 2010;**65**:1955-1958. DOI: 10.1093/jac/dkq257

[23] Singh R, Sahore S, Kaur P, Rani A, Ray P. Penetration barrier contributes to bacterial biofilm-associated resistance against only select antibiotics, and exhibits genus-, strain- and antibiotic-specific differences. Pathogens and Disease. 2016;**74**:1-20. DOI: 10.1093/femspd/ftw056

[24] Anutrakunchai C, Bolscher JGM, Krom BP, Kanthawong S, Chareonsudjai S, Taweechaisupapong S. Impact of nutritional stress on drug susceptibility and biofilm structures of *Burkholderia pseudomallei* and *Burkholderia thailandensis* grown in static and microfluidic systems. PLoS One. 2018;**13**:e0194946. DOI: 10.1371/journal.pone.0194946

[25] Hoiby N, Bjarnsholt T, Givskov M, Molin S, Ciofu O. Antibiotic resistance of bacterial biofilms. International Journal of Antimicrobial Agents. 2010;**35**:322-332. DOI: 10.1016/j.ijantimicag.2009.12.011

[26] Deepigaa M. Antibacterial resistance of bacteria in biofilms. Research Journal of Pharmacy and Technology. 2017;**10**:4019-4023. DOI: 10.5958/0974-360X.2017.00728.4

[27] Brooun A, Liu S, Lewis K. A dose-response study of antibiotic

resistance in *Pseudomonas aeruginosa* biofilms. Antimicrobial Agents and Chemotherapy. 2000;**44**:640-646. DOI: 10.1128/AAC.44.3.640-646.2000

[28] Lewis K. Persister cells, dormancy and infectious disease. Nature Reviews. Microbiology. 2007;**5**:48-56. DOI: 10.1038/nrmicro1557

[29] Van Mellaert L, Shahrooei M, Hofmans D, Van Eldere J. Immunoprophylaxisandimmunotherapy of *Staphylococcus epidermidis* infections: Challenges and prospects. Expert Review of Vaccines. 2012;**11**:319-334. DOI: 10.1586/erv.11.190

[30] Qin ZQ, Ou YZ, Yang L, Zhu YL, Tolker-Nielsen T, Molin S, et al. Role of autolysin-mediated DNA release in biofilm formation of *Staphylococcus epidermidis*. Microbiology. 2007;**153**:2083-2092. DOI: 10.1099/mic.0.2007/006031-0

[31] Arora S, Uhlemann AC, Lowy FD, Hook M. A novel MSCRAMM subfamily in coagulase negative *Staphylococcal* species. Frontiers in Microbiology. 2016;**7**:1-9. DOI: 10.3389/fmicb.2016.00540

[32] Marcello Abbondio M, Fois I, Longheu C, Azara E, Tola S. Biofilm production, quorum sensing system and analysis of virulence factors of *Staphylococcus epidermidis* collected from sheep milk samples. Small Ruminant Research. 2019;**174**:83-87. DOI: 10.1016/j.smallrumres.2019.03.017

[33] Pei L, Palma M, Nilsson M, Guss B, Flock JI. Functional studies of a fibrinogen binding protein from *Staphylococcus epidermidis*. Infection and Immunity. 1999;**67**:4525-4530

[34] Rennermalm A, Nilsson M, Flock JI. The fibrinogen binding protein of *Staphylococcus epidermidis* is a target for opsonic antibodies. Infection and Immunity. 2004;**72**:3081-3083. DOI: 10.1128/IAI.72.5.3081-3083.2004

[35] Maira-Litrán T, Kropec A, Goldmann DA, Pier GB. Comparative opsonic and protective activities of *Staphylococcus aureus* conjugate vaccines containing native or deacetylated staphylococcal poly-N-acetyl-β-(1-6)-glucosamine. Infection and Immunity. 2005;**73**:6752-6762. DOI: 10.1128/IAI.73.10.6752-6762.2005

[36] Gomes F, Leite B, Teixeira P, Oliveira R. Strategies to control *Staphylococcus epidermidis* biofilms. Science Against Microbial Pathogens. 2011;**2011**:842-852

[37] Gomes F, Teixeira P, Cerca N, Ceir H, Oliveira R. Virulence gene expression by *Staphylococcus epidermidis* biofilm cells exposed to antibiotics. Microbial Drug Resistance. 2011;**17**:191-196. DOI: 10.1089/mdr.2010.0149

[38] Sun DQ, Accavitti MA, Bryers JD. Inhibition of biofilm formation by monoclonal antibodies against *Staphylococcus epidermidis* RP62A accumulation-associated protein. Clinical and Diagnostic Laboratory Immunology. 2005;**12**:93-100. DOI: 10.1128/CDLI.12.1.93-100.2005

[39] Lönn-Stensrud J, Benneche T, Scheie AA. Furanones and thiophenones in control of *Staphylococcus epidermidis* biofilm infections? Science and Technology Against Microbial Pathogens. 2011;**2011**:155-159. DOI: 10.1142/9789814354868_0030

[40] Leite B, Gomes F, Pilar T, Clovis S, Elisabeth P, Oliveira R. *Staphylococcus epidermidis* biofilms control by N-acetylcysteine and rifampicin. American Journal of Therapeutics. 2013;**20**:322-328. DOI: 10.1097/MJT.0b013e318209e17b

[41] Kaplan JB, Velliyagounder K, Ragunath C, Rohde H, Dietrich M,

Knobloch J, et al. Genes involved in the synthesis and degradation of matrix polysaccharide in *Actinobacillus actinomycetemcomitans* and *Actinobacillus pleuropneumoniae* biofilms. Journal of Bacteriology. 2004;**186**:8213-8220. DOI: 10.1128/JB.186.24.8213-8220.2004

Chapter 8

Combating Biofilm and Quorum Sensing: A New Strategy to Fight Infections

Cynthia Amaning Danquah, Samuel Osei-Djarbeng, Theresah Appiah, Yaw Duah Boakye and Francis Adu

Abstract

Biofilms are structured aggregates of bacterial cells that are embedded in self-produced extracellular polymeric substances. Various pathogens initiate a disease process by creating organized biofilms that enhance their ability to adhere, replicate to accumulate, and express their virulence potential. Quorum sensing, which refers to the bacterial cell-to-cell communication resulting from production and response to *N*-acyl homoserine lactone signal molecules, also plays an important role in virulence and biofilm formation. Attenuation of microorganisms' virulence such that they fail to adapt to the hosts' environment could be a new strategic fight against pathogens. Thus, agents or products that possess anti-biofilm formation and/or anti-quorum sensing activities could go a long way to manage microbial infections. The incidence of microbial resistance can be reduced by the use of anti-biofilm formation and anti-quorum sensing agents.

Keywords: biofilm, quorum sensing, bacteria, acyl homoserine lactone

1. Introduction

Biofilm is a population of cells growing on a surface and enclosed in an exopolysaccharide matrix [1]. The physiology, structure and chemistry of the biofilm vary with the nature of its resident microbes and local environment [2].

Most important feature among biofilms is that their structural integrity critically depends upon the extracellular matrix produced by their constituent cells. They are notoriously difficult to eradicate and are a source of many recalcitrant infections [2]. Biofilms are associated with serious health issues stemming from persistent infections due to the contamination of medical devices (intravenous and urinary catheters), artificial implants and drinking water pollution among others [3].

Intercellular signaling, often referred to as quorum sensing (QS), has been shown to be involved in biofilm development [4]. Quorum sensing relies on small, secreted signaling molecules; much like hormones in higher organisms, to initiate coordinated responses across a population and it contributes to behaviors that enable microbes to resist antimicrobial compounds [5]. Quorum sensing signaling activation can lead to antimicrobial resistance of the pathogens, thus increasing the therapy difficulty of diseases [4].

IntechOpen

The key concern about biofilms is their contribution to the development of resistance against antimicrobial agents, and with the on-going emergence of antibiotic-resistant pathogens, there is a current need for development of alternative therapeutic strategies [6].

An anti-virulence approach by which quorum sensing is impeded could be a viable means to manipulate bacterial processes, especially pathogenic traits that are harmful to human and animal health and agricultural productivity [7]. Further research into the identification and development of chemical compounds and enzymes that facilitate quorum-sensing inhibition (QSI) by targeting signaling molecules, signal biogenesis, or signal detection are required [7]. Anti-QS agents can abolish the QS signaling and prevent the biofilm formation, therefore reducing bacterial virulence without causing drug-resistant to the pathogens, suggesting that anti-QS agents could be potential alternatives for antibiotics [8]. An effective clinical strategy for treating bacterial diseases in the near future will be to combine anti-QS agents with conventional antibiotics since this can significantly improve the efficacy of therapeutic drugs and decrease the cost of human healthcare [9].

2. Microbial biodiversity in biofilm systems

Biofilms are mixed microbial cultures normally consisting predominantly of prokaryotes with some eukaryotes. Thus, in addition to microbial cells, the surrounding environment contains a range of macromolecular products in which exopolysaccharide secreted by the cells is the dominant macromolecular component, while the water content is probably about 90–97% [10, 11]. Secreted products also include enzymes and other proteins, bacteriocins, and low mass solutes and nucleic acid released through cell lysis. The lysis may occur either naturally with cell aging or through the action of phage and bacteriocins.

Opportunistic pathogens, viruses, parasitic protozoa, toxin releasing algae and fungi and enteric bacteria e.g. *Escherichia coli, Klebsiella pneumoniae, Klebsiella oxytoca, Enterobacter cloacae, Enterobacter agglomerans, Helicobacter pylori, Shigella spp., Campylobacter spp., Salmonella spp., Clostridium perfringens, Enterococcus faecium, Enterococcus faecalis* and environmental pathogenic bacteria like *Legionella pneumophila, Pseudomonas aeruginosa, Pseudomonas fluorescens, Aeromonas hydrophila, Aeromonas caviae, Mycobacterium avium, Mycobacterium xenopi* etc. are associated with biofilms present in drinking water [12, 13].

Biofilms present complex assemblies of microorganisms attached to surfaces. They are dynamic structures in which various metabolic activities and interactions between the component cells occur [10]. Studies on microorganisms and biofilm formation have revealed diverse complex social behavior including cooperation in foraging, building, reproduction, dispersion and communication among microorganisms [14]. The organisms within a biofilm setup may include a single or diverse species of microorganisms. In the biofilm, bacteria can share nutrients and are sheltered from harmful factors in the environment, such as desiccation, antibiotics, and a host body's immune system.

Bacteria, fungi, viruses, protozoa and cyanobacteria that are common pathogens are all involved in biofilm formation [15].

2.1 Bacterial biofilms

About 99.9% of all bacteria live in biofilm communities [16]. A biofilm usually begins to form when a free-swimming bacterium attaches to a surface. Pathogenic organisms are found on most food items including seafoods and biofilm forming pathogens are found on such seafoods as crabs [17], pacific oysters [18], shrimps [19] etc. Public health and clinical microbiologists recognize that biofilms are present everywhere in nature and are responsible for a number of human infections. Infectious caused by microbial communities include urinary tract infections, middle-ear infections, dental plaque, gingivitis, endocarditis, cystic fibrosis. Biofilms on persistent indwelling devices such as catheter, contact lenses, heart valves and joint prostheses are also responsible for many recurrent infections [20, 21]. Biofilms on indwelling medical devices may be composed of Gram-positive or Gram-negative bacteria. Bacteria commonly isolated from these devices include the Gram-positive *Enterococcus faecalis, Staphylococcus aureus, Staphylococcus epidermidis*, and *Streptococcus viridans;* and the Gram-negative *Escherichia coli, Klebsiella pneumoniae, Proteus mirabilis,* and *Pseudomonas aeruginosa* [22]. It has been shown that virtually all indwelling central venous catheters are colonized by microorganisms embedded in a biofilm matrix. Among these *S. epidermidis* and *S. aureus* are commonly present on cardiovascular devices [23], causing about 40–50% of infections related to heart valve [14].

The organisms that form biofilms on medical devices originate from patient's skin microflora, exogenous microflora from health-care personnel, or contaminated infusates. Biofilms associated with catheters may initially be composed of single species, but with the passage of time they become multi-specie communities. Some urinary tract and bloodstream infections are also caused by biofilm-associated indwelling medical devices with 50–70% of infections related to catheter [12]. Chronic infections, inflammation and tissue damage caused by many strains of single species are often found in polymicrobial communities [24].

Bacteria that reside in a biofilm community usually will not grow when cultured, a situation normally referred to as "viable, but not culturable". The reason is that to change to the planktonic state from a biofilm-producing phenotype, bacteria require complex and specific environmental and signaling factors that are not available in a culture plate [25]. This therefore suggests that analyzing biofilm samples for bacterial infective agents during infections may show negative results and the real cause of the infections may not be detected if culturing is the only investigative procedure.

2.2 Fungal biofilms

Many medically important fungi produce biofilms and they include *Candida*, *Aspergillus*, *Cryptococcus*, *Trichosporon*, *Coccidioides*, and *Pneumocystis*. *Candida albicans* biofilms are primarily made up of yeast-form and hyphal cells, both of which are required for biofilm formation [26]. The formation of *Candida albicans* biofilm follows a sequential process that involves adherence to a substrate (either abiotic or mucosal surface), proliferation of yeast cells over the surface, and induction of hyphal formation [27]. As the biofilm matures extracellular matrix (ECM) accumulates. Many other Candida spp. form ECM-containing biofilms but do not produce true hyphae and they include *Candida tropicalis*, *Candida parapsilosis*, and *Candida glabrata* [28]. Aspergillus biofilms can form both on abiotic and biotic surfaces and the initial colonizing cells that adhere to the substrate are conidia. Mycelia (the hyphal form) develop as the biofilm matures [29]. *Aspergillus fumigatus* produces two forms of biofilm infections: Aspergilloma and Aspergillosis. Aspergilloma infections present an intertwined ball of hyphae while aspergillosis infections present individual separated hyphae [30].

Trichosporon asahii forms biofilms comprised of yeast and hyphal cells embedded in matrix, as do those of *Coccidioides immitis*. *Cryptococcus neoformans* forms biofilms consisting of yeast cells on many abiotic substrates [31]. Although *Cryptococcus neoformans* forms hyphae in the course of mating, no hyphae have been observed in *Cryptococcus neoformans* biofilms. Similarly, *Pneumocystis species* do not produce hyphal structures as part of their biofilms [32]. Hyphal formation is therefore, not a uniform feature of fungal biofilms.

2.3 Protozoan biofilms

Free-living protozoans are single celled eukaryotic organisms and are divided into amoebae, flagellates and ciliates. All the three protozoan groups have been found in fresh water biofilms. Although many different species are found in association with biofilms, their level of association differs. The protozoans *Cyclospora cayetanensis, Cryptosporidium spp., and Toxoplasma gondii* have all been found in biofilm communities [22].

2.4 Virus involvement in biofilms

Viruses are obligatory intracellular parasites and are found in communities where cells in which they live are found. Viruses are, thus, found in biofilms communities associated with the bacteria, fungi and protozoa they infect.

Many phages may produce polysaccharases or polysaccharide lyases. Some phages are also known to produce enzymes that degrade the poly-Q-glutamic acid capsule of *Bacillus* spp. [33]. Various structures including extracellular polymers and heterologous microbial cells may impede viral access to the bacterial cell surface. Phage may carry on their surfaces enzymes that degrade bacterial polysaccharides including those of biofilm structures. These enzymes are very specific and seldom act on more than a few closely related polysaccharide structures [34]. Numerous phages have been isolated which induce enzymes capable of degrading the exopolysaccharide of various Gram-negative bacterial genera. These include phage for biofilm-forming bacteria. It has been observed that the extracellular matrix of the biofilms does not protect the bacterial cells from infection with phage T4 [35].

Many biofilms possess an open architecture with water-filled channels, which would allow the phage access to the biofilm interior [36]. As biofilms age and cells die and slough off, potential new viral receptor sites may become available. As bacteria excel at adapting to differing nutrient conditions, changes to the host cell surface could be expected with either loss or gain of possible phage receptors. A further factor which might influence phage retention within biofilms lies in the role of hydrophobic and electrostatic interactions. In the interaction of a coliphage with both hydrophobic and hydrophilic membranes, a critical factor in the retention of the phage was its iso-electric point [37].

In complex biofilms in natural environments, eukaryotic algae may also be present [38]. Under these circumstances algal cell lysis through viral action is also possible as many viruses for algal species have now been isolated and identified [39].

3. Biofilms in respiratory tract infections

It is becoming progressively more accepted that biofilm formation is an important cause of morbidity in respiratory tract infections [40]. Biofilms may be involved in some respiratory infections, including ventilator-associated pneumonia, bronchiectasis, bronchitis, cystic fibrosis and upper respiratory airway infections [41].

3.1 Upper respiratory tract infections

Infectious diseases that affect the upper respiratory tract include otitis media, sinusitis, tonsillitis, adenoiditis, pharyngotonsillitis, adenoiditis and chronic rhinosinusitis [42]. In otitis media, infections may be as a result of both respiratory viruses and bacteria such as non-capsulated *Haemophilus influenza, Streptococcus pneumonia*, *Streptococcus pyogenes, Moraxella catarrhalis* and *Staphylococcus aureus*, triggering the appearance of polymicrobial biofilms [43].

The most cited reason for childhood visits to physicians is otitis media with effusion (OME) and is again one of the most reasons for antibiotic therapy in children. Even though OME is regarded as a sterile inflammatory process, current data using a chinchilla model suggest that viable bacteria are present in intricate communities referred to as mucosal biofilms [44]. It is interesting to know that intracellular *Haemophilus influenzae*, *Streptococcus pneumoniae*, *Staphylococcus aureus* and *Moraxella catarrhalis in situ* are found in adenoids from children going through adenoidectomy for the treatment of hypertrophic adenoids or chronic otitis media using Fluorescence in situ hybridization [45]. *Haemophilus influenzae* and intracellular *S. pneumoniae* have also been in middle ear mucosal biopsies in children with chronic otitis media [46].

Biofilms were seen in the sinus tissues of 72% of patients affected by chronic rhinosinusitis and the cultured organisms identified included *H. influenzae* (28%), *P. aeruginosa* (22%), *S. aureus* (50%), and fungi (22%). The presence of bacterial biofilms was linked to persistent mucosal inflammation after endoscopic sinus surgery [47]. Assessment of some chronic infections in the upper respiratory tract including recurrent tonsillitis and chronic rhinosinusitis in human clinical specimens suggests that both attachment and aggregated bacteria are present [48]. For instance, electron microscopy and culture were used to show that biofilms were associated with the mucosal epithelium of tonsils in 73% of tonsils removed for tonsillitis and 75% of those tonsils removed due to hypertrophic tonsils alone [49]. Calo *et al.* [42] found bacterial biofilms in recurrent and chronic infectious diseases of the upper respiratory tract (adenoiditis, tonsillitis, and chronic rhinosinusitis) and concluded that biofilms formation plays a role in upper airway infections.

3.2 Tissue related infections

3.2.1 Cystic fibrosis (CF)

Cystic fibrosis (CF) is a protracted disease of the lower respiratory tract. The most frequent serious clinical complication in CF today is chronic endobronchial infection with *Pseudomonas aeruginosa*. *Pseudomonas aeruginosa* is a microorganism characterized by the capacity to produce large amounts of alginate and developed as a biofilm where micro-colonies of bacteria embedded in a matrix of alginate attack the lower respiratory tract [42]. Cystic fibrosis occurs as a result of a mutation in the CF transmembrane conductance regulator gene that encodes a cyclic AMP-regulated chloride ion channel. The mutation causes defective ion transport across epithelial cell surfaces in the upper airways, interfering with the removal of particles and microbial cells trapped in the overlying mucus and causing increased susceptibility to bacterial infection. Therefore, the airways of patients with CF are almost always infected with different bacterial species, but *P. aeruginosa* infection causes the greatest problem of morbidity and mortality [43]. *Pseudomonas aeruginosa* is the most common bacterial species that causes respiratory tract infection in CF patients and can be seen in about half of all cases and in up to 70% of adults [44]. Other pathogens such as *Staphylococcus aureus*, *Achromobacter xylosoxidans*, *Burkholderia cepacia* complex and *Stenotrophomonas maltophilia* have also been found to cause CF and are linked to biofilm formation [45].

3.2.2 *Cystic fibrosis with chronic lung infections*

A major difficulty in this type of infection is contamination of lower respiratory secretions with the normal oropharyngeal flora, particularly as members of the normal flora (e.g. *Haemophilus influenzae*, *Staphylococcus aureus*, *Streptococcus pneumonia* and *Moraxella catarrhalis*) are common lung pathogens in CF [46, 47]. The incidence of bacterial lung infections in CF is high because the mucoid polysaccharidic material that accumulates on the respiratory epithelium due to the fact that impaired mucociliary removal in the bronchi of such patients favors biofilm formation. The capacity of *Pseudomonas aeruginosa* to form biofilms is believed to be the primary reason for its survival in the CF lung, despite a high inflammatory response and intensive antibiotic treatment [48]. Chronic airway infections cause an increase deterioration of lung tissue, a decline in pulmonary function and, finally, respiratory failure and death in cystic fibrosis (CF) patients [49].

3.2.3 *Chronic obstructive pulmonary disease (COPD)*

The role of biofilms in patients with COPD has not been directly validated but has been hypothesized considering the evidence showing that the respiratory tracts of these patients are frequently colonized by pathogens. Murphy and Kirkham [50] have recently confirmed that biofilms do play a role in COPD where they identified major outer membrane proteins of Non-typeable *H. influenzae* during its growth as a biofilm. Even if direct proof of biofilm formation *in vivo* is lacking, biofilms may reasonably be considered to be involved in the vicious cycle of infection/inflammation leading to disease development in patients with COPD [51].

3.2.4 *Non-cystic fibrosis bronchiectasis*

In bronchiectasis not due to cystic fibrosis, infections result in changes in the muscular and elastic components of the bronchial wall, which become distorted and expanded. Airways gradually become unable to clear mucus, leading to serious lung infections, which in turn cause more damage to the bronchi [52]. Recently biofilm formation has been demonstrated *in vivo* and is assumed to play a significant role in the pathophysiological cascade of the disease [53]. Bacterial biofilm formation by *Pseudomonas aeruginosa* or *Klebsiella pneumoniae* is common in bronchiectasis and could be an essential factor that makes infections in bronchiectasis obstinate. Other pathogens such as *Prevotella sp*., *Veillonella sp*. and *Neisseria sp*. have also been identified recently in patients with bronchiectasis to form biofilms [54].

3.2.5 *Bronchitis*

Prolonged bacterial bronchitis may be caused by chronic infections of the respiratory tract. In children especially, the condition appears to be secondary to impaired mucociliary removal that produces an environment favorable for bacteria to become established, usually in the form of biofilms. The most commonly involved bacteria include *Haemophilus influenzae* (30–70%), *Moraxella catarrhalis and Streptococcus pneumonia* [55].

3.2.6 *Diffuse pan-bronchiolitis*

Diffuse pan-bronchiolitis (DPB) is an unusual inflammatory lung disease of unknown etiology found in adult Japanese patients. With this disease, chronic endobronchial infection with *Pseudomonas aeruginosa* biofilms leading to respiratory failure is common. It is a severe, progressive form of bronchiolitis (Inflammation and congestion in the bronchioles of the lung) [56].

3.3 Device-related infections

In device-related infections such as ventilator-associated pneumonia (VAP), biofilms result in microbial persistence and reduced response to treatment. Biofilm formation within the first 24 h after intubation has been reported in 95% of endotracheal tubes [57]. Pathogens in both endotracheal tube biofilm and secretions accrued within the airways/endotracheal tubes in 56 to 70% of patients with VAP have been reported. *Pseudomonas aeruginosa* and *Acinetobacter baumannii* are the most common bacteria that colonize these devices [57].

3.4 Biofilm forming organisms associated with respiratory tract infections

This section presents the role of biofilms in respiratory tract infections, with specific emphasis on the biofilms formed by *Pseudomonas*, *Staphylococcus*, and *Haemophilus*, the primary pathogens associated with respiratory tract infections [58] although additional important pathogens, including *Streptococcus pneumoniae*, *Bordetella* and *Mycobacterium* species do play a role [59].

3.4.1 Biofilms formed by Pseudomonas aeruginosa

Pseudomonas aeruginosa is a recognized common pathogen in respiratory tract infections although other members of the genus *Pseudomonas* are able to form biofilms [7]. Respiratory infections caused by *P. aeruginosa* are a major globally clinical issue, especially for patients with chronic pulmonary disorders, such as those with cystic fibrosis (CF), non-CF bronchiectasis, severe chronic obstructive pulmonary disease (COPD) and ventilator-associated pneumonia [60]. This bacterium is a difficult opportunistic pathogen that readily forms biofilms on most surfaces [5]. The intricate steps of biofilm formation by *P. aeruginosa* are considered to be a developmental process. The stages of *P. aeruginosa* biofilm formation can be seen by several strategies. One easy technique is the scanning electron microscope (SEM) of *P. aeruginosa* grown on glass surfaces or tracheal explants. Biofilms form when planktonic *P. aeruginosa* bacteria get attached to a surface using adhesins such as type IV fimbriae and flagella, and begin to colonize. In this regard type IV fimbriae and flagella *P. aeruginosa* mutants are severely compromised in initiation of biofilm formation [58, 61]. Additionally, the process of surface translocation mediated by type IV fimbriae (twitching motility) is essential for initiation of biofilm formation by *P. aeruginosa* [58]. Most probable, twitching motility confers synchronized cell movement along the surface as well as cell–cell communications that lead to the formation of micro-colonies. The coordination of events for the initiation and formation of biofilms requires cell–cell interactions that are mediated by quorum sensing [62]. Following this, the micro-colonies mature into distinctive three-dimensional structures that pose the most severe scenario for clinical treatment. This structure is typically trapped in a matrix material that may be composed of protein, polysaccharide, or nucleic acid. Nonetheless, it has been proposed guluronic and mannuronic acids [63] are the major constituents of the biofilm matrix [64]. Recent data also suggest that DNA also contributes to this matrix [60].

3.4.2 Biofilms formed by Staphylococcus species

The adherence of *Staphylococcus* directly to an implanted device (intravascular catheters, prosthetic devices, and other indwelling medical devices) or indirectly via host proteins is the first step in the development of a biofilm. This is followed by a buildup of multilayered cell clusters on the polymer surface [65]. When *Staphylococcus* bacteria get within 50 nm of a surface, they adhere through hydrophobic interactions, van der Waal's forces, and when present, fimbriae and pili also contribute to its adhesion [66]. A biofilm-associated protein (Bap) is reported to contribute to *S. aureus* biofilm formation. The second phase of *Staphylococcus* biofilm formation is the accumulation of complex cell clusters mediated by intercellular adhesion. A 140 kDa extracellular protein, known as the accumulation associated protein (AAP), appears responsible for accumulative growth on polymer substances [67]. It has been hypothesized that AAP is involved in anchoring the polysaccharide adhesion PIA (polysaccharide intercellular adhesion) to the cell surface [63]. The extracellular polysaccharide adhesion antigen PIA is a well-described polysaccharide antigen that is linked to cellular aggregation or clustering. Lastly, the generation of a slime glycocalyx is believed to be the climaxing event in the staphylococcal biofilm developmental process. This slime layer is not essential for surface colonization and appears variable between strains. However, when present, the slime layer protects the bacteria from host defenses and some antibiotics. As in *P. aeruginosa*, organization of complex communities within *Staphylococcus* biofilms is a coordinated effort and requires cell–cell communication [68].

3.4.3 Biofilms formed by Haemophilus influenzae

Non-typeable *H. influenzae* (NTHI) strains are members of the normal human nasopharyngeal flora, as well as frequent opportunistic pathogens of both the upper and lower respiratory tracts. It is an important cause of otitis media in children and lower respiratory tract infection in adults with chronic obstructive pulmonary disease (COPD). Recently, it has been shown that NTHI can form biofilms both *in vitro* and *in vivo* [69]. Considerable diversity in the ability of NTHi isolates to form biofilms has also been reported. A NTHi pilus defective strain was reduced three-to four fold in biofilm formation compared with its isogenic parental NTHi isolate, signifying a role of the pilus in biofilm development. Although this is the case for other gram-negative bacteria [70], nonetheless, it is quite clear that NTHi strains have the ability to form biofilms both *in vitro* and *in vivo* [69]. Earlier studies of cell envelopes during growth of *H. influenzae* as a biofilm established an increased abundance of a ~30 kDa protein [58], peroxiredoxin-glutaredoxin (PGdx) [71], that is expressed by *H. influenzae* during biofilm growth and this probably contributes to its persistence in the upper respiratory tract infections.

3.4.4 Biofilms formed by other microorganisms

Streptococcus pneumoniae: *Streptococcus pneumoniae* is a frequent colonizer of the human nasopharynx and a significant human respiratory pathogen that causes a variety of diseases such as community-acquired pneumonia and otitis media in children [72]. Colonizing pneumococci form well-ordered biofilm communities in the nasopharyngeal environment, but the exact role of biofilms and their interaction with the host during colonization and disease is not yet explicit [73]. However, investigators have speculated that pneumococci form biofilms in the nasopharynx *in vivo* [74]. Recently, pneumococci have been reported for the first time to form highly structured biofilms during colonization of the murine nasopharynx [75]. Mice were also inoculated intranasally with the pneumococcal strain EF3030, a clinical isolate known to be non-invasive and an efficient colonizer in murine models, and found to form biofilms [76].

Bordetella species: *Bordetellae* are respiratory pathogens that infect both humans and animals. *Bordetella bronchiseptica* causes asymptomatic and long-term to life-long infections in animal nasopharynges while the human pathogen, *B. pertussis* is the etiological agent of the acute disease whooping cough in infants and young children. One proposed hypothesis to explain the survival and continued persistence of *Bordetella spp*. in the mammalian nasopharynx is that these organisms produce surface-adherent communities known as biofilms [77]. Researchers have recently established the ability of the three classical *Bordetella species* (*Bordetella pertussis, Bordetella bronchiseptica,* and *Bordetella parapertussis*) to form biofilms on abiotic surfaces [78]. It is assumed that *Bordetella* biofilm formation may play a role in the pathogenic cycle, precisely in persistence within the nasopharynx [79]. The capacity to form biofilms in mice suggests a role for *Bordetella* mode of existence during human infections. Clusters and tangles (reminiscent of biofilms) of *Bordetella pertussis* adherent to ciliated cells in explant cultures and tissue biopsies of pertussis patients have been documented [79]. As reported for other biofilm-forming organisms, extracellular DNA and exopolysaccharide are vital for biofilm formation by *Bordetella bronchiseptica*. The observation of biofilm-like structures *in vivo* in the nasal epithelium of *Bordetella bronchiseptica* infected mice showed that these communities expressed a polysaccharide essential for *in vivo* biofilm development [75, 76]. In *Bordetella*, BvgAS-regulated factors, including the filamentous hemagglutinin and adenylate cyclase, may also contribute to biofilm formation [79].

Mycobacterium species: Mycobacterial infections have been shown to form biofilms, most notably *Mycobacterium tuberculosis*, which under the conducive environments, can self-assemble. Among the non-tuberculous mycobacteria, *Mycobacterium avium* complex (MAC) and the rapidly growing mycobacteria, including *Mycobacterium abscessus* complex, have been reported to produce biofilms either *in vitro* or in environmental reservoirs [80], but *in vivo* conditions have not been investigated. *Mycobacterium abscessus* complex is an evolving threat to patients with cystic fibrosis [81], that become infected at an early stage and worsens clinically as the persistent infection results in inflammation and tissue damage.

4. Quorum sensing

In the control of microbial infections, two strategies are normally envisaged; killing the organisms or attenuation of the organisms' virulence such that they fail to adapt to the host environment. The former approach is what is generally favored; the latter lacks specific targets for rational drug design. It has, however, been realized that Gram-negative bacteria use small molecules known as acyl homoserine lactones to regulate the production of secondary metabolites and virulence factors, and this could offer a novel target to address the strategy of attenuating the organisms' virulence thereby impairing their adaptation to the host system. Recent research has highlighted the importance of cell-to-cell interactions or communications, referred to as Quorum Sensing (QS), in microorganisms. Many bacterial species employ a complex mechanistic communication system to transmit information among themselves. Bacteria can act in response to a variety of chemical signals produced by the same species along with others produced by other species, and this provides a way for intraspecies and interspecies cross-communication and interruption of signals. The ability of bacteria to dispatch, pull together, and process information allow them to act as "multicellular" organisms and enhance their survival in complex environments [82].

Any mechanism capable of disrupting QS signals can be used to reduce survival of the microorganism thereby preventing or reducing virulence in the host environment. Such methods of interruption of the QS include:

- Disruption of biosynthesis of signal molecules,
- Application of QS antagonists (e.g. use of extracts from higher plants and algae and other chemical compounds),
- Chemical inactivation of quorum sensing signals,
- Biodegradation of signal molecule.

Agents capable of inhibiting the growth of microorganisms or disrupting the quorum sensing mechanisms of the microorganisms or interrupting the biofilm formation may be useful in the fight against microbial pathogenicity.

4.1 Anti-quorum sensing activity

It has now become apparent that different types of microorganisms have evolved the ability to recognize and act in response to the presence of other microorganisms in their neighborhood. Most Gram-negative bacteria produce and respond to *N*-acyl homoserine lactone (AHLs) signal molecules to regulate production of secondary metabolites in order to monitor their own population density. These molecules, at a threshold population density, act together with cellular receptors and elicit the expression of target genes such as those involved in virulence, antimicrobial production, motility and swarming, sporulation, bioluminescence and biofilm formation. The concept of quorum sensing (QS) was initially described in *Vibrio fischeri*, a luminescent marine bacterium. It was observed that the organisms express genes controlling light emission (the luciferase enzyme) when in symbiotic association with its hosts, the squid [83]. At low population densities (i.e. free-living in seawater) *Vibrio fischeri* does not express luciferase and so is non-luminescent. However, when cultured in the laboratory to high cell densities, they express bioluminescence with a blue-green light. They do not emit light unless they detect a concentration high enough of their own AHL. These organisms usually form symbiotic relationships with some fish and squid species such as *Euprymna scolopes*. *Euprymna scolopes* appears bioluminescent in dark surroundings because of high-population of the cells (*Vibrio fischeri*) in a specialized light organ. *Euprymna scolopes,* in return, offers nutrients to the *Vibrio fischeri* population. The QS system originally identified in *Vibrios* involved two genes, luxl and luxR. The Luxl codes for an enzyme, which synthesizes 3-oxo-C6-homoserine lactone (an auto-inducer as they are produced by the same cells whose metabolism they regulate) [82].

The unpleasant side effects of antibiotics (such as ototoxicity and nephrotoxicity associated with the aminoglycosides) have led to preference for preventive rather than curative approach towards fighting infectious diseases. Inhibition of quorum sensing activity has been hypothesized as one approach that can be useful in preventing bacterial infection. It could provide an additional approach to antibiotic mediated bactericidal or bacteriostatic activity thereby reducing the risk of successful establishment of infections or resistance development in the bacteria. This is supported by the protective effect of QS inhibition demonstrated in animal infection models. A simple animal infection model on QS was launched in *Caenorhabditis elegans*, a nematode that feeds on bacteria. When fed on opportunistic pathogens such as *P. aeruginosa*, the worm was mostly destroyed within a short time after taking in the bacteria; presumably annihilated by the actions of cyanide and phenazines produced by the bacteria [84]. However, in instances where the worms ingested *P. aeruginosa* with mutations in the QS-controlling systems, they were not killed but were rather sustained on the bacteria. This model highlights the involvement of QS-regulated virulence factors in pathogenicity of *Pseudomonas aeruginosa*. It is obvious from such models that interruption of the QS apparatus of bacteria by plant extracts or other chemical compounds may offer a novel and an exciting approach to fight the existing problems associated with antimicrobial chemotherapy.

Many bacteria produce AHL molecules in response to QS and so could be used as biomonitor organisms in screening of compounds for anti-QS activity. Such bacteria include *Chromobacterium violaceum, Erwinia carotovora* and *Pseudomonas aeruginosa*.

5. Medicinal plants with biofilm inhibition activity

Natural products have been identified to inhibit biofilm formation in microorganisms. The exact mechanism for most of the agents is yet to be elucidated. Medicinal plants have been identified as rich source of bioactive compounds that have the capability of interfering with biofilm formation but most of these studies are still in the early stages of drug development. The anti-biofilm effects of medicinal plants have been proposed to be due to the inhibition of formation of polymer matrix, suppression of cell adhesion and attachment, interruption of extracellular matrix formation and reduction in virulence factors production and activation, thereby blocking QS network and biofilm development [85].

Medicinal plants belonging to various plant families reported to have biofilm inhibitory activity are listed in **Table 1**; the part of the plant (leaves, fruits, stem

Plant name	Family	Part used	Solvent	Biofilm inhibition activity	Reference
Punica granatum L	Lythraceae	Fruit	Methanol	Inhibit biofilm formation in *E. coli* by 70% at 150 μg/mL	[86]
Salvia fruticosa Mill.	Lamiaceae	Aerial parts	Ethanol	Inhibit biofilm formation by 60.9% at 0.78 mg/mL	[87]
Vaccinium corymbosum L	Ericaceae	Fruit	Decoction	Reducing 47% MRSA biofilm viable counts. 12.5 mg/mL	[88]
Commelina benghalensis L.	Commelinaceae	Whole plant	Distilled water	Inhibited the biofilm formation at 250 μg/mL	[89]
Curcuma longa L.	Zingiberaceae	Rhizome	Aqueous	Removed 30 to 40% of biofilm at 5–0.63 μg/mL	

Plant name	Family	Part used	Solvent	Biofilm inhibition activity	Reference
Euphorbia hirta L.	Euphorbiaceae	Aerial parts	Methanol	Biofilm inhibition and eradication activity against *P. aeruginosa* observed at 0.25 and 0.5 mg/ml, respectively	[90]
Terminalia bellirica (Gaertn.) Roxb	Combretaceae	Dried fruit	Ethanol	Inhibition biofilm formation by 89.8 and 92.2% at 125 and 250 μg/mL, respectively	[91]
Azadirachta indica A. Juss	Meliaceae	Leaf	Distilled water	Reduced biofilm completely by 35% at 5% w/v	[92]
Commiphora leptophloeos (Mart.) J.B. Gillet	Burseraceae	Stem bark	Distilled water	Inhibition of cell adhesion above 80% at 4.0 mg/mL	[93]
Bauhinia acuruana (Moric)	Fabaceae	Fruit	Distilled water	Inhibition of biofilm formation was determined to be 77.8 ± 5.0% at 4.0 mg/mL	
Camellia sinensis (L.) Kuntze	Theaceae	Leaves	Ethanol	Inhibited the cell adhesion by 78.7% 0.5%w/v	[94]

Table 1.
Medicinal plants with anti-biofilm activity.

bark, rhizome) used, the various solvents used for extraction and their ability to inhibit cell adhesion or to eradicate biofilm formed by different pathogens have been mentioned.

6. Conclusion

Combatting biofilm and quorum sensing is a good strategy to reduce microbial pathogenicity and thus fight infections. This can be achieved by finding effective agents that can inhibit biofilm formation and disrupt quorum sensing mechanisms. Natural products particularly medicinal plants are a rich source of bioactive compounds that have served as useful leads in the development of drugs. Rigorous evaluation of medicinal plants can therefore lead to novel anti-biofilm and anti-quorum sensing agents.

Author details

Cynthia Amaning Danquah[1*], Samuel Osei-Djarbeng[2], Theresah Appiah[3], Yaw Duah Boakye[3] and Francis Adu[3]

1 Department of Pharmacology, Faculty of Pharmacy and Pharmaceutical Sciences, Kwame Nkrumah University of Science and Technology, Kumasi, Ghana

2 Department of Pharmaceutical Sciences, Kumasi Technical University, Kumasi, Ghana

3 Department of Pharmaceutical Microbiology, Faculty of Pharmacy and Pharmaceutical Sciences, Kwame Nkrumah University of Science and Technology, Kumasi, Ghana

*Address all correspondence to: cadanquah.pharm@knust.edu.gh

References

[1] Davey ME, O'Toole GA. Microbial biofilms: From ecology to molecular genetics. Microbiology and Molecular Biology Reviews. 2000;**64**:847-867

[2] Donlan RM. Biofilms: Microbial life on surfaces. Emerging Infectious Diseases. 2002;**8**:881-890

[3] Bjarnsholt T. The role of bacterial biofilms in chronic infections. APMIS. 2013;**121**:1-58

[4] Sifri CD. *Quorum sensing*: Bacteria Talk Sense. Clinical Infectious Diseases. 2008;**47**:1070-1076

[5] Davies DG, Parsek MR, Pearson JP, Iglewski BH, Costerton JW, Greenberg EP. The involvement of cell-to-cell signals in the development of a bacterial biofilm. Science. 1998;**280**:295-298

[6] Lewis K. Riddle of biofilm resistance. Antimicrobial Agents and Chemotherapy. 2001;**45**:999-1007

[7] Costerton JW, Stewart PS, Greenberg EP. Bacterial biofilms: A common cause of persistent infections. Science. 1999;**284**:1318-1322

[8] Hentzer M, Givskov M. Pharmacological inhibition of quorum sensing for the treatment of chronic bacterial infections. Journal of Clinical Investigation. 2003;**119**:1300-1307

[9] Cheesman MJ, Ilanko A, Blonk B, Cock IE. Developing new antimicrobial therapies: Are synergistic combinations of plant extracts/compounds with conventional antibiotics the solution? Pharmacognosy Reviews. 2017;**11**:57-72

[10] Sutherland IW, Hughes KA, Skillman LC, Tait K. The interaction of phage and biofilms. FEMS Microbiology Letters. 2004;**232**:2321-2326

[11] Zhang XQ, Bishop PL, Kupferle MJ. Measurement of polysaccharides and proteins in biofilm extracellular polymers. Water Science and Technology. 1998;**37**:345-348

[12] Rodney MD. Biofilm formation: A clinically relevant microbiological process. Clinical Infectious Diseases. 2001;**33**:1387-1392

[13] Costerton JW. Structure of biofilms. In: Geesey GG, Lewandowski Z, Flemming HC, editors. Biofouling and Biocorrosion in Industrial Water Systems. CRC Press, USA; 1994. ISBN 087371 928 X

[14] Crespi BJ. The evolution of social behaviour in microorganisms. Trends in Ecology & Evolution. 2001;**16**:178-183

[15] Trasneem U, Yasin N, Qasim M, Nisa I, Shah F, Rasheed U, et al. Biofilm producing bacteria: A serious threat to public health in developing countries. Journal of Food Science and Nutrition. 2018;**1**:25-31

[16] Adetunji VO, Isola OT. Crystal violet binding assay for assessment of biofilm formation by *Listeria monocytogenes* and *Listeria spp*. on Wood, steel and glass surfaces. Global Veterinaria. 2011;**6**:6-10

[17] Reguera G, Kolter R. Virulence and the environment: A novel role for *Vibrio cholerae* toxin-coregulated pili in biofilm formation on chitin. Journal of Bacteriology. 2005;**187**:3551-3555

[18] Alisha M, Aagesen AM, Sureerat P, Yi-Cheng S, Häsea CC. Persistence of *Vibrio parahaemolyticus* in the Pacific oyster, *Crassostrea gigas*, is a multifactorial process involving pili and flagella but not type III secretion systems or phase variation. Applied and Environmental Microbiology. 2013;**79**:3303-3305

[19] Norhana MNW, Poole SE, Deeth HC, Dykes GA. The effects of

temperature, chlorine and acids on the survival of *Listeria* and *Salmonella* strains associated with uncooked shrimp carapace and cooked shrimp flesh. Food Microbiology. 2010;**27**:250-256

[20] Donlan RM. Biofilms and device-associated infections. Emerging Infectious Diseases. 2001;7:277-281

[21] Delle-Bovi RJ, Smits A, Pylypiw HM. Rapid method for the determination of total monosaccharide in *Enterobacter cloacae* strains using Fourier transform infrared spectroscopy. American Journal of Analytical Chemistry. 2011;**2**:212-216

[22] Stickler DJ. Bacterial biofilms and the encrustation of urethral catheters. Biofouling. 1996;**9**:293-305

[23] Otto M. *Staphylococcus epidermidis*-the "accidental" pathogen. Nature Reviews Microbiology. 2009;**7**:555-567

[24] Hall-Stoodley L, Stoodley P. Evolving concepts in biofilm infections. Cellular Microbiology. 2009;**11**:1034-1043

[25] Healy DY, Leid GJ, Sanderson RA, Hunsaker DH. Biofilms with fungi in chronic rhinosinusitis. Otolaryngology–Head and Neck Surgery. 2008;**138**:641-647

[26] Fanning S, Mitchell AP. Fungal Biofilms. PLoS Pathogens. 2012;**8**:e1002585

[27] Finkel JS, Mitchell AP. Genetic control of *Candida albicans* biofilm development. Nature Reviews Microbiology. 2011;**9**:109-118

[28] Silva S, Negri M, Henriques M, Oliveira R, Williams DW, Azeredo J. Adherence and biofilm formation of non-*Candida albicans* Candida species. Trends in Microbiology. 2011;**19**:241-247

[29] Mowat E, Williams C, Jones B, McChlery S, Ramage G. The characteristics of *Aspergillus fumigatus* mycetoma development: Is this a biofilm? Medical Mycology. 2009;**47**:S120-S126

[30] Loussert C, Schmitt C, Prevost MC, Balloy V, Fadel E, Philippe B, et al. *In vivo* biofilm composition of *Aspergillus fumigatus*. Cellular Microbiology. 2010;**12**:405-410

[31] Martinez LR, Casadevall A. *Cryptococcus neoformans* biofilm formation depends on surface support and carbon source and reduces fungal cell susceptibility to heat, cold, and UV light. Applied and Environmental Microbiology. 2007;**73**:4592-4601

[32] Cushion MT, Collins MS, Linke MJ. Biofilm formation by *Pneumocystis spp*. Eukaryotic Cell. 2009;**8**:197-206

[33] Kimura K, Itoh Y. Characterization of poly-Q-glutamate hydrolase encoded by a bacteriophage genome: Possible role in phage infection of *Bacillus subtilis* encapsulated with poly-Q-glutamate. Applied and Environmental Microbiology. 2003;**69**:2491-2497

[34] Sutherland IW. Polysaccharases for microbial polysaccharides. Carbohydrate Polymers. 1999;**116**:319-328

[35] Doolittle MM, Cooney JJ, Caldwell DE. Lytic infection of *Escherichia coli* biofilms by bacteriophage T4. Canadian Journal of Microbiology. 1995;**41**:12-18

[36] Wood SR, Kirkham J, Marsh PD, Shore RC, Nattress B, Robinson C. Architecture of intact natural human plaque biofilms studied by confocal laser scanning microscopy. Journal of Dental Research. 2000;**79**:21-27

[37] Van Voorthuizen EM, Ashbolt NJ, Schafer AI. Role of hydrophobic and electrostatic interactions for initial enteric virus retention by MF membranes. Journal of Membrane Science. 2001;**194**:69-79

[38] Van Etten JL, Lane LC, Meints RH. Viruses and virus-like particles of eukaryotic algae. Microbiological Reviews. 1991;**55**:586-620

[39] Wilson WH, Tarran GA, Schroder D, Cox M, Oke J, Malin G. Isolation of viruses responsible for the demise of *Emiliania huxleyi* bloom in the English Channel. Journal of the Marine Biological Association. 2002;**82**:369-377

[40] Anderson MJ, Patrick J, Parks MLP. A mucosal model to study microbial biofilm development and anti-biofilm therapeutics. Journal of Microbiological Methods. 2012;**92**:201-208

[41] Blasi F, Page C, Maria G, Pallecchi L, Gabriella M, Rogliani P, et al. The effect of N-acetylcysteine on biofilms : Implications for the treatment of respiratory tract infections. Respiratory Medicine. 2016;**117**:190-197

[42] Calo L, Passàli GC, Galli J, Fadda G, Paludetti G. Role of biofilms in chronic inflammatory diseases of the upper airways. Advances in Oto-Rhino-Laryngology. 2011;**72**:93-96

[43] Hamilos DL. Host-microbial interactions in patients with chronic rhinosinusitis. Journal of Allergy and Clinical Immunology. 2014;**133**(3):640-653

[44] Ehrlich GD, Veeh R, Wang XJ, Costerton W, Hayes JD, Hu FZ, et al. Mucosal biofilm formation on middle-ear mucosa in the chinchilla model of otitis media. JAMA. 2002;**287**:1710-1715

[45] Forsgren J, Samuelson A, Ahlin A, Jonasson J, Rynnel-Dagöö B, Lindberg A. *Haemophilus influenzae* resides and multiplies intracellularly in human adenoid tissue as demonstrated by *in situ* hybridization and bacterial viability assay. Infection and Immunity. 1994;**62**:673-679

[46] Coates H, Thornton R, Langlands J, Filion P, Keil AD, Vijayasekaran S, et al. The role of chronic infection in children with otitis media with effusion: Evidence for intracellular persistence of bacteria. Otolaryngology-Head and Neck Surgery. 2008;**138**:778-781

[47] Goddard AF, Staudinger BJ, Dowd SE, Joshi-Datar A, Wolcott RD, Aitken ML, et al. Direct sampling of cystic fibrosis lungs indicates that DNA-based analyses of upper-airway specimens can misrepresent lung microbiota. Proceedings of the National Academy of Sciences of the USA. 2012;**109**:13769-13774

[48] Koch C, Hoiby N. Pathogenesis of cystic fibrosis. Lancet. 1993;**341**: 165-1069

[49] Sibley CD, Rabin H, Surette MG. Cystic fibrosis: A polymicrobial infectious disease. Future Microbiology. 2006;**1**:53-61

[50] Murphy TF, Kirkham C. Biofilm formation by non-typeable *Haemophilus influenzae*: Strain variability, outer membrane antigen expression and role of pili. FEMS Microbiology Letters. 2002;**2**:81-89

[51] Chalmers JD, Aliberti S, Blasi F. Management of bronchiectasis in adults. European Respiratory Journal. 2015;**45**:1446-1462

[52] Rogers GB, van der Gast CJ, Serisier DJ. Predominant pathogen competition and core microbiota divergence in chronic airway infection. International Society for Microbial Ecology. 2014;**9**:217-225

[53] Priftis KN, Litt D, Manglani S, Anthracopoulos MB, Thickett K, Tzanakaki G, et al. Bacterial bronchitis caused by *Streptococcus pneumoniae* and non-typable *Haemophilus influenzae* in

children: The impact of vaccination. Chest. 2013;**143**:152-157

[54] Folkesson A, Jelsbak L, Yang L, Johansen HK, Ciofu O, Høiby N, et al. Adaptation of *Pseudomonas aeruginosa* to the cystic fibrosis airway: An evolutionary perspective. Nature Reviews Microbiology. 2012;**10**:841-851

[55] Mietto C, Pinciroli R, Patel N, Berra L. Ventilator associated pneumonia: Evolving definitions and preventive strategies. Respiratory Care. 2013;**58**:990-1007

[56] Jackson K, Keyser R, Wozniak DJ. The role of biofilms in airway disease. Thieme. 2003;**24**:663-670

[57] Lipuma J. The changing microbial epidemiology in cystic fibrosis. Clinical Microbiology Reviews. 2010;**23**:299-323

[58] O'Toole GA. Initiation of biofilm formation in *Pseudomonas fluorescens* WCS365 proceeds via multiple convergent signalling pathways: A genetic analysis. Molecular Microbiology. 1998;**28**:449-461

[59] Grimwood K, Kyd JM, Owen SJ, Massa HM, Cripps AW. Vaccination against respiratory *Pseudomonas aeruginosa* infection. Human Vaccines & Immunotherapeutics. 2014;**11**:14-20

[60] Whitchurch CB, Tolker-Nielsen T, Ragas PC, Mattick JS. Extracellular DNA required for bacterial biofilm formation. Science. 2002;**295**:1487

[61] Evans LR, Linker A. Production and characterization of the slime polysaccharide of *Pseudomonas aeruginosa*. Journal of Bacteriology. 1973;**116**:915-924

[62] Davies DG, Chakrabarty AM, Geesey GG. Exopolysaccharide production in biofilms: Substratum activation of alginate gene expression by *Pseudomonas aeruginosa*. Applied and Environmental Microbiology. 1993;**59**:1181-1186

[63] von Eiff C, Peters G, Heilmann C. Pathogenesis of infections due to coagulase-negative *staphylococci*. Lancet Infectious Diseases. 2002;**2**:677-685

[64] Schierholz JM, Beuth J. Implant infections: A haven for opportunistic bacteria. Journal of Hospital Infection. 2001;**49**:87-93

[65] Hussain M, Herrmann M, von Eiff C, Perdreau-Remington F, Peters G. A 140-kilodalton extracellular protein is essential for the accumulation of *Staphylococcus epidermidis* strains on surfaces. Infection and Immunity. 1997;**65**:519-524

[66] Balaban N, Goldkorn T, Gov Y, Hirshberg M, Koyfman N, Matthews HR, et al. Regulation of *Staphylococcus aureus* pathogenesis via target of RNAIII-activating protein (TRAP). Journal of Biological Chemistry. 2001;**276**:2658-2667

[67] Jurcisek JA, Bakaletz LO. Biofilms formed by non-typeable *Haemophilus influenzae in vivo* contain both double-stranded DNA and type IV pilin protein. Molecular Biology of Pathogens. 2007;**189**:3868-3875

[68] O'Toole GA, Kolter R. Flagellar and twitching motility are necessary for *Pseudomonas aeruginosa* biofilm development. Molecular Microbiology. 1998;**30**:295-304

[69] Hoiby N, Johansen HK, Moser C, Song Z, Ciofu O, Kharazmi A. *Pseudomonas aeruginosa* and the *in vitro* and *in vivo* biofilm mode of growth. Microbes and Infection. 2001;**3**:23-35

[70] Hoa M, Tomovic S, Nistico L, Hall-Stoodley L, Stoodley P, Sachdeva L, et al. Identification of adenoid biofilms with middle ear pathogens in otitis-prone children utilizing SEM

and FISH. International Journal of Pediatric Otorhinolaryngology. 2009;**73**:1242-1248

[71] Chao Y, Marks LR, Pettigrew MM, Hakansson AP. *Streptococcus pneumoniae* biofilm formation and dispersion during colonization and disease. Frontiers in Cellular and Infection Microbiology. 2015;**4**:194

[72] Sanchez CJ, Shivshankar P, Stol K, Trakhtenbroit S, Sullam PM, Sauer K, et al. The pneumococcal serine-rich repeat protein is an intra-species bacterial adhesin that promotes bacterial aggregation *in vivo* and in biofilms. PLoS Pathogens. 2010;**6**:e1001044

[73] Marks LR, Parameswaran GI, Hakansson AP. Pneumococcal interactions with epithelial cells are crucial for optimal biofilm formation and colonization *in vitro* and *in vivo*. Infection and Immunity. 2012;**80**:2744-2760

[74] Palaniappan R, Singh S, Singh UP, Sakthivel SK, Ades EW, Briles DE, et al. Differential PsaA-PspA-, PspC-, and PdB-specific immune responses in a mouse model of pneumococcal carriage. Infection and Immunity. 2005;**73**:1006-1013

[75] Sloan GP, Love CF, Sukumar N, Mishra M, Deora R. The *Bordetella* bps polysaccharide is critical for biofilm development in the mouse respiratory tract. Jorunal of Bacteriology. 2007;**189**:8270-8276

[76] Conover MS, Mishra M, Deora R. Extracellular DNA is essential for maintaining *Bordetella* biofilm integrity on abiotic surfaces and in the upper respiratory tract of mice. PLoS One. 2011;**6**:e16861

[77] Paddock CD, Sanden GN, Cherry JD, Gal AA, Langston C, Tatti KM, et al. Pathology and pathogenesis of fatal *Bordetella pertussis* infection in infants. Respiratory Medicine. 2008;**47**:328-338

[78] Parise G, Mishra M, Itoh Y, Romeo T, Deora R. Role of a putative polysaccharide locus in *Bordetella* biofilm development. Journal of Bacteriology. 2007;**189**:750-760

[79] Serra DO, Conover MS, Arnal L, Sloan GP, Rodriguez ME, Yantorno OM, et al. FHA-mediated cell-substrate and cell-cell adhesions are critical for *Bordetella pertussis* biofilm formation on abiotic surfaces and in the mouse nose and the trachea. PLoS One. 2011;**6**:e28811

[80] Falkinham JO. Surrounded by mycobacteria: Non-tuberculous mycobacteria in the human environment. Journal of Applied Microbiology. 2009;**107**:356-367

[81] Leung JM, Olivier KN. Non-tuberculous mycobacteria: The changing epidemiology and treatment challenges in cystic fibrosis. Current Opinion in Pulmonary Medicine. 2013;**19**:662-669

[82] Swift S, Downie JA, Whitehead WA. Quorum sensing as a population-density-dependent determinant of bacteria physiology. Advances in Microbial Physiology. 2001;**45**:199-200

[83] Wagner VE, Bushnell D, Passador L, Brooks AI, Iglewski BH. Microarray analysis of *Pseudomonas aeruginosa*. Quorum-sensing regulons: Effects of growth phase and environment. Journal of Bacteriology. 2003;**185**:2080-2095

[84] Nealson KH, Platt T, Hastings JW. Cellular control of the synthesis and activity of the bacterial luminescent system. Journal of Bacteriology. 1970;**104**:313-322

[85] Lu L, Hu W, Tian Z, Yuan D, Yi G, Zhou Y, et al. Developing natural

products as potential anti-biofilm agents. Chinese Medicine. 2019;**14**:11

[86] Bakkiyaraj D, Nandhini J, Malathy B, Pandian S. The anti-biofilm potential of pomegranate (*Punica granatum* L.) extract against human bacterial and fungal pathogens. Biofouling. 2013;**29**:929-937

[87] Al-Bakri A, Othman G, Afifi F. Determination of the antibiofilm, antiadhesive, and anti-MRSA activities of seven *Salvia* species. Pharmacognosy Magazine. 2010;**6**:2640-2670

[88] Silva S, Costa E, Costa M, Pereira M, Pereira J, Soares J, et al. Aqueous extracts of *Vaccinium corymbosum* as inhibitors of *Staphylococcus aureus*. Food Control. 2015;**51**:314-320

[89] Chusri S, Phatthalung P, Voravuthikunchai S. Anti-biofilm activity of Quercus infectoria G. Olivier against methicillin-resistant *Staphylococcus aureus*. Letters in Applied Microbiology. 2012;**54**:511-517

[90] Perumal S, Mahmud R. Chemical analysis, inhibition of biofilm formation and biofilm eradication potential of *Euphorbia hirta* L. against clinical isolates and standard strains. BMC Complementary and Alternative Medicine. 2013;**13**:346

[91] Yadav S. Antibiofilm formation activity of *Terminalia bellerica* plant extract against clinical isolates of *Streptococcus mutans* and *Streptococcus sobrinus* implication in oral hygiene. International Journal of Pharmaceutical & Biological Archive. 2012;**3**:6

[92] Syed H, Khalid A, Sikander KS, Nazia B, Shahana U. Detection of *Mycobacterium smegmatis* biofilm and its control by natural agents. International Journal of Current Microbiology and Applied Sciences. 2014;**3**:801-812

[93] Trentin D, Giordani R, Zimmer K, da Silva A, da Silva M, Correia MT, et al. Potential of medicinal plants from the Brazilian semi-arid region (Caatinga) against *Staphylococcus epidermidis* planktonic and biofilm lifestyles. Journal of Ethnopharmacology. 2011;**137**:327-335

[94] Limsong J, Benjavongkulchai E, Kuvatanasuchati J. Inhibitory effect of some herbal extracts on adherence of *Streptococcus mutans*. Journal of Ethnopharmacology. 2004;**92**:281-289

Section 3

Biofilm and Infection

Chapter 9

Antibiotic Resistance in Biofilm

Sadık Dincer, Fatima Masume Uslu and Anil Delik

Abstract

Biofilms can be found on several living and nonliving surfaces, which are formed by a group of microorganisms, complex assembly of proteins, polysaccharides, and DNAs in an extracellular polymeric matrix. By forming a biofilm, bacteria protect themselves from host defense, disinfectants, and antibiotics. Bacteria inside biofilm are much more resistant to antimicrobial agents than planktonic forms since bacteria that are unresisting to antimicrobial agents in any way can turn resistant after forming a biofilm. Low penetration of antibiotics into the biofilm, slow reproduction, and the existence of adaptive stress response constitute the multiphased defense of the bacterium. This antibiotic resistance, which is provided by biofilm, makes the treatments, which use effective antibiotic doses on the bacterium in planktonic shape, difficult. Biofilm formation potential of bacteria appears as an important virulence factor in ensuring the colonization on the living tissues or medical devices and makes the treatment difficult. The aim of this chapter is to overview the current knowledge of antimicrobial resistance mechanisms in biofilms.

Keywords: biofilm, antibiotic resistance, bacteria, antimicrobial agents

1. Introduction

Bacteria can grow in biofilms on a wide variety of surfaces and attach to inert or alive surfaces, including tissues, industrial surfaces, and artificial devices, such as catheters, intrauterine contraceptive devices, and prosthetic medical devices, implants, cardiac valves, dental materials, and contact lenses [1, 2]. Biofilm growth confers several advantages to bacteria, including protective against hostile environments conditions such as osmotic stress, metal toxicity, and antibiotic exposure.

Biofilm-associated drug resistance and tolerance play a major role in the pathogenesis of many subacute and chronic bacterial diseases and their recalcitrance to antibiotic treatment, especially in medical device-related infections.

The definition of biofilm has been made with the development of new techniques for the direct examination of biofilms over the last four decades. Initially, a biofilm was defined as the composition of bacterial communities bound to coated surfaces in a glycocalyx matrix; subsequently, the correct definition of biofilm was made not only by considering its easily observable properties, such as cells irreversibly attached to a surface or interface embedded in an extracellular polymeric matrix material, but also by taking into account other physiological properties of these organisms such as altered growth rate and different gene expression [3].

A biofilm can be described as a microbially derived sessile community characterized by cells. These cells are irreversibly attached to a surface or interface or to each other, are inserted in a matrix of extracellular polymeric substances (EPSs) that they have produced, and exhibit an altered phenotype in terms of growth rate and gene transcription [4].

EPSs consist of proteins, cellulose, alginates, extracellular teichoic acid, poly-N-acetyl, and other organic compounds [4, 5] and play a critical role in the formation of glucosamine, lipids, nucleic acids, phospholipids, polysaccharides, and extracellular DNA (eDNA) and in physical interactions [4].

The stages that occur during the biofilm development are the initial attachment of the planktonic cell to the surface, followed by cell differentiation, EPS secretion, maturation, and dispersion of biofilm [6]. It can be summarized in three main stages: irreversible adhesion to the surface, being followed by bacterial division and production of the extracellular matrix, and, finally, disassembly of the matrix and dispersion of bacteria [2]. Quorum Sensing (QS) is one of the regulatory mechanisms that plays an important role in coordinating biofilm formation in many species but QS may not be the primary regulatory mechanism and serves as a checkpoint during the development of biofilm [6].

2. Causes of antibiotic failure in biofilm

Antibiotic resistance is the acquired ability of a microorganism to resist the effect of an antimicrobial agent and is associated with inheritable antibiotic resistance. On the other hand, antibiotic tolerance is a transient and nonheritable phenotype defined by the physiological state of biofilm cell populations. Also it can be provided by biofilm-specific characteristics that limit drug diffusion and activity [7]. For an antimicrobial agent to act on biofilm-forming microorganisms, it must overcome some factors, such as an increased number of resistant mutants, high cell density, molecular exchanges, substance delivery, efflux pump, and persistent cells.

2.1 Antibiotic penetration

Antibiotic molecules ought to penetrate throughout the biofilm matrix to impact the covered cells. The extracellular polymeric matrix influences the amount of the molecule, which is transferred to the inner layer of biofilm and interacts with an antibiotic agent, so it provides an anti-spread barrier for an antimicrobial agent. Biofilm EPS confers a physical barrier containing numerous anionic and cationic molecules such as proteins, glycoproteins, and glycolipid that can bind charged antimicrobial agents and provide shelter for microorganisms [8]. For example in *Pseudomonas aeruginosa* biofilms, Pel exopolysaccharides, an EPS component is able to spread cationic antibiotics such as aminoglycosides and, thus, provides tolerance to these molecules [9].

The adsorption sites of the matrix also limit the transportation of antimicrobial substances. Glycocalyx layer, component of EPS, can accumulate antibacterial molecule up to 25% of its weight and serve as an adherent for exoenzymes [10].

It is commonly accepted that in written materials lowered antibiotic penetration toward the EPS layer does not adequately clarify the risen resistance of microorganisms forming biofilm against most antimicrobial agents. The act of lowered antibiotic penetration in developing biofilm is not clear due to the fact that even antibiotics, which quickly disperse the biofilm, do not lead to notable cell death. It is suggested that reduction of antibiotics penetration might provide time for an adaptive phenotypic response, which can probably reduce susceptibility [11].

2.2 Accumulation of antibiotic-degrading enzymes in the matrix

The microorganisms that form biofilm are able to collect high amounts of β-lactamases in the biofilm matrix as a defense mechanism.

When *P. aeruginosa* biofilm matrix accumulates β-lactamases, it can lead to increased hydrolysis of antibiotics, such as imipenem and ceftazidime. It is demonstrated that *P. aeruginosa* PAO1-J32 biofilms have shown high promoter (ampC β-lactamases) activity, which is determined by scanning confocal laser photomicrographs [12]. Also, while ampicillin cannot reach the deeper layers of *Klebsiella pneumoniae* biofilms associated with β-lactamase activity, deletion of β-lactamase increases the amount of ampicillin that reaches the deep layer [13].

2.3 DNA in biofilm matrix

Extracellular DNA (eDNA) is a significant and common component ingredient of the bacterial biofilm matrix. The eDNA can be obtained endogenously without quorum sensing-mediated release, from the outer membrane or from the cell integrity-degraded biofilm microorganisms [14]. DNA can increase biofilm resistance to certain antimicrobial agents [15].

One of the mechanisms by which the DNA increases biofilm resistance is that it causes changes in outer membrane because DNA is an anionic molecule; it is able to chelate cations, such as magnesium ions and cause a lowering Mg^{2+} concentration in membrane. Magnesium restriction in *P. aeruginosa* and *Salmonella enterica* serovar Typhimurium is an environmental signal that induces energizing of the two-component systems PhoPQ and PmrAB to provide antimicrobial resistance [16].

These signal molecules are responsible for the rearrangement of the PA3552-3559 operon. The operon encodes to protein having enzymatic activity that attaches aminoarabinose to Lipid A part of the lipopolysaccharide layer, so it provides resistance against cationic peptide and aminoglycoside [17].

A polyamine, spermidine, localized to the outer membrane contributes to saving the cell from aminoglycosides and cationic peptides that are antimicrobial agents by lowering outer membrane penetrability for these positively charged molecules. Spermidine synthesis is another resistance mechanism induced by eDNA-associated cation restriction in *P. aeruginosa* [18].

Playing a physical role in defense against antibiotics, eDNA has also provided horizontal transfer of antibiotic resistance genes between microorganism cells forming biofilm [14].

2.4 Growth rate, stress response, and persistent cells

During growth in biofilm structures, physiological heterogeneity happens due to the occurrence of oxygen and other nutrients gradient in biofilms. This gradient is created because cells that are close to the surface of the biofilm consume obtainable nutrient sources and oxygen before the nutrients disperse into depth of the biofilm [19]. Nutrient and oxygen concentration gradients develop and cause bacterial populations that display different growth rates [20]. The effect of many antibiotics depends on growth. Because most antibiotics aim at some kind of produced macromolecule, it is unexpected that these agents will have much impact on the microorganisms in biofilm that limit macromolecular production, so conventional antibiotics are usually less affected against metabolically inactive or slow-growing cells.

In biofilms, a small subpopulation of bacteria can be reversibly transformed into slowly growing cells. These cells are known as persistent or dormant cells. Persistent cells are generated stochastically or under endogenous stress (e.g., oxidative stress and exposure to antibiotics) and are highly resistant to being killed by antibiotics [21, 22]. When these cells are compared with active and rapidly growing bacteria, lower metabolism rate makes these cells less susceptible to antibiotics. High levels of persistent cells are seen in chronic urinary tract infections and the lungs of patients with cystic fibrosis, especially, when the penetration of the immune system components is limited. The dormant phenotype is characterized by down-regulation of functions, such as energy production and biosynthesis.

Persistent formation is enhanced by toxin/antitoxin (TA) systems induced by environmental factors or DNA damage. TA systems do the following: (i) inhibition of protein synthesis by phosphorylation of the elongation factor, Ef-Tu (e.g., HipBA), translation inhibition and subsequent tolerance to antibiotics; (ii) expressing the TA modules (e.g., TisB toxin forming an anion channel in the membrane) leading to a decrease in PMF and ATP levels; and (iii) breakdown of mRNA (e.g., RelE and MazF toxins) and inhibition of translation. Prolonged treatment with aminoglycosides and RNA polymerase inhibitor rifampicin may prevent persistent resuscitation with synergistic effects with TA systems [23]. It is suggested that fluoroquinolones can induce TisB toxin by causing DNA damage in *Escherichia coli* [24]. In biofilms, many TA systems are associated with multidrug-tolerant persistent cells. However, this tolerance is limited to specific antibiotics and TA [25].

Bacteria are equipped with a range of stress responses that make them possible to deal with environmental change, such as oxidative stress, unexpected temperature changes, low water activity, deprivation, and DNA damage [26]. These adaptive responses serve to enhance bacterial survivability. Adaptive stress responses can influence antimicrobial susceptibility since these responses impact on many of the same cellular components and processes that are aimed by antimicrobials [27].

Heterogeneity in the biofilm is one of the causes of the stress response [26]. Cells within hypoxic zones have decreased metabolic activity and are in a state like stationary phase [28]. It is known that many of the stress responses result in bacterial cells entering stationary phase.

Nutrient starvation also induces (p)ppGpp production, which mediates a global stress response known as the stringent response. The stringent response and (p)ppGpp signaling contribute to multidrug tolerance in *P. aeruginosa* biofilms. It is shown that ofloxacin, gentamicin, meropenem, and colistin killing increased upon inactivation of the stringent response [29]. Nutrient starvation also induced ofloxacin tolerance in *E. coli* K-12 biofilm through mechanisms dependent on the stringent and SOS response [30].

2.5 Quorum sensing

Despite their self-sufficiency, bacteria interact with neighbors to accomplish collective activities, such as bioluminescence production, biofilm development, and exoenzyme secretion. This cooperation occurs through a mechanism: quorum sensing (QS) [31]. Quorum sensing (QS) is cell-to-cell communication at the molecular level controlled by chemical signaling molecules called autoinducers (AIs) [32]. Due to QS, bacteria can recognize the population density by measuring the accumulation of signaling molecules that are secreted from members of the community. The accumulation of the signal in the extracellular environment is adequate to activate the response only when the population density is high [33].

Recent studies indicate that in many bacterial species, activation of QS happens in the formed biofilm activating the maturation and disassembly of the biofilm. The initial adhesion step seems not suitable for the accumulation of signal molecules. Then, with the next steps, the attached bacteria are divided and form microcolonies, population density rises, and so signal molecules can reach adequate levels to activate the maturation and disassembly of the biofilm in a coordinate manner (**Figure 1**). The time nutrients and other resources become limited and waste products accumulate, biofilm dispersion is imperative to provide bacteria to escape and colonize new niches [33].

P. aeruginosa harbors two complete AHL circuits, *lasI/lasR* and *rhlI/rhlR*, the *lasI/R* circuit being hierarchically positioned upstream, of the *rhlI/R* circuit. It is reported that *las*-mediated QS inhibits the production of exopolysaccharide, Pel, which builds the biofilm matrix [34].

Another element controlled by QS in *P. aeruginosa* biofilm development is rhamnolipids production [35]. These biosurfactant rhamnolipids caused bacterial detachment of *Pseudomonas* biofilms or even biofilms produced by other microorganisms (*Bordetella bronchiseptica* and *C. albicans*) [36].

In *Staphylococcus aureus*, Agr is a QS regulation system [37]. It was demonstrated in *S. aureus* that a specific class of secreted peptides (phenol-soluble modulins, PSMs) that have surfactant-like properties mediates the main impact of Agr in biofilm dispersion. PSM operons transcription is under strict control by AgrA and agr mutants lack PSM production [38]. Also, it is shown that by analysis of biofilm tridimensional structure with confocal laser scanning microscopy, PSMs impacted the biofilm volume, thickness, roughness, and channel formation.

2.6 Efflux pumps

Efflux pumps are membrane proteins that are related to the export of harmful substances from within the bacterial cell into the external environment. They are found in all species of bacteria, and efflux pump genes can be found in bacterial chromosomes or mobile genetic elements, such as plasmids. A wide array of substrates, such as antibiotics, detergents, dyes, toxins, and waste metabolites are extruded by efflux pumps [39].

There are five known different classes of bacterial efflux pumps, which are the major facilitator superfamily (MF), the small multidrug resistance family (SMR), the ATP-binding cassette family (ABC), the resistance nodulation-division family (RND), and the multidrug and toxic compound extrusion family (MATE) [40]. To carry out the antimicrobial agent flow, the ABC family system hydrolyzes the ATP

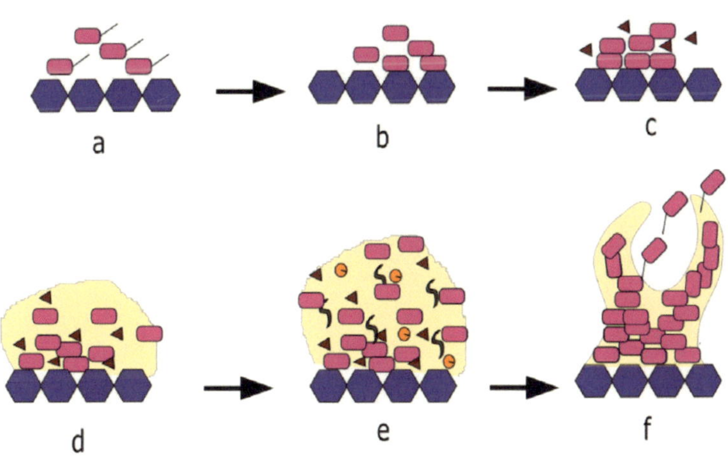

Figure 1.
Stages of bacterial biofilm formation ((a) Planktonic cells adhere to a surface, (b) Initial attachment; cell proliferates to form a monolayer over the surface, (c–d) Increase in cell numbers results in the synthesis of elevated levels of autoinducers and EPS, (e) A mature biofilm with increased resistance to hostile environmental factors, (f) Dispersion of bacteria, ◀: Autoinducers, ʕ: eDNA, ●: Enzymes).

while the MF family, the MATE family, and the RND family act as secondary carriers and catalyze the drug ion antiproton [41].

Efflux pumps play a role in the natural resistance to antibiotics in some pathogens. These pumps also cause acquired resistance by overexpression and contribute to other resistance mechanisms [20]. Overproduction of the efflux pump can lead to multidrug resistance. Bacterial efflux pumps perform multidrug resistance (MDR) phenotype [42]. The efflux pump slows down the diffusion of hydrophilic solutes by downregulating the "porin" production in several pathogenic bacteria, such as *E. coli*, *Enterobacter aerogenes,* and *Klebsiella pneumonia*, thereby decreasing the transmembrane diffusion of lipophilic solutes [43].

Some multidrug efflux pumps contribute significantly to biofilm formation and this mechanism can be used to help bacteria overcome attacks from several classes of antibiotics. Extremely reduced biofilm formation has been reported for mutant *E. coli* that does not have the various genes associated with efflux pumps [44].

Upregulation of some efflux pumps (MexAB-OprM and MexCD-OprJ) in resistant *P. aeruginosa* biofilms has been observed in the presence of azithromycin [45]. It is reported that flow pump PA 1874-1877 is associated with biofilm-specific resistance to antibiotics. When these genes are mutated, lower resistance to aminoglycosides and fluoroquinolones is seen in biofilm conditions [46].

Efflux pumps may play different roles in biofilm formation; several studies proposed that efflux of EPSs and QS molecules to facilitate biofilm matrix formation and regulate QS, respectively, lead to indirect regulation of genes involved in biofilm formation and influence aggregation by promoting or preventing adhesion to surfaces and other cells [39].

2.7 Genetic diversity

Genetic diversity provides bacterial adaptation, evolution, and survival in hostile environments. Biofilms are considered as a reservoir of genetic diversity. In biofilms, the emergence and spread of antibiotic resistance genes increase with horizontal gene transfer (HGT). HGT can happen through the transfer of plasmids among microorganism cells in a biofilm by conjugation. Actually, studies that are practiced by certain researchers have demonstrated that plasmid moving among bacterial cells might be more effective in biofilms than planktonic cells and that probably arises from proximity of microorganism cells in planktonic shape. In addition, some bacteria have the ability to pick up DNA from the biofilm matrix. The highly hydrated matrix provides favorable conditions for natural transformation [47]. Incidence of antibiotic resistance gene cassettes is determined more than 100-fold higher in biofilms than in planktonic cells [48].

Mutation frequency can be another factor that increases antibiotic resistance or tolerance. There is proof in the literature that cells in biofilms accumulate mutations at a higher rate than planktonic cells and these mutations may contribute to increase of antibiotic resistance [16]. Some bacteria have ability to pick up DNA from the biofilm matrix in addition.

2.8 Multispecies interactions

Many laboratory studies about biofilm associated with antibiotic resistance and tolerance mechanisms have focused on monospecies biofilms in the literature and this issue is becoming progressively apparent. Interactions among microorganisms that are different species in a biofilm can alter the general antimicrobial resistance of the population. When we regard that many infections are polymicrobial, these interactions may be considered clinically important [49].

Studies are showing that antimicrobial resistance in multispecies biofilms is much higher than that in monospecies biofilms in available literature. For instance, it is determined that in vivo *P. aeruginosa* growing in a monospecies biofilm is twice more vulnerable to gentamicin antibiotic than that growing in multispecies biofilm consisting of *S. aureus, Enterococcus faecalis,* and *Finegoldia magna*. The molecular mechanism that underlies this multispecies biofilm model, which increases gentamicin tolerance, is not known [50].

A clinically important model of multispecies biofilm infection includes *Moraxella catarrhalis* and *Streptococcus pneumoniae*. These bacteria play a role in the pathogenesis of otitis media, a biofilm-mediated infection that may be multi microbial. When antibiotic therapy is required, otitis media is commonly treated by amoxicillin. However, in stubborn cases, second-line treatments, such as amoxicillin-associated β-lactamase inhibitor or azithromycin, are applied. It is determined that in the biofilm consisting of two species, *M. catarrhalis* produces β-lactamase that provides resistance of *S. pneumoniae* against amoxicillin. Reciprocally, *S. pneumoniae* protects *M. catarrhalis* from azithromycin with an unknown mechanism [51].

Interactions between different microorganisms and their effects on biofilm susceptibility to antibiotics have also been examined in polymicrobial biofilms. *C. albicans*, an opportunistic fungal pathogen, and *S. aureus* have a high resistance to vancomycin in a dual species. In a biofilm that is composed of *C. albicans* and *S. aureus*, *S. aureus* is associated with the fungal hyphae via the *C. albicans* Als3p adhesin and becomes covered with biofilm matrix probably derived from *C. albicans* [52].

Owing to the fungal matrix component, β-1,3-glucan, which is thought to act as a barrier to vancomycin diffusion into the biofilm, Staphylococcal resistance to vancomycin is increased in polymicrobial biofilms formed with *C. albicans* [53].

In polymicrobial biofilms, molecular basis may increase antibiotic resistance. In a study focused on *P. aeruginosa* and *S. maltophilia* two-species biofilm, it is determined that an intercellular signaling molecule that is secreted by *S. maltophilia* is sensed by the two-component sensor, BptS in *P. aeruginosa*, inducing upregulation of the PmrA-regulated PA3552_3559 and PA4773_4775 genes. These two operon gene products provide resistance to polymyxins, which is a cationic antimicrobial peptide. Actually, *P. aeruginosa* cultured in a biofilm with *S. maltophilia* have reduced vulnerability to polymyxin B and colistin compared to *P. aeruginosa,* single-species biofilms [54].

3. Approaches aimed at overcoming biofilm resistance

Biofilm infections can be treated and dispersed by the mixture of traditional antibiotics and substances called biofilm disrupting. The dissolution of biofilm is the first step in the ability of the host organism's immune system to remove microbial pathogens [55]. The combined antibiotics with the biofilm-dispersing medicines can bring a promising outcome. Most biofilm-dispersing medicines do not kill the pathogenic cells when they are used alone. For instance, patulin was analyzed with the aim of acyl-homoserine lactone removal in *P. aeruginosa*, but it had no effect on the existence of *P. aeruginosa* cells in a given biofilm. Although only patulin had no effect on the *P. aeruginosa,* the combination of patulin with antibiotic tobramycin was more effective and caused serious killing of the bacterial cells [56]. Another study showed that the mixture of the quorum controlling compounds with the antibiotic tigecycline increased the susceptibility of *S. aureus* fourfold compared to tigecycline alone [57]. Furthermore, the treatment of *S. aureus* with the mixture of cis-2-decenoic acid and ciprofloxacin is improved from 11 to 87% compared to antibiotic alone.

Considering the rising number of antibiotic-resistant pathogens, QS inhibitors can be used as a mixture with the remaining sensitive antibiotics to complement their effects. These molecules mainly act by suppressing the QS system, and their practice with antibiotics leads to effective cure at much lower dosages of the drug than necessary, which may result in reduced therapeutic costs. These combinations can be beneficial in the cure of chronic infections, such as chronic urinary tract, cystic fibrosis, or prosthetic infections and biofilms are a barrier to antibiotic diffusion in these chronic diseases.

There is an urgent need for new methods in the cure of biofilm-associated infections. For instance, cyclic di-GMP (c-di-GMP) is a commonly protected prokaryotic second messenger signal molecule necessary for biofilm development [58]. New inhibitors of diguanylate cyclase enzymes were identified by using in silico screening, and they tested them successfully in vitro. Inhibitors of flow pumps can also be recommended to complement the effect of antimicrobial agent and needed to be tested in vivo.

The choice of antimicrobial agents also seems to be significant because some of them may act as agonists for biofilm formation and some may disrupt it. The usage and dosages of novel antibiotics should be checked and clinically synthesized antibiotics should be tested at impactful concentrations by considering their distribution in biofilms and the detrimental effects of signaling molecules. Other compounds act as key enzymes in the biosynthesis of these signaling molecules and play a role in regulating virulence factor production and biofilm formation. A ligand-based strategy will allow the identification of new inhibitors in the future.

Better usage of the new active molecules can be supported by understanding mechanisms of antimicrobial agents activity as well as the molecular mechanisms associated with biofilm formation and recalcitrance [5].

4. Conclusion

Biofilm infections are highly resistant to antibiotics and physical treatments and it is known that there are many strategies that support biofilm antibiotic resistance and tolerance, such as persistent cells, adaptive responses, and limited antibiotic penetration. It is also known that the underlying mechanisms of antibiotic tolerance and resistance in biofilms have a genetic basis in many cases.

In human diseases, highly organized bacterial cells gradually induce immune responses to form biofilms responsible for chronic infections that lead to tissue damage and permanent pathology. Therefore, the formation of biofilm is considered a critical concern in health care services.

Exploring promising cure methods for biofilm-associated infections is an urgent task. Few innovative and effective antibiotic strategies have been tried, such as dispersion of biofilms, antibiotic combinations with quorum sensing inhibitors, and a mixture of all these new techniques. Although the mentioned anti-biofilm strategies are important research areas, they are still in infancy and have not undergone clinical research and entered the commercial market. We hope that new anti-biofilm molecules based on finding universal substances that do not harm cells and synergistic with commonly used antibiotics will be available in the near future.

Conflict of interest

The authors declare no conflict of interest.

Author details

Sadık Dincer*, Fatima Masume Uslu and Anil Delik
Cukurova University, Adana, Turkey

*Address all correspondence to: sdincer@cu.edu.tr

References

[1] Reg Bott T. Industrial biofouling. In: Biofilms in Industry. Edgbaston, UK: Elsevier Inc.; 2011. pp. 181-201. DOI: 10.1016/ B978-0-444-53224-4.10007-5

[2] Jamal M, Ahmad W, Andleeb S, Jalil F, Imran M, Nawaz MA, et al. Bacterial biofilm and associated infections. Journal of the Chinese Medical Association. 2018;**81**(1):7-11. DOI: 10.1016/j.jcma.2017.07.012

[3] Gebreyohannes G, Nyerere A, Bii C, Sbhatu DB. Challenges of intervention, treatment, and antibiotic resistance of biofilm-forming microorganisms. Heliyon. 2019;**5**:e02192. DOI: 10.1016/j.heliyon.2019.e02192

[4] Flemming HC, Wingender J, Szewzyk U, Steinberg P, Rice SA, Kjelleberg S. Biofilms: An emergent form of bacterial life. Nature Reviews. Microbiology. 2016;**14**:563-575. DOI: 10.1038/nrmicro.2016.94

[5] Jolivet-Gougeon A, Bonnaure-Mallet M. Biofilms as a mechanism of bacterial resistance. Drug Discovery Today: Technologies. 2014;**11**:49-56. DOI: 10.1016/j.ddtec.2014.02.003

[6] Mangwani N, Kumari S, Das S. Bacterial biofilms and quorum sensing: Fidelity in bioremediation technology. Biotechnology and Genetic Engineering Reviews. 2016;**32**(1-2):43-73

[7] Hathroubi S, Mekni MA, Domenico P, Nguyen D, Jacques M. Biofilms: Microbial shelters against antibiotics. Microbial Drug Resistance. 2017;**23**(2):147-156. DOI: 10.1089/mdr.2016.0087

[8] Nadell CD, Drescher K, Wingreen NS, Bassler BL. Extracellular matrix structure governs invasion resistance in bacterial biofilms. The ISME Journal. 2015;**9**:1700-1709. DOI: 10.1038/ismej.2014.246

[9] Colvin KM, Gordon VD, Murakami K, Borlee BR, Wozniak DJ, Wong GC, et al. The pel polysaccharide can serve a structural and protective role in the biofilm matrix of *Pseudomonas aeruginosa*. PLoS Pathogens. 2011;**7**(1):e1001264. DOI: 10.1371/journal.ppat.1001264

[10] Sugano M, Morisaki H, Negishi Y, Endo-Takahashi Y, Kuwata H, Miyazaki T, et al. Potential effect of cationic liposomes on interactions with oral bacterial cells and biofilms. Journal of Liposome Research. 2016;**26**(2):156-162. DOI: 10.3109/08982104.2015.1063648

[11] Tseng BS, Zhang W, Harrison JJ, Quach TP, Song JL, Penterman J, et al. The extracellular matrix protects *Pseudomonas aeruginosa* biofilms by limiting the penetration of tobramycin. Environmental Microbiology. 2013;**15**(10):2865-2878. DOI: 10.1111/1462-2920.12155

[12] Bagge N, Hentzer M, Andersen JB, Ciofu O, Givskov M, Hoiby N. Dynamics and spatial distribution of beta-lactamase expression in *Pseudomonas aeruginosa* biofilms. Antimicrobial Agents and Chemotherapy. 2004;**48**(4):1168-1174. DOI: 10.1128/AAC.48.4.1168-1174.2004

[13] Anderl JN, Franklin MJ, Stewart PS. Role of antibiotic penetration limitation in *Klebsiella pneumoniae* biofilm resistance to ampicillin and ciprofloxacin. Antimicrobial Agents and Chemotherapy. 2000;**44**(7):1818-1824. DOI: 10.1128/AAC.44.7.1818-1824.2000

[14] Hall CW, Mah TF. Molecular mechanisms of biofilm based antibiotic resistance and tolerance in pathogenic bacteria. FEMS Microbiology Reviews. 2017;**41**(3):276-301. DOI: 10.1093/femsre/fux010

[15] Chiang WC, Nilsson M, Jensen PØ, Høiby N, Nielsen TE, Givskov M, et al. Shields against aminoglycosides in *Pseudomonas aeruginosa* biofilms. Antimicrobial Agents and Chemotherapy. 2013;**57**(5):2352-2361. DOI: 10.1128/AAC.00001-13

[16] Wilton M, Charron-Mazenod L, Moore R, Lewenza S. Extracellular DNA acidifies biofilms and induces aminoglycoside resistance in *Pseudomonas aeruginosa*. Antimicrobial Agents and Chemotherapy. 2016;**60**(1):544-553. DOI: 10.1128/AAC.01650-15

[17] Lewenza S. Extracellular DNA-induced antimicrobial peptide resistance mechanisms in *Pseudomonas aeruginosa*. Frontiers in Microbiology. 2013;**4**(21):1-6. DOI: 10.3389/fmicb.2013.00021

[18] Johnson L, Mulcahy H, Kanevets U, Shi Y, Lewenza S. Surface-localized spermidine protects the *Pseudomonas aeruginosa* outer membrane from antibiotic treatment and oxidative stress. Journal of Bacteriology. 2011;**194**(4):813-826. DOI: 10.1128/JB.05230-11

[19] Stewart PS, Franklin MJ. Physiological heterogeneity in biofilms. Nature Reviews. Microbiology. 2008;**6**:199-210. DOI: 10.1038/nrmicro1838

[20] Blanco P, Hernando-Amado S, Reales-Calderon JA, Corona F, Lira F, Alcalde-Rico M, et al. Bacterial multidrug efflux pumps: Much more than antibiotic resistance determinants. Microorganisms. 2016;(1):1-19. DOI: 10.3390/microorganisms4010014

[21] Öner ET. Microbial production of extracellular polysaccharides from biomass. In: Pretreatment Techniques for Biofuels and Biorefineries. Berlin, Heidelberg: Springer; 2013. pp. 35-56. DOI: 10.1007/978-3-642-32735-3_2

[22] Germain E, Roghanian M, Gerdes K, Maisonneuve E. Stochastic induction of persistent cells by HipA through (p)ppGpp-mediated activation of mRNA endonucleases. Proceedings of the National Academy of Sciences of the United States of America. DOI: 10.1073/pnas.1423536112

[23] Keren I, Mulcahy LR, Lewis K. Persistent eradication: Lessons from the world of natural products. In: Hopwood DA, editor. Natural Product Biosynthesis by Microorganisms and Plants, Part C. Springer; 2012. 449 pp. DOI: 10.1016/B978-0-12-404634-4.00019-X

[24] Dorr T, Vulic M, Lewis K. Ciprofloxacin causes persistent formation by inducing the TisB toxin in *Escherichia coli*. PLoS Biology. 2010;**8**(2):e1000317. DOI: 10.1371/journal.pbio.1000317

[25] Harrison JJ, Wade WD, Akierman S, Vacchi-Suzzi C, Stremick CA, Turner RJ. The chromosomal toxin gene yafQ is a determinant of multidrug tolerance for *Escherichia coli* growing in a biofilm. Antimicrobial Agents and Chemotherapy. 2009;**53**(6):2253-2258. DOI: 10.1128/AAC.00043-09

[26] Stewart PS. Mechanisms of antibiotic resistance in bacterial biofilms. International Journal of Medical Microbiology. 2002;**292**(2):107-113. DOI: 10.1078/1438-4221-00196

[27] Poole K. Stress responses as determinants of antimicrobial resistance in Gram-negative bacteria. Trends in Microbiology. 2012;**20**(5):227-234. DOI: 10.1016/j.tim.2012.02.004

[28] Stewart PS, Zhang T, Xu R. Reaction–diffusion theory explains hypoxia and heterogeneous growth within microbial biofilms associated with chronic infections. npj Biofilms and Microbiomes. 2016;**2**:16012. DOI: 10.1038/npjbiofilms.2016.12

[29] Nguyen D, Joshi-Datar A, Lepine F, Bauerle E, Olakanmi O, Beer K, et al. Active starvation responses mediate antibiotic tolerance in biofilms and nutrient-limited bacteria. Science. 2011;**334**(6058):982-986. DOI: 10.1126/science.1211037

[30] Zheng Z, Stewart PS. Growth limitation of *Staphylococcus epidermidis* in biofilms contributes to rifampin tolerance. Biofilms. 2004;**1**(1):31-35. DOI: 10.1017/S1479050503001042

[31] Rutherford ST, Bassler BL. Bacterial quorum sensing: Its role in virulence and possibilities for its control. Cold Spring Harbor Perspectives in Medicine. 2012;**2**(11):a012427. DOI: 10.1101/cshperspect.a012427

[32] Bhardwaj AK, Vinothkumar K, Rajpara N. Bacterial quorum sensing inhibitors: Attractive alternatives for control of infectious pathogens showing multiple drug resistance. Recent Patents on Anti-Infective Drug Discovery. 2013;**8**(1):68-83. DOI: 10.2174/1574891X11308010012

[33] Solano C, Echeverz M, Lasa I. Biofilm dispersion and quorum sensing. Current Opinion in Microbiology. 2014;**18**:96-104. DOI: 10.1016/j.mib.2014.02.008

[34] Ueda A, Wood TK. Connecting quorum sensing, c-di-GMP, Pel polysaccharide, and biofilm formation in *Pseudomonas aeruginosa* through tyrosine phosphatase TpbA (PA3885). PLoS Pathogens. 2009;**5**(6):e1000483. DOI: 10.1371/journal.ppat.1000483

[35] Diggle SP, Winzer K, Chhabra SR, Chhabra SR, Worrall KE, Camara M, et al. The *Pseudomonas aeruginosa* quinolone signal molecule overcomes the cell density dependency of the quorum sensing hierarchy, regulates rhl dependent genes at the onset of stationary phase and can be produced in the absence of LasR. Molecular Microbiology. 2003;**50**(1):29-43. DOI: 10.1046/j.1365-2958.2003.03672.x

[36] Singh N, Pemmaraju SC, Pruthi PA, Cameotra SS, Pruthi V. *Candida* biofilm disrupting ability of di-rhamnolipid (RL-2) produced from *Pseudomonas aeruginosa* DSVP20. Applied Biochemistry and Biotechnology. 2013;**169**(8):2374-2391. DOI: 10.1007/s12010-013-0149-7

[37] Boles BR, Horswill AR. Agr-mediated dispersal of *Staphylococcus aureus* biofilms. PLoS Pathogens. 2008;**4**(4):e1000052. DOI: 10.1371/journal.ppat.1000052

[38] Periasamy S, Joo H-S, Duong AC, Bach THL, Tan VY, Chatterjee SS, et al. How *Staphylococcus aureus* biofilms develop their characteristic structure. Proceedings of the National Academy of Sciences of the United States of America. 2012;**109**(4):1281-1286. DOI: 10.1073/pnas.1115006109

[39] Alav I, Sutton JM, Rahman KM. Role of bacterial efflux pumps in biofilm formation. The Journal of Antimicrobial Chemotherapy. 2018;**73**(8):2003-2020. DOI: 10.1093/jac/dky042

[40] Singh S, Singh SK, Chowdhury I, Singh R. Understanding the mechanism of bacterial biofilms resistance to antimicrobial agents. The Open Microbiology Journal. 2017;**11**(1):53-62. DOI: 10.2174/1874285801711010053

[41] Poole K. Efflux-mediated antimicrobial resistance. The Journal of Antimicrobial Chemotherapy. 2005;**56**(1):20-51. DOI: 10.1093/jacdki171

[42] Bolla JM, Alibert-Franco S, Handzlik J, et al. Strategies for bypassing the membrane barrier in multidrug resistant gram-negative bacteria. FEBS Letters. 2011;**585**(11):1682-1690. DOI: 10.1016/j.febslet.2011.04.054

[43] Pagès JM, James CE, Winterhalter M. The porin and the permeating antibiotic: A selective diffusion barrier in Gram-negative bacteria. Nature Reviews. Microbiology. 2008;**6**:893-903. DOI: 10.1038/nrmicro1994

[44] Matsumura K, Furukawa S, Ogihara H, Morinaga Y. Roles of multidrug efflux pumps on the biofilm formation of *Escherichia coli* K-12. Biocontrol Science. 2011;**16**(2):69-72. DOI: 10.4265/bio.16.69

[45] Gillis RJ, White KG, Choi KH, Wagner VE, Schweizer HP, Iglewski BH. Molecular basis of azithromycin-resistant *Pseudomonas aeruginosa* biofilms. Antimicrobial Agents and Chemotherapy. 2012;**158**:2975-2986. DOI: 10.1128/AAC.49.9.3858-3867.2005

[46] Zhang L, Mah TF. Involvement of a novel efflux system in biofilm-specific resistance to antibiotics. Journal of Bacteriology. 2008;**190**(13):4447-4452. DOI: 10.1128/JB.01655-07

[47] Madsen JS, Burmølle M, Hansen LH, Sørensen SJ. The interconnection between biofilm formation and horizontal gene transfer. FEMS Immunology and Medical Microbiology. DOI: 10.1111/j.1574-695X.2012.00960.x

[48] Strugeon E, Tilloy V, Ploy M-C, da Re S. The stringent response promotes antibiotic resistance dissemination by regulating integron integrase expression in biofilms. MBio. 2016;**7**(4):e00868-16. DOI: 10.1128/mBio.00868-16

[49] Harriott MM, Noverr MC. Importance of *Candida*-bacterial polymicrobial biofilms in disease. Trends in Microbiology. 2011;**19**(11):557-563. DOI: 10.1016/j.tim.2011.07.004

[50] Dalton T, Dowd SE, Wolcott RD, Sun Y, Watters C, Griswold JA, et al. An in vivo polymicrobial biofilm wound infection model to study interspecies interactions. PLoS One. 2011;**6**(11):e27317. DOI: 10.1371/journal.pone.0027317

[51] Perez AC, Pang B, King LB, Tan L, Murrah KA, Reimche JL, et al. Residence of *Streptococcus pneumoniae* and *Moraxella catarrhalis* within polymicrobial biofilm promotes antibiotic resistance and bacterial persistence in vivo. Pathogens and Disease. 2014;**70**(3):280-288. DOI: 10.1111/2049-632X.12129

[52] Peters BM, Ovchinnikova ES, Krom BP, Schlecht LM, Zhou H, Hoyer LL, et al. *Staphylococcus aureus* adherence to *Candida albicans* hyphae is mediated by the hyphal adhesin Als3p. Microbiology. 2012;**158**(Pt 12):2975-2986. DOI: 10.1099/mic.0.062109-0

[53] Kong EF, Tsui C, Kucharíková S, Andes D, Van Dijck P, Jabra-Rizk MA. Commensal protection of *Staphylococcus aureus* against antimicrobials by *Candida albicans* biofilm matrix. MBio. 2016;**7**(5):e01365-16 DOI: 10.1128/mBio.01365-16

[54] Ryan RP, Fouhy Y, Garcia BF, Watt SA, Niehaus K, Yang L, et al. Interspecies signalling via the *Stenotrophomonas maltophilia* diffusible signal factor influences biofilm formation and polymyxin tolerance in *Pseudomonas aeruginosa*. Molecular Microbiology. 2008;**68**(1):75-86. DOI: 10.1111/j.1365-2958.2008.06132.x

[55] Romilly C, Lays C, Tomasini A, Caldelari I, Benito Y, Hammann P, et al. A non-coding RNA promotes bacterial persistence and decreases virulence by regulating a regulator in *Staphylococcus aureus*. PLoS Pathogens. 2014;**10**(3):e1003979. DOI: 10.1371/journal.ppat.1003979

[56] Rasmussen TB, Skindersoe ME, Bjarnsholt T, Phipps RK, Christensen KB, Jensen PO. Identity and effects of quorum-sensing inhibitors produced by *Penicillium* species. Microbiology. 2005;**151**(5):1325-1340. DOI: 10.1099/mic.0.27715-0

[57] Simonetti O, Cirioni O, Mocchegiani F, Cacciatore I, Silvestri C, Baldassarre L, et al. The efficacy of the quorum sensing inhibitor FS8 and tigecycline in preventing prosthesis biofilm in an animal model of staphylococcal infection. International Journal of Molecular Sciences. 2013;**14**(8):16321-16332. DOI: 10.3390/ijms140816321

[58] Sambanthamoorthy K, Luo C, Pattabiraman N, Feng X, Koestler B, Waters CM, et al. Identification of small molecules inhibiting diguanylate cyclases to control bacterial biofilm development. Biofouling. 2014;**30**(1):17-28. DOI: 10.1080/08927014.2013.832224

Chapter 10

Microbial Biofilms

Princy Choudhary, Sangeeta Singh and Vishnu Agarwal

Abstract

Biofilms are the aggregation of microbial cells, which are associated with the surface in almost an irreversible manner. It exists in variety of forms like dental plaque, pond scum, or the slimy build up in sink. Biofilm formation involves sequence of steps like conditioning, attachment, metabolism, and detachment. Biofilm consists of water channels, EPS (Exopolysaccharide), and eDNA (Environmental DNA), which plays an important role in nutrient circulation, its development, and structure stabilization. Resistance of planktonic bacteria against antimicrobial agents gets increased on the formation of biofilm, which may be the presence of diffusive barrier EPS or neutralizing enzyme, cells undergoing starvation, or due to spore formation. There are numerous factors, which affects biofilm formation such as substratum effects, conditioning film on substratum, hydrodynamics, characteristics of the aqueous medium, cell characteristics, and environmental factors. Biofilm can cause industrial, medical, and household damage and is a reason for loss of billions of dollars every year. Development of biofilm on catheters, medical implants, and devices is a major cause of infections and diseases in humans. Examples include Plaque, Native Valve Endocarditis, Otitis media, Prostatitis, Cystic fibrosis, Periodontitis, Osteomyelitis, and many more.

Keywords: biofilm, EPS, microbes, medical implants, resistance

1. Introduction

Biofilms are the aggregation of microbial cells, which are associated with the surface in almost an irreversible manner, i.e. cannot be removed by gently rising [1]. They are attached with a biotic or abiotic surface integrated into the matrix that they have produced [2]. An accustomed biofilm provides favorable conditions for genetic material mobility between the cells and has a defined architecture. It is also reported that these surface-associated microorganisms possess definite phenotype with reference to growth rate and gene transcription [1].

The credit of discovery of microbial biofilm can be given to Van Leeuwenhoek who, with his simple microscope first observed the microorganisms on tooth surface [1].

2. Biofilm formation

A biofilm may be composed of one microbial species or many microbial species found on a variety of living or nonliving surfaces. However, mixed species biofilms

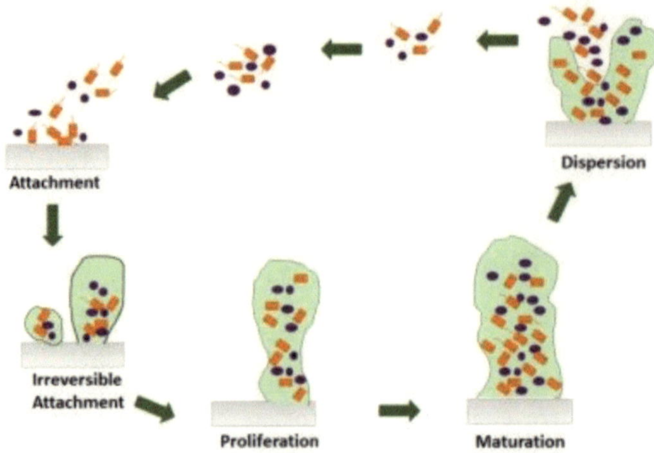

Figure 1.
Stages of biofilm development.

form the majority in most of the environments and single species biofilms host the surface of medical implants and hence being the reason of infections.

The initiation of biofilm formation have some requirements as the bacteria must be capable of attaching itself to and moving on the surface, detecting their cell density and ultimately to form a 3-D mesh of cells enclosed by exo-polysaccharide [3]. There is also an important role of cell membrane proteins, extracellular polysaccharides and signaling molecules [2] (**Figure 1**).

2.1 Biofilm formation steps

Step1. Attachment: Conditioning layer is formed which have a loose collection of carbohydrates and proteins which gets unite with minerals in hard water. It attracts the microbial cells to get attached with the surface.

Step2. Irreversible attachment: As soon as conditioning layer formed, electrical charge accumulates on the surface which attracts the bacteria having opposite charge that result in irreversible attachment of microbial cells. The charges are sufficiently weak that microorganisms could be easily removed by the mild cleanser and sanitizers.

Step3. Proliferation: In this phase, bacteria get attached to the surface as well as with each other by secreting EPS (an extracellular polymeric substance) that entraps the cells within a glue-like matrix.

Step4. Maturation: The biofilm environment consists of the nutrient-rich layer which supports the rapid growth of microorganisms. Complex diffusion channels are present in a mature biofilm to transport nutrients, oxygen and other components required for bacterial growth and removes waste products and dead cells [4, 5].

Step5. Dispersion: It is the process of dispersal of biofilm in which actively growing cells gradually sheds daughter cells [1]. Because as long as fresh nutrients are kept providing, biofilm continues to grow and when they get nutrient deprived, they return to their planktonic mode by detaching themselves from the surface [3]. This process probably happens to allow bacterial cells to get sufficient nutrients [2]. There is also a possibility of the detachment process to be species-specific as *Pseudomonas fluorescence* recolonizes surface after approx. 5 hours, *Vibrio harveyi* after 2 hours and *Vibrio parahaemolyticus* after 4 hours [1].

3. The composition of biofilm

Biofilm is primarily composed of bacterial micro-colonies which are non-randomly distributed in a shaped matrix or glycocalyx [6]. Mostly, these micro-colonies are rod-like or mushroom-shaped or they can have one or more types of bacteria. Based on bacteria type, the composition of micro-colonies contains 10–25% (by volume) of microbial cells and 79–90% (by volume) of the matrix [2, 6]. Extensive bacterial growth assists in the rapid formation of visible layers of microbes accompanied by excretion of EPS in an abundant amount [6]. At bottom of most of the biofilms, a dense layer of microorganism is bound together in polysaccharide matrix with other organic and inorganic components. The successive layer is highly irregular and loose and may extend into surrounding medium [6].

3.1 Water channels

These are present in between the micro-colonies which act as the simple circulatory system for distributing nutrients and receiving harmful metabolites [2].

3.2 EPS

Exopolysaccharide which is produced by the bacteria, are the major component of a biofilm. It constitutes about 50–90% of the total organic matter in a biofilm [6]. It is mainly composed of polysaccharides, some of which may neutral or polyanionic in case of Gram-negative bacteria or cationic as in case of Gram-positive bacteria. The anionic property of polysaccharide is confirmed by the presence of uronic acids (such as D-glucuronic, D-galacturonic, and mannuronic acids) or ketal-linked pyruvate. This anionic property plays an important role in the association of divalent cations like calcium and magnesium that have been shown to provide greater binding force in developed biofilm by cross-linking with polymer strands [1]. Along with the polysaccharide (which constitutes 1–2% of EPS), EPS also contains proteins [<1–2% (including enzymes)], DNA (<1%), RNA (<1%) as well as some lipids and humic substances [7].

3.3 eDNA

The microbial genetics and the environment in which bacteria grows are the determining factors for the composition of a biofilm. *Pseudomonas aeruginosa, Streptococcus intermedius*, *Enterococcus faecalis* and *Staphylococcus* are the species in which eDNA was initially observed.

One of the common mechanism by which eDNA is released is Autolysis. Released eDNA plays an important role in the development of the biofilm, biofilm structure stabilization as well as in gene transfer mechanisms. This genetic transfer is responsible for spreading of virulence and antibiotic resistance genes in circulating strains exposed to the selective pressure of medical treatment. *Streptococcus pneumonia* and related *Streptococci* are a good example of this [8].

4. Drug resistance and biofilm

In a biofilm, rendering biofilm becomes ten to thousand times less prone to several antimicrobial agents than the same planktonic culture grown bacterium. As an example, it has been seen that there is an increase of 600-fold concentration in sodium hypochlorite (an oxidizing biocide that is counted in most effective antibacterial drugs) for killing biofilm cells of *Staphylococcus aureus* as compared with its planktonic form [9]. Moreover, as compared to planktonic form, bacteria in biofilms shows a discrete physiology like reduced metabolic rate and enhanced cell to cell communication which helps in developing resistance to antibiotics or reduce their effects [10]. In the attempt to describe the resistance of biofilms to antibiotics, three assumptions have been made:

1. Slow or partial diffusion of antibiotics into inner layers of biofilm. This is due to EPS matrix which has biofilm entrenched bacteria, act as a diffusive barrier [2].

2. In the biofilm microenvironment, some microbial cells fall into a state of slow growth or starvation due to nutrient limitation or accumulation of harmful metabolites. These are not vulnerable to many antimicrobial agents [2, 11].

3. The differentiation of a bacterial subpopulation resembles the process of spore formation. It has a distinctive and highly resistance phenotype (a biologically programmed response to bacterial sessile life form) that protects them from antibacterial effects [2].

Presence of neutralizing enzymes also contributes to the antibiotic resistance in the biofilm. These proteinaceous enzymes degrade or inactivate antibiotics by mechanisms like hydrolysis and modification of antimicrobials by different biochemical reactions [7].

Although, intensive and insistent treatment of antibiotic is effective in reducing the biofilm and controlling the exacerbations of chronic biofilm infections but are not able to eliminate biofilm infections it is possibly because the minimal concentration of antibiotic (required to eliminate a mature biofilm) is challenging to reach *in vivo*. Hence, if a bacterial biofilm infection is established, it becomes much difficult to eradicate [12].

Experimental studies suggested that in most of the cases antibiotic treatment alone is not sufficient to eliminate infections of biofilm [12]. In a study, a nanoparticle called ciprofloxacin-loaded poly (lactic-co-glycolic acid), that were functionalized with DNase I, were prepared to observe their antibiofilm activity against *P. aeruginosa* biofilms. It has been found that they release ciprofloxacin in a controlled manner, as well as they effectively target and disassemble the biofilm by degrading the extracellular DNA that stabilizes the EPS [10]. Biofilm combination therapy is usually recommended for treating biofilm infections as this is found to be substantially better than antibiotic monotherapy [12].

5. Factors affecting biofilm formation

A number of factors such as substratum effects, hydrodynamics and various properties of cell surface play an important role in microbial attachment [1].

5.1 Substratum effects

As the surface roughness increases microbial colonization increases because as the roughness increases, surface area increases and shear forces get diminished. And considering extent and rate of attachment, it has been seen that microorganisms get attached to more rapidly to hydrophobic and nonpolar surfaces as Teflon and other plastics rather than to glass and other materials having hydrophilic properties.

5.2 Conditioning films forming on the substratum

When a material surface gets exposed to any aqueous medium, it gets immediately coated with polymers from that surface or become conditioned. The coating or film is found to be organic in nature formed within minutes of exposure. The nature of these films is found to be quite different for surfaces exposed in the human host. As an example, "acquired pellicle," a proteinaceous conditioning film, develops on tooth enamel surface. A pellicle is composed of glycoprotein, lysozymes, phosphoproteins, albumin, lipids and gingival crevice fluid. Oral cavity bacteria get adhered within hours of exposure to this pellicle conditioned surface.

5.3 Hydrodynamics

The hydrodynamic flow layer is the zone of negligible flow which is found at the immediately adjacent to the substratum/liquid interface. The flow velocity of this zone is negligible and its thickness is inversely proportional to the linear velocity. Substantial mixing or turbulence is the main characteristics shown by the region outside the boundary layer. The hydrodynamic boundary layer can considerably affect the interaction between cells and substratum. The velocity characteristic of the liquid governs the association of cells with the submerged surfaces. At, very low linear velocities, the cells must navigate through the hydrodynamic boundary layer, and cell size and cell motility govern its association with the surface. The boundary layer decreases, as the velocity increases and cells will be exposed to progressively larger turbulence and mixing. Therefore, higher linear velocities would be supposed to form a more rapid association with the surface, at least until velocities become high enough to apply abundant shear forces on the attaching cells, that results in detachment of these cells [1].

5.4 Characteristics of the aqueous medium

Characteristics of the aqueous medium such as temperature, pH, nutrient level and ionic strength possibly play an important role in attachment of microbes with the substratum. As an example, it has been found that the attachment of *Pseudomonas fluorescens* to glass surface is affected by an increase in the concentration of several cations (sodium, calcium, lanthanum, ferric iron), perhaps by reducing the repulsive forces between the negatively charged bacterial cells and the glass surfaces.

5.5 Properties of the cells

The rate and extent of adherence of microbes depends on the properties of cells like cell surface hydrophobicity, as hydrophobic interactions tend to increase with an increasing nonpolar nature of one or both involved surfaces and adhesion increases with increase in hydrophobicity, presence of fimbriae and flagella as fimbriae contribute to cell surface hydrophobicity probably by overcoming the initial electrostatic repulsion barrier that exists between the cell and substratum and production of EPS. EPS might be hydrophobic, although mostly they are both hydrophilic and hydrophobic. Numerous bacterial EPS have the backbone of 1,3- or 1,4-β-linked hexose residues and tend to be less deformable, more rigid and inadequately soluble or insoluble in specific cases although other EPS molecules may be water soluble. Researches also showed that different organisms produce different amounts of EPS and the amount of EPS increases with age of the biofilm. Antimicrobial resistance properties in the biofilm are possibly mediated by the EPS by impeding the mass transport of antibiotics through the biofilm, which might be by binding directly to these agents [1]. EPS formation is an essential part of biofilm formation as studies on *Staphylococcus epidermidis* have shown that if genes responsible for the synthesis of EPS matrix are inactivated then bacteria lose the ability to form biofilm [2].

5.6 Environmental factors

Different environmental factors affect the biofilm formation; listed below:

5.6.1 Availability of certain nutrients

It has been shown by studies on *Listeria monocytogenes* that an optimum level of phosphate is very important for biofilm formation and gets stimulated by the presence of carbohydrates mannose and trehalose.

5.6.2 Presence of oxygen

Presence of oxygen regulates Biofilm formation in *Escherichia coli*. In the absence of sufficient oxygen supply biofilm does not form as bacteria could not adhere to the substrate surface.

5.6.3 Environmental pH

Environmental pH effects were observed by studying on *Vibrio cholerae*. Optimal pH for multiplication of *V. cholerae* is 8.2 and below pH 7 i.e., in acidic environment the bacteria lose their ability to form biofilm as they lose mobility.

On the other hand, bacteria like *S. epidermidis* and *E. coli* do not need an alkaline environment for multiplying hence they easily form a biofilm on urethral catheters where urine pH is acidic.

5.6.4 Temperature

When temperature was kept high, *L. monocytogenes* did not form biofilm as the bacteria wasn't able to adhere itself to the substrate surface [2].

6. Diseases due to biofilm

Besides infecting the industrial pipelines, waste water channels, oral cavity, ventilators, catheters, and medical implants, they are a major cause of human diseases [11]. Infections and diseases in humans are mostly due to development of biofilm on or within indwelling implants or devices such as contact lenses, bio prosthetic and mechanical heart valves, pacemakers, intra-arterial and intravenous catheters, central venous catheters, peritoneal dialysis catheters, urinary catheters, joint prosthesis, voice prosthesis, penile prosthesis, ureteral stents, biliary stents, endotracheal tubes, nephrostomy tubes, intrauterine contraceptive devices (IUDs) [13, 14]. A biofilm may be composed of gram-positive or gram-negative microorganisms which may arise from the skin of a patient, health worker, tap water or any other environmental source [5].

Biofilm growth usually was seen in the lungs of cystic fibrosis patients causing chronic bronchopneumonia, in the middle ear in patients with chronic and secretory otitis media, in chronic rhino sinusitis, in chronic osteomyelitis and in chronic

Gram-positive microorganisms	Site of infections and diseases
Acidogenic gram-positive cocci (e.g. *Streptococcus*)	Dental caries
Gram-positive cocci (e.g. *Staphylococci*)	Musculoskeletal infections
Group A *Streptococci*	Necrotizing fasciitis
Viridans Group *Streptococci*	Native valve endocarditis
S. epidermidis and *S. aureus*	Sutures, exit sites and arteriovenous shunts
S. epidermidis, E. faecalis	Urinary catheter cystitis
S. epidermidis, S. aureus, Corynebacterium species, *Micrococcus* species, *Enterococcus* species, *Candida albicans*, Group B *Streptococci*	IUDs
C. albicans, S. epidermidis	Hickman catheter
S. epidermidis, S. aureus, E. faecalis, C. albicans	Central venous catheter
Viridans Streptococci, Enterococci	Mechanical heart valves
Hemolytic Streptococci, Enterococci	Orthopedic devices
S. epidermidis, S. aureus,	Penile prosthesis
Gram-negative microorganisms	**The site of infections and diseases**
Nontypable strains of *Haemophilus influenzae*	Otitis media
E. coli (enteric bacteria)	Biliary tract infection, bacterial prostatitis
P. aeruginosa and Burkholderia cepacia	Cystic fibrosis pneumonia
Pseudomonas pseudomallei	Melioidosis nosocomial infections
Klebsiella pneumoniae, Proteus mirabilis	Urinary catheter cystitis
K. pneumoniae, P. aeruginosa	Central venous catheter
Proteus mirabilis, Bacteroides species, *P. aeruginosa, E. coli*	Orthopedic devices

Table 1.
Different infections and involved microorganisms [11].

wounds [15]. Infections and then diseases occur because of these two reasons: (a) Implantation of any medical device cause tissue damage which attracts platelets and fibrin accumulation at the site of the attachment. The damaged tissue aids in colonizing the microorganisms [13]. (b) Drug resistance and inflammation in host might get stimulated by biofilm formation which results in sustained infections [16] (**Table 1**).

7. Biofilm on some common medical devices

7.1 Central venous catheter biofilms

Commonly found organisms on catheter biofilm are *S. epidermidis, S. aureus, K. pneumoniae, C. albicans, P. aeruginosa,* and *E. faecalis.* These might get emerged from patient's skin microflora, exogenous microflora from health-care personnel, or infected infusates. It has been reported that inner lumen of long-term catheters (30 days) and an external surface of short-term catheters (<10 days) has more biofilm formation. Microbial growth may depend on the nature of fluid delivered through a central venous catheter, as it has been seen that gram-negative microorganisms grow well in the intravenous fluid than gram-positive organisms [17].

Many studies have been done to control or avoid biofilm formation in these devices. Few remarkable results are:

- It has been found in a research that microbial colonies of the left arterial catheter can be eliminated by addition of sodium metabisulfite to the dextrose-heparin flush.

- Less colonization was seen on catheters coated with minocycline and rifampin than those coated with chlorhexidine and silver sulfadiazine [5].

7.2 Mechanical heart valve biofilms

Microorganisms like *S. epidermidis*, *S. aureus, Streptococcus* species, Gram-negative bacilli, diphtheroids, *Enterococci and Candida* species develop biofilm on the components of mechanical heart valves and surrounded heart tissues, which lead to a condition called prosthetic valve endocarditis. Also, it more often develops on the tissue surrounding the prosthesis or on the sewing cuff fabric that attaches a device to the tissue than on the valve itself. The source of the microorganism somehow tells its identity as, if it gets originate from an invasive process like dental work then it possibly belongs to *Streptococcus* species or it also might get originated during surgery (early endocarditis, mainly due to *S. epidermidis*) or from an indwelling medical device.

To prevent initial attachment of the microbes, anti-microbial agents are provided during valve replacement or any invasive process like dental work. It has also been found out that less inflammation was caused when silver coated sewing cuff of St. Jude mechanical heart valve was implanted than an uncoated one [5, 17].

7.3 Urinary catheter biofilms

Organisms which develop biofilm on these devices are *S. epidermidis*, *E. faecalis*, *E. coli*, *Proteus mirabilis*, *P. aeruginosa*, *K. pneumonia* and other Gram-negative organisms [17]. These catheters are tubular latex or silicone devices that are inserted via urethra into the bladder. It may be of an open system in which catheter drains into an open collection center or close system in which it vacates into a securely fastened bag. In open system, catheter gets quickly contaminated and chances of UTI (Urinary Tract Infection) are much more than in closed system. The chances of microbes to develop biofilm and hence causing UTI is more as long as the catheter remains on its place as it has been found out that approximately 10 to 50% of the patients undergoing short-term catheterization (up to 7 days) and around all the patients undergoing long-term catheterization (>30 days) gets infected with UTI [5].

It has been shown in studies that hydrophobicity of both organism and surface is responsible factors for microbial attachment on the catheter as a wide range of microbial colonies are found to be attached on the catheter's surface which displays both hydrophobic and hydrophilic regions [17]. Bacterial attachment is also enhanced by an increase in urinary pH and ionic strength by divalent cations (Mg and Ca). Urease is produced by some of the organisms of this biofilm which is responsible for hydrolyzing the urea to ammonium hydroxide. As a result, pH at the biofilm-urine interface gets higher, which causes precipitation of minerals such as struvite and hydroxyapatite. These biofilms having mineral components form encrustations which can completely block the catheter's inner lumen [5].

Several approaches have been done to control biofilm formation on urinary catheters like the use of antimicrobial ointments and lubricants, bladder instillation,

DOI: http://dx.doi.org/10.5772/intechopen.90790

antimicrobial agents in collection bags, impregnation of catheters by silver oxides like antimicrobial agents or systemic antibiotics. Also, biofilm of many Gram-negative microorganisms can be reduced by exposing to mandelic acid in combination with lactic acid [17].

7.4 Contact lenses biofilms

Microbes get readily attached to the surface of both type of contact lenses i.e. soft contact lenses and hard contact lenses (differentiated according to the material used, design, wear schedule and frequency of disposal). Nature of substrate, water content, polymer composition, electrolyte concentration and type of bacterial strains governs the degree of adherence of microbes to the lenses. The storage case of a lens has been implicated as the primary source of contamination [5].

Staphylococcus, *Serratia* and *Pseudomonas* are some most common bacterial species obtained in contact lenses. *Staphylococci* are found affiliated with contact lens induced peripheral ulcer, blepharitis and conjunctivitis while *Serratia* and *Pseudomonas* species known to contribute in corneal inflammation and infection [18].

7.5 Intrauterine devices

The tail part of IUDs which is made up of a plastic microfilament surrounded by nylon sheath is possibly the primary source of infection. Microorganisms that contaminate IUDs are *Lactobacillus plantarum*, *S. epidermidis*, *C. albicans*, *S. aureus*, species of *Corynebacterium*, *Enterococcus* species [5].

8. Some common biofilm infections

8.1 Dental biofilms

Dental biofilms, commonly known as plaque are the most studied biofilm in human. It involves hundreds of species of bacteria. Some significant microbes include *Porphyromonas gingivalis, Bacteroides forsythus*, *Actinobacillus actinomycetemcomitans*, *Treponema denticola*, and a number of *Streptococci* including *Streptococcus mutans* [11].

After a good oral wash or dental cleaning, the tooth enamel acquires a coating called as pellicle which is composed of various proteins and glycoproteins of host origin. Then with the help of adhesion molecules and pilli, first *Streptococci* then *Actinomycetes* colonizes the teeth surface. Bacterial cells start interacting with each other on the pellicle and a number of *Streptococci* and related organisms starts synthesizing insoluble glucan via glucan binding protein. After few successive colonization with few more organisms, demineralization of tooth enamel starts (which leads to caries) by the acids which are produced by fermentation of the dietary sucrose and other carbohydrates [11].

8.2 Native valve endocarditis (NVE)

This condition arises due to the interaction between bacteria, vascular endothelium and generally of mitral, aortic, tricuspid and pulmonary valves of the heart. The organisms responsible for these conditions are species of *Streptococcus*, *Staphylococcus*, *Pneumococci*, *Candida*, *Aspergillus,* and some Gram-negative bacteria, which get access to the blood stream via the oropharynx, gastrointestinal, and urinary tract. When the intact endothelium gets damaged, microbes adhered to it and as a result nonbacterial endocarditis (NBTE) develops at the site of injury and thrombus (accumulation of platelets, fibrin, and red blood cells) formed [13]. Fibronectin which has been found as a thrombotic lesion of the heart valve can simultaneously bind to fibrin, collagen, human cell and bacteria. Fibronectin receptors are found in many bacterial species like *Staphylococcus and Streptococcus* [5].

Many antibiotic therapies are suggested depending on the organisms involved as Penicillin is recommended for normal treatment of *Streptococcal endocarditis* and for synergistic killing gentamycin may be supplemented. Fluconazole can successfully terminate the effect of *Candida endocarditis* [5].

8.3 Otitis media

It is a condition of chronic ear infection caused due to inflammation of mucoperiosteal lining [5]. In the middle ear cavity, fluid gets accumulated which ultimately affects speech development and learning capability of the patient. However, its complete etiology is still under research [7]. Various organisms responsible for otitis media include *S. pneumoniae, H. influenzae, Moraxella catarrhalis, S. epidermidis, P. aeruginosa,* etc. As due to limited penetration of antibiotic, its low concentration is present in middle ear fluid, hence strong antibiotics like amoxicillin, cefaclor, erythromycin, and clarithromycin are needed for combating otitis media [5].

8.4 Chronic bacterial prostatitis

Prostatitis is the inflammation of the prostate gland which possibly occurs due to the microorganisms that have ascended from the urethra or by the reflux of infected urine into prostatic ducts which vacates into the posterior urethra. Once the microbe gets entered in the prostatic duct, they start multiplying rapidly and can form sporadic micro-colonies and biofilms which gets adhered to the epithelial cells of the system of ducts. Microbes responsible for this infection are *E. coli, P. aeruginosa*, species of *Klebsiella*, *Proteus, Serratia*, *Bacteroides,* etc. [5].

8.5 Cystic fibrosis

Cystic Fibrosis is a chronic bacterial infection of intrapulmonary airways with *P. aeruginosa* [19]. Its consequences include thickening of mucus in many body systems which results in impaired mucociliary clearance of microorganisms and chronic infection in lungs. The infection gets punctuated by acute aggravation of disease and inflammation which will lead to lung failure and premature death [20]. According to the genetic etiology, one out of more than 1500 potential mutations in the cystic fibrosis transmembrane conductance regulator (CFTR) gene results in its malfunction, as a result sodium absorption is inhibited through epithelial sodium channel. And due to hyper-absorption of water, airway surface liquid gets depleted, mucociliary clearance get depleted and inhaled bacteria are allowed to remain within airway [20, 21].

Microscopic studies of sputum samples and lung tissue section have shown the presence of biofilm or micro-colonies in the airways. These biofilms are able to grow larger than 100 μm in diameter [22]. Some common cystic fibrosis pathogens include *S. aureus, H. influenzae* and ultimately predominant one *P. aeruginosa* [20, 21]. *P. aeruginosa* has some adaptive mechanisms which make it survive and persist for several decades in CF patient's respiratory tract. Biofilm adaptation of *P. aeruginosa* makes it resistant to antibiotic therapy and inflammatory defense mechanism. This also makes it survive in different conditions like whether it is aerobic respiratory zone or the conductive zone of the lungs which have anaerobic sputum or the paranasal sinuses where mucus too has a lower concentration of oxygen [15].

Early antimicrobial treatment (i.e. during early colonization period of microbes) for preventing chronic infection of *P. aeruginosa* may give a possibility for successful treatment of cystic fibrosis, as this chronic infection may postpone for several years by giving an early treatment with ciprofloxacin and colistin [5, 22].

8.6 Periodontitis

Periodontitis is the infection of supporting tissues of teeth, gums (gingiva) and periodontal tissues (gingiva, alveolar bone, and periodontal ligament). Its chronic form may lead to exfoliation of teeth. The primary site of periodontitis is sub gingival crevice which is the channel between the tooth root and the gum. Organisms responsible for this infection are *Fusobacterium nucleatum*, *Peptostreptococcus micros*, *Eubacterium timidum*, *E. brachy*, *Pseudomonas anerobicus,* and predominate one *P. gingivalis.* They can easily colonize the surface of the oral cavity which helps them in invading mucosal cells, altering calcium flux in epithelial cells and in releasing toxins. As a result, plaque (a climax biofilm community) is formed within 2–3 weeks. Calculus or tartar is the mineralized plaque which acts as a resistance against the antimicrobial activity of saliva in protecting tooth enamel, as a consequence of which dental carries and periodontal diseases occurs [5].

Dental plaque or biofilm cannot be eliminated, only their pathogenic nature can be minimized by minimizing the bioburden and effectively maintain a normal oral flora via oral hygiene methods [6, 23].

8.7 Osteomyelitis

Osteomyelitis is an inflammatory bone disorder characterized by infection in bone/bone marrow which leads to necrosis and bone destruction [24, 25]. When complex multi-resistant biofilm has established, treatment of osteomyelitis becomes more challenging. Due to increased bacterial resistance to antibiotics in biofilm mode, they cause persistent infections. It has been found that in more than 50% osteomyelitis cases, causative organisms are *S. aureus* and *S. epidermidis* [24].

Although, endoprostheses which are found to be an increasingly common source of infection, surgically implanted devices or other implants like orthopedic internal fixation devices also represents a remarkable risk factor for the development of osteomyelitis. Stainless steel, titanium, titanium alloys are most commonly used materials in implants in which stainless steel is found to be associated with greater infection rate as compared to titanium. A possible reason of this is might be that soft tissues get firmly adhered to a titanium-implant surface while a fibrous capsule is formed enclosing a liquid filled space around the steel implants. This unvascularized space is less accessible to host defense mechanisms where bacteria can multiply and freely spread. Studies showed that *S. aureus* and *S. epidermidis* adhesion to the surface can be reduced by the use of coatings based on human proteins such as albumin or human serum. Coatings of poly(1-lysine)-grafted-poly(ethylene glycol) (PLL-g-PEG) when extensively studied for use in biomedical applications, it has been found to be highly effective in reducing the absorption of blood serum, blood plasma and single proteins like fibrinogen and albumin. Fibroblast and osteoblast cell adhesion get remarkably reduced by spreading of metal oxide surface coated with PLL-g-PEG in comparison to uncoated surfaces [25].

Author details

Princy Choudhary[1], Sangeeta Singh[1*] and Vishnu Agarwal[2]

1 Department of Applied Science, Indian Institute of Information Technology Allahabad, Prayagraj, UP, India

2 Department of Biotechnology, Motilal Nehru National Institute of Technology, Allahabad, Prayagraj, UP, India

*Address all correspondence to: sangeeta@iiita.ac.in

References

[1] Donlan RM. Biofilms: Microbial life on surfaces. Emerging Infectious Diseases. 2002;**8**(9):881-890

[2] Marić S, Vraneš J. Characteristics and significance of microbial biofilm formation. Periodicum Biologorum. 2007;**109**:115-121

[3] O'Toole G, Kaplan HB, Kolter R. Biofilm formation as microbial development. Annual Reviews in Microbiology. 2000;**54**(1):49-79

[4] Abu Bakar M et al. Chronic tonsillitis and biofilms: A brief overview of treatment modalities. Journal of Inflammation Research. 2018;**11**:329

[5] Kokare CR et al. Biofilm: Importance and applications. Indian Journal of Biotechnology. 2009;**8**(2):159-168

[6] Saini R et al. Dental plaque: A complex biofilm. Pravara Medical Review. 2015;**7**(1):9-14

[7] Jamal M et al. Bacterial biofilm: Its composition, formation and role in human infections. RRJMB. 2015;**4**:1-14

[8] Montanaro L et al. Extracellular DNA in biofilms. The International Journal of Artificial Organs. 2011;**34**(9):824-831

[9] Davies D. Understanding biofilm resistance to antibacterial agents. Nature Reviews Drug Discovery. 2003;**2**(2):114-122

[10] Baelo A et al. Disassembling bacterial extracellular matrix with DNase-coated nanoparticles to enhance antibiotic delivery in biofilm infections. Journal of Controlled Release. 2015;**209**:150-158

[11] Aparna MS, Yadav S. Biofilms: Microbes and disease. Brazilian Journal of Infectious Diseases. 2008;**12**(6):526-530

[12] Wu H et al. Strategies for combating bacterial biofilm infections. International Journal of Oral Science. 2015;**7**(1):1-7

[13] Percival SL et al. Introduction to biofilms. In: Biofilms and Veterinary Medicine. Berlin Heidelberg: Springer; 2011. pp. 41-68

[14] Costerton JW et al. Bacterial biofilms in nature and disease. Annual Reviews in Microbiology. 1987;**41**(1):435-464

[15] Høiby N et al. The clinical impact of bacterial biofilms. International Journal of Oral Science. 2011;**3**(2):55

[16] Chen L, Wen Y-M. The role of bacterial biofilm in persistent infections and control strategies. International Journal of Oral Science. 2011;**3**(2):66

[17] Donlan RM. Biofilms and device-associated infections. Emerging Infectious Diseases. 2001;**7**(2):277

[18] Wu YT-Y et al. Contact lens hygiene compliance and lens case contamination: A review. Contact Lens & Anterior Eye. 2015;**38**(5):307-316

[19] Matsui H et al. A physical linkage between cystic fibrosis airway surface dehydration and *Pseudomonas aeruginosa* biofilms. Proceedings of the National Academy of Sciences. 2006;**103**(48):18131-18136

[20] Woo JKK et al. Biofilm dispersal cells of a cystic fibrosis *Pseudomonas aeruginosa* isolate exhibit variability in functional traits likely to contribute to persistent infection. FEMS Immunology and Medical Microbiology. 2012;**66**(2):251-264

[21] Davies JC, Bilton D. Bugs, biofilms, and resistance in cystic fibrosis. Respiratory Care. 2009;**54**(5):628-640

[22] Anderson GG. *Pseudomonas aeruginosa* Biofilm Formation in the CF Lung and its Implications for Therapy. Rijeka: INTECH Open Access Publisher; 2012

[23] Khuller N. The biofilm concept and its role in prevention of periodontal disease. Revista de Clínica e Pesquisa Odontológica. 2009;**5**(1):53-57

[24] Gomes D, Pereira M, Bettencourt AF. Osteomyelitis: An overview of antimicrobial therapy. Brazilian Journal of Pharmaceutical Sciences. 2013;**49**(1):13-27

[25] Roy M et al. Pathophysiology and Pathogenesis of Osteomyelitis. Rijeka: INTECH Open Access Publisher; 2012

Chapter 11

Essential Oils as an Innovative Approach against Biofilm of Multidrug-Resistant *Staphylococcus aureus*

Victor Alves Carneiro, Ramaiana Soares Melo, Antônio Mateus Gomes Pereira, Águida Maria Albuquerque Azevedo, Maria Nágila Carneiro Matos, Rafaela Mesquita Bastos Cavalcante, Renan Rhonalty Rocha, Vinícius de Queiroz Albuquerque, Jesús Alberto Pérez Guerrero and Francisco Eduardo Aragão Catunda Junior

Abstract

Staphylococcus aureus is one of the most common pathogens that cause recurrent, chronic, and biofilm-related diseases. Biofilms are the major form of bacterial structures capable of secreting polysaccharides that provide intrinsic protection against environmental stress like high concentrations of antibiotics. This, along with the emergence of multidrug-resistant strains, has made *S. aureus* infections a worldwide problem as a result of the inefficiency of the conventional medications. Plant essential oils (EOs) are an important source for drug discovery and pharmaceutical development due to their diverse biological activities, such as antimicrobial agents. The EOs' microbicide action is extensively reported at the scientific literature and frequently associated with bioactive molecules, such as aldehydes and terpenes. However, the ability of some EOs to inhibit biofilm formation has been poorly explored and it is still unclear how they could be applied in specific treatments against well-known infections. Therefore, this chapter will address virulence factors and biofilm formation of *S. aureus*, as well as bioprospecting of essential oil as a promising source in the search for new bioactive compounds employed in the fight against this microorganism.

Keywords: antibiofilm activity, biofilm-related diseases, essential oil, natural products, *Staphylococcus aureus*

1. Introduction

The emergence of multidrug-resistant (MDR) bacteria is correlated with selective pressure caused by the indiscriminate use of antibiotics, which reduces therapeutic options available [1]. Consequently, it leads to a serious public health problem frequently associated with increase of healthcare costs and high morbimortality rates [2]. One worldwide recognized bacterial pathogen with the ability to develop severe clinical conditions such as pneumonia and septicemia is *Staphylococcus aureus* [3]. Historically, this bacterium has shown a great ability to become resistant to several antibiotics [4]. Furthermore, *S. aureus* has a highlighted ability to build surface-associated bacterial communities, called biofilm, being one of the most determinant factors for the development of chronic infections, and it is the major cause of treatment failure [5–7].

Recently, the use of natural compounds, such as EOs obtained from different parts of the plants, is receiving attention for their biological activities, including antioxidant, anti-inflammatory, and anticancer effect [8]. Moreover, EOs have been frequently mentioned on scientific literature as a promising antimicrobial agent, being effective against a wide range of pathogenic bacteria and yeast [9, 10]. Thus, this chapter will present a comprehensive overview about general features of *S. aureus*, including virulence factors, antibiotic resistance, and biofilm formation. Additionally, it will introduce the EOs used as potential therapeutic approaches against biofilm of multidrug-resistant *S. aureus*.

2. *Staphylococcus aureus*

2.1 Clinical relevance and virulence factors

Member of the *Micrococcaceae* family, *S. aureus* is Gram-positive cocci-shape arranged in a grape-like cluster. The cells are anaerobic facultative and catalase-positive with approximately 0.5–1.5 μm in diameter. Overall, 52 species have been described in the staphylococcal genus, *S. aureus* being, by far, the member most clinically relevant [11]. *S. aureus* genome has been completely sequenced and three main components were observed: conserved genes, variable genes, and mobile genetic elements (MGE). More than 97% of the *S. aureus* genome is composed of highly conserved genes found in all staphylococcal strains. More than 700 genes are variable and their distribution defines different lineages [12, 13]. Apart from the core genes, there are several numbers of MGE acquired by horizontal gene transfer by bacteriophages, transposons, and plasmids that contribute to genome plasticity and evolution, such as the antibiotics resistance and virulence gene dissemination [14].

Widely disseminated in nature, *S. aureus* is a commensal component of human cutaneous and mucosal microbiota as well as an adaptive pathogen that leads to numerous invasive and, sometimes, fatal infections [15, 16]. This microorganism can be easily spread by the hands or expelled from the respiratory tract. About 30% of the population is colonized by *S. aureus*, and this increases to 60% when it involves the healthcare environment, implying in either cases high risk of further infection [17]. As a pathogen, this bacterium causes various suppurative diseases, such as boils, carbuncles, folliculitis, and scalded-skin syndrome [18]. Additionally, the lymphatic system and bloodstream contributed to bacterial spread to other parts of the body causing osteomyelitis, medical device infection, endocarditis, and pneumonia [19]. Furthermore, the presence of a variety of antimicrobial resistance mechanisms in some strains leads to treatment failure, increasing healthcare costs and risk of death [20].

Bacterial sepsis confirmed by blood cultures in pediatric hospitals, Gram-positive bacteria (62%) were involved in most of the infection cases. Among them, the major reported strains were *S. aureus* (15%), followed by *Staphylococcus* coagulase negative (11%) and *Streptococcus pneumoniae* (10%) [21]. In addition, serious cases of high virulence profile community-associated *S. aureus* (CA-MRSA) infections have been reported globally in recent decades [22, 23]. In Taiwan, for instance, 423 cases of CA-MRSA infections were reported in children, and most of them were associated to bone, joint, and deep soft tissue infections and pneumonia [24]. Despite each disease profile, the staphylococcal species is frequently correlated with both community- and hospital-acquired infections, and it has steadily increased [25, 26]. Thus, it is necessary to look for new therapeutic alternatives to minimize this public health problem [27].

S. aureus can survive in its hosts as a commensal bacterium for a long time; however, it can also be considered one of the most relevant human pathogens [28]. This bacterium has mechanisms to evade the host's immune response through production of a variety of virulence factors, such as adhesins, exotoxins, and hydrolytic enzymes (e.g., coagulase, catalase, and staphylokinase), as summarized in **Figure 1** [29, 30].

The bacterial adherence to extracellular matrix cells in the host is one of the most important steps for colonization. It is mediated mainly by surface-anchored proteins classified as MSCRAMM (microbial surface components recognizing adhesive matrix molecules) family. Among them, two fibronectin binding proteins, FnbA and FnbB, contribute considerably to epithelial tissue colonization in various pathological manifestations and medical device-related infections [31]. Other cell surface protein related in adhesion mechanisms are named clumping factor A and B (ClfA and ClfB). The first one has a highlighted ability to interact with soluble proteins, fibrinogen, and fibrin, present in blood plasma. These surface components aid the microorganism to interact with plasma protein-coated biomaterials and, consequently, make possible the colonization and biofilm formation on medical devices [32]. The ClfB is frequently associated to nasal colonization due to high affinity to cornified envelope of the nostrils, which promotes the formation of skin abscesses by binding to the protein loricrin [33]. It is worth mentioning genes capable of encoding proteins on the cell surface, *cna* (collagen adhesin), *ebp* (elastin-binding protein), *bbp* (bone sialoprotein-binding protein), and *eno* (laminin-binding protein), closely related to pathogenesis of implant infections caused by *S. aureus* [34–36].

2.2 Antibiotic resistance and biofilm formation

Historically, infections caused by MDR *S. aureus* strains have been often reported worldwide. This microorganism has a notable ability to acquire systems of antibiotics inactivation. Production of reduced-affinity penicillin binding,

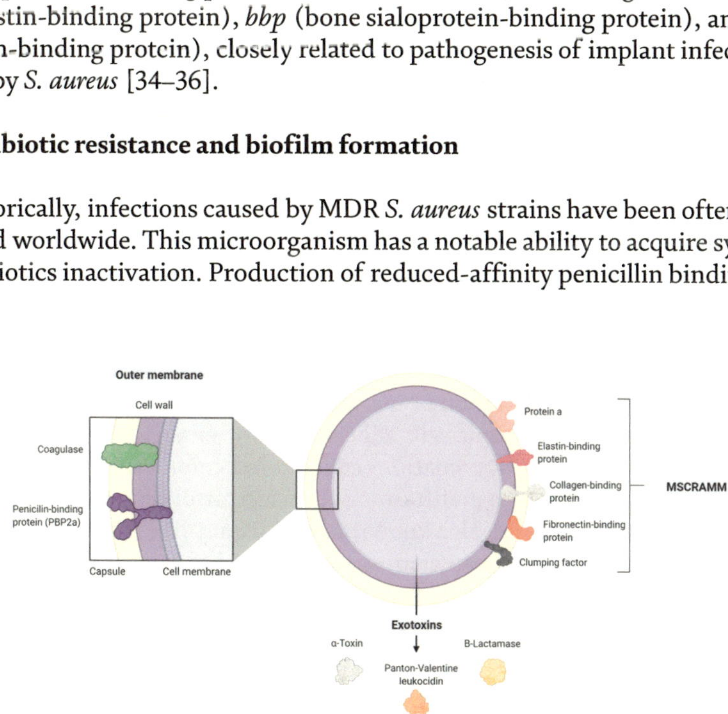

Figure 1.
Illustration of virulence factors produced by S. aureus.

ribosomal active site methylation, and efflux pumps that remove the antibiotic from the bacterial cell are the most cited mechanisms of antibiotic resistance developed by *S. aureus* cells [37]. The isolation of antibiotic-resistant strains began after the introduction of penicillin and methicillin into clinical practice, when resistant lineages, known as methicillin-resistant *S. aureus* (MRSA), were reported in 1950s and 1960s, respectively [38]. This resistance profile is mediated by mecA and mecC genes, which encode penicillin-binding protein 2a (PBP2a) in cell-wall synthesis. Those mec gene complexes are carried in an MGE known as the staphylococcal cassette chromosome mec (SCCmec), which can be acquired by horizontal gene transfer among related species [39].

Subsequently, vancomycin was used as an alternative to cases of MRSA infection [40]. However, the constant use of this antibiotic leads to the emergence of vancomycin-resistant *S. aureus* (VRSA) strains, first detected in 2002 [41]. Due to the fact that VRSA strains are generally also resistant to teicoplanin, the use of other abbreviations has been suggested: GISA (*S. aureus* of intermediate sensitivity to glycopeptides) and GRSA (*S. aureus* glycopeptide resistant) [42]. Moreover, some strains presented a relevant phenomenon known as heterogeneous resistance (heteroresistance) to vancomycin, where they have a mechanism of tolerance against this antibiotic. These strains, called hVISA, display a vancomycin-susceptible profile by microdilution assay; however, some individual cells into bacterial community might exhibit VRSA features [43].

Furthermore, the ability of some microorganisms to form cellular agglomerates, such as biofilms, contributes way more for antibiotic resistance. In summary, biofilm is a three-dimensional community of microorganisms covered and embedded in a self-produced matrix of extracellular polymeric substances (EPS) [44]. Such multicellular structure provides intrinsic protection for biofilm-embedded cells against hostile environments, for instance extreme temperature and pH, high salinity and pressure, poor nutrients, and antibiotics [45–47]. Microorganisms that grow on biofilms often exhibited different physiology profile from planktonic cells, especially in terms of their response to antibiotic treatment [48]. Although biofilm lifestyle can arise from a single cell, differential environmental conditions throughout the community can potentiate the development of distinct subpopulations. Gradients in oxygen, nutrients, and electron acceptors can cause heterogeneous gene expression throughout a biofilm. This communication between these bacterial cells, called quorum sensing, mediated the genes expression and activate virulence factors [49].

S. aureus has a great capacity to form biofilms on human body tissues and medical devices, increasing the risk of invasive infections [50]. It is estimated that *S. aureus* causes about 40–50% of prosthetic heart valve infections, 50–70% of catheter biofilm infections, and 87% of bloodstream infections [24]. The main stages of biofilm formation consist of four sequential steps: attachment, formation of microcolonies, accumulation or maturation, and detachment or dispersal (**Figure 2**) [51]. Firstly, planktonic cells adhere to biotic or abiotic surfaces and further proliferate into sticky aggregations called microcolonies. The EPS produced by bacterial cells during biofilm maturation serves as scaffold for establishing this three-dimensional architecture, also known as mushroom-like structures. Upon reaching a specific cell density, a mechanism is triggered to initiate EPS degradation that releases cells embedded into biofilm to disperse and reinitiate the biofilm formation at distal sites [7].

S. aureus shows a variety of adhesins that mediate attachment to host factors, essential for biofilm formation [48]. These proteins are surface-associated by different means, such as ionic or hydrophobic interactions, such as autolysin, SERAM (secretable expanded repertoire adhesive molecules) proteins, membrane-spanning proteins, and the polysaccharide intercellular adhesin (PIA) [52]. It is worth to

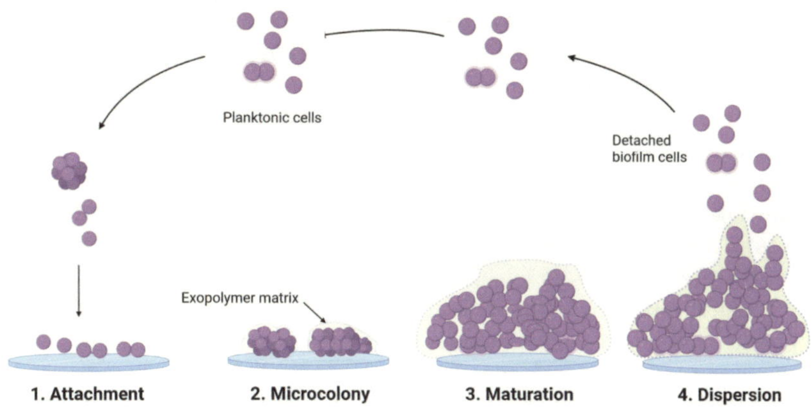

Figure 2.
General steps of S. aureus *biofilm formation.*

highlight that the presence of the ica gene located in the icaADBC (intercellular adhesion [ica]) operon works like a genetic determinant that contributes for biofilm establishment [53]. This genetic element can mediate the production of an extracellular mucopolysaccharide composed mainly of N-acetylglucosamine, which is associated to adhesion and colonization of several surfaces [54, 55].

Thus, several steps regarding biofilm formation of *S. aureus* might be considered as target to antibiofilm approaches. As many conventional antimicrobial agents have no satisfactory effect against mature biofilms, EOs already used for hundreds of years as a natural medicine to combat a variety of infections became a great antimicrobial alternative. The EOs are made up of various compounds, and it is further believed that this makes it difficult to develop bacterial resistance compared to antibiotics that have only one target action, making it attractive to fight MDR biofilm-forming bacteria [56]. Then, such attributes qualify the EOs as an important product from natural source to be explored by pharmaceutical industry [57, 58].

3. Essential oils

3.1 General aspect

Essential oils are compounds obtained from the secondary metabolism of the plants. They are characterized as complex mixtures of volatile compounds abundant in aromatic plants found in different parts of the plant, including leaves, flowers, stem, roots, seeds, and fruits [59]. There is a diversity of these substances described in the literature in commercial use, such as in perfumes, pharmaceuticals, cosmetics, insecticides, and food additives [60].

Generally, they are oily-looking liquids at room temperature of complex mixtures of volatile lipophilic substances, usually with pleasant scent. In water, EOs have a limited solubility, which allows their separation by steam or water distillation. Other methods to obtain EOs include cold-press extraction used for citrus peels, separation by solubility using organic solvents, and through supercritical fluid extraction [61]. They are usually unstable against environmental factors such as light, temperature, water activity, and salinity, affecting their constitution, contributing to the appearance of chemotypes with particular compositions. Depending on the technique used in the course of a separation, reactions such as ester hydrolysis, autoxidation, and rearrangements may occur, leading to the formation of artifacts and modifying their biological activity [62].

Compounds included in the EOs are produced in the cytoplasm and plastids of plant cells through the action of terpene synthase enzymes (TPSs), in which they use substrates from two pathways involved in the synthesis of terpenes: the mevalonate (MVA) and the methyl-eritritol phosphate (MEP) pathways [63]. They are localized and stored in complex secretory structures, such as glands, secretory cavities, hairs or trichomes, epidermal cells, internal secretory cells, and the secretory pockets [64]. Many of these substances are now known to be directly involved in the defense or attraction mechanisms of plants and often show interesting biological activities [65].

Despite containing two or three main components at a level of 20–70%, EOs are very complex mixtures of substances. In general, the majority components are formed by terpenes and phenylpropanoids [66]. In the very first definitions of EOs, these were frequently identified with terpenes, principally mono- and sesquiterpenes. Other substances have also been identified as alcohols, aldehydes, ketones, phenols, esters, ethers, oxides, peroxides, furans, organic acids, lactones, coumarins, sulfur compounds, anthraquinones, and alkaloids [61]. In the mixture, such compounds come in different concentrations. Usually, one of them is the majority compound, with others in lower grades and some in very low quantities (trace). EOs are composed of volatile hydrocarbons, and they may contain oxygen, sulfur, and halogens (rare) in their chemical structure [67]. In a reduced number of species, the predominant components are aromatic molecules, and these include thyme (thymol and carvacrol), peppermint (menthol), and anise (anethol) [68].

3.2 Antimicrobial and antibiofilm potential

Humans have used EOs for thousands of years, not only as aromatic extracts and for beauty care and culinary uses, but also in folk medicine, due to their many different pharmacological activities, such as antiseptic, anti-inflammatory, and analgesic properties [65]. Some of the EOs, and their components, have demonstrated the relevant antimicrobial potential against a wide range of microbial pathogens [69]. Additionally, Gram-positive bacteria, such as *S. aureus*, seem to be much more susceptible to EOs than Gram-negative cells, probably due to cellular surface constitution. Gram-positive has only the inner membrane and a cell wall that allows hydrophobic molecules to easily penetrate into the cells. For instance, phenolic compounds have a dose-dependent effect, at low concentrations they interfere with enzymes involved in energy production, and at high amounts they can denature proteins [70, 71].

The broad-spectrum activity of EOs is related to the diverse chemical reactions of aldehydes, phenolic compounds, and terpenes, synthesized from secondary metabolism by different plant parts [10]. The EO action is attributed to the ability of their constituents to interact with the cell membrane and consequently disturb the microbial integrity, leading to cell death [72]. However, EO bioactive components can have several cellular targets, and they are mainly associated with cytoplasmic coagulation, inhibition of ATP-production enzymes, alteration in ion transport, cell-wall damage, and bacterial membrane destruction (**Figure 3**) [73].

The emergence of MRD pathogens has caused an interesting shift from the onerous development of novel classes of antibiotics to the more straightforward application of synergism or combinatory therapy in the hope of reviving the efficacy and effectiveness of existing antibiotics [74]. Several studies have demonstrated that there was synergetic effect when two or more EOs are mixed together. Moreover, there are also reports of synergistic activity of EOs when used in combination with well-known antibiotics. When blended with other antimicrobial agents, the

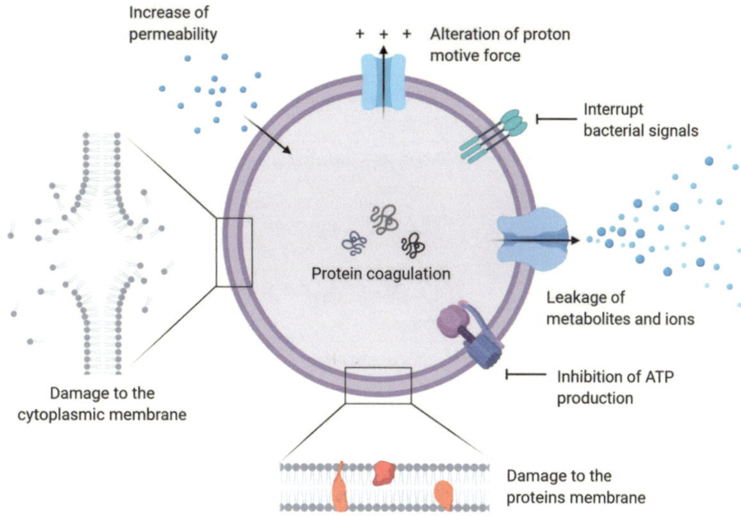

Figure 3.
General antibacterial mechanisms of essential oils.

constituents of EOs can unlock the cell membrane channels, thus opening the passage of antimicrobial agents to reach their target sites [75].

The capacity of some EOs to inhibit biofilm formation has been less explored; however, some reports suggested their utilization as potent inhibitor of virulence factors and biofilm formation [76]. So far, a plethora of potential antibiofilm agents, mainly inspired by natural products, has been developed and shown great promise in either facilitating the dispersion of preformed biofilms or inhibiting the formation of new biofilms *in vitro* [77]. In contrast to conventional antibiotics, the recently developed antibiofilm molecules do not directly affect bacterial survival and thus the expectation is that resistance to these molecules will not readily occur [77].

Table 1 shows some studies based on the *S. aureus* antibiofilm activity of OEs extracted from several plant sources. The EO action on biofilm inhibition and dispersal can be related to reactivity, hydrophobicity, and the diffusion rate of the EO in the matrix, as well as the biofilm composition and structure [78]. Same studies correlated sublethal concentrations of EOs with their capability to inhibit the first steps of biofilm formation. The main constituents of OE can act by several ways to disturb the biofilm development, such as blockage of the quorum-sense system, inhibition of the flagellar gene transcription, or through interference with bacterial motility [71].

Antibiofilm agents can have different therapeutic applications depending on their effects on the biofilm: compounds that interfere with biofilm formation could be exploited in the prophylaxis of implant surgery or for the coatings in medical devices, whereas agents able to disperse biofilm structure could be administered in combination with conventional antibiotics for the treatment of biofilm-associated infections [96]. Despite the growing number of new potent EO-based antibiofilm compounds described, there is still a great challenge in the development of antibiofilm drugs. Once the EO compounds, which has such activity, discovered so far need further optimizations to improve potency for it become one clinical candidate for such approach. Other EO features such as stability, volatility, encapsulation, and optimal dosage should be considered for the development of EO-based antibiofilm drugs. However, it is expected that in the coming years some of these compounds would be translated into antibiofilm drugs.

Scientific name	Plant part	Major chemical compounds	Resistant phenotype	Ref.
Artemisia absinthium	Aerial	1,8 Cineole, methyl chavicol, camphor	MRSA, MRSAMupR	[79]
Artemisia dracunculus	Aerial	Methyl chavicol, methyl eugenol	MRSA, MRSAMupR	[79]
Artemisia longifolia	Aerial	Camphor, 1,8-cineole	MRSA, MRSAMupR	[79]
Artemisia frigida	Aerial	1,8-cineole, methyl chavicol, camphor	MRSA, MRSAMupR	[79]
Cinnamomum zeylancium	Bark and leaves	Cinnamaldehyde	MRSA	[80]
Cymbopogon citratus	Fruit	Ethanolic compounds	MSSA MRSAMupR	[80]
Cymbopogon nardus	Leaves	Eugenol, cinnamaldehyde, citral, geraniol	MSSA, MRSA	[80, 81]
Eucalyptus globulus	Aerial	Eucalyptol, [+] spathulenol, α-pinene	MRSA, MSSA	[82]
Lippia alba	Aerial	Geranial, neral	MSSA	[83]
Mentha piperita	Aerial	Menthol, menthone, menthyl acetate	MRSA, MSSA	[84]
Melaleuca alternifolia	Aerial	α-Terpineol, terpinen-4-ol	MSSA	[85]
Myrtus communis	Leaves	Eugenol, α-terpineol, γ-terpinene	MSSA	[86]
Ocimum gratissimum	Leaves	Eugenol, 1,8-cineole	MRSA, MSSA	[87]
Origanum vulgare	Leaves and Arial	1-Terpineol, sabinene, γ-terpinene	MRSA, MSSA	[88, 89]
Piper nigrum	Leaves	Limonene, sabinene, β-pinene	MSSA	[90]
Rosmarinus officinalis	Leaves and flower	1,8-cineol, camphor, α-pinene	MSSA	[91]
Satureja hortensis	Leaves	β-cubebene, limonene, α-pinene	MSSA, MRSA	[92]
Satureja montana	Leaves	Carvacrol, p-cymene, δ-terpinene	MRSA	[93]
Syzygium aromaticum	Aerial	Eugenol, caryophyllene	MSSA	[90]
Thymus vulgaris	Aerial	p-Cymene, γ-terpinene	MRSA, MSSA	[82, 94]
Thymus daenensis	Aerial	Carvacrol, γ-terpinene	MSSA	[92]
Zataria multiflora	Leaves	Thymol, carvacrol, rho-cymene	MRSA, MSSA	[95]

Table 1.
Summarized antibiofilm activity of EOs against MDR S. aureus.

4. Conclusion

Due to the emergence of multiresistant strains and biofilm formation, there is an urgent need to find effective alternatives against *S. aureus*. Thus, EOs became a promising alternative for treatment and prophylaxis of infections caused by *S. aureus*. Many EOs have proven to be effective antimicrobials and antibiofilm, opening the possibility of using EOs in clinical formulations alone or in synergy with already known antibiotics. However, further research is needed to better understand the interactions between the steps of biofilms formation with the EOs and their constituents separately, as well. In addition, more acute studies in relation to volatility and solubility should be done in order to increase the essential oils' antimicrobial potential as a pharmacological product.

Conflict of interest

The authors declare no conflict of interest.

Author details

Victor Alves Carneiro[1,2*], Ramaiana Soares Melo[1], Antônio Mateus Gomes Pereira[1], Águida Maria Albuquerque Azevedo[1], Maria Nágila Carneiro Matos[2], Rafaela Mesquita Bastos Cavalcante[2], Renan Rhonalty Rocha[2], Vinícius de Queiroz Albuquerque[1], Jesús Alberto Pérez Guerrero[2] and Francisco Eduardo Aragão Catunda Junior[3]

1 Center of Bioprospection and Experimentation Molecular Applied [NUBEM], University Center INTA—UNINTA, Sobral, Ceará, Brazil

2 Laboratory of Biofilms and Antimicrobial Agents [LaBAM], Federal University of Ceará, Sobral, Ceará, Brazil

3 Center for Exact, Natural and Technological Sciences, State University of the Tocantina Region of Maranhão, Imperatriz, Maranhão, Brazil

*Address all correspondence to: viktorcarneiro@gmail.com

References

[1] Karam G, Chastre J, Wilcox MH, Vincent JL. Antibiotic strategies in the era of multidrug resistance. Critical Care. 2016;**20**(1):1-9

[2] Van DD, Paterson DL. Multidrug-resistant bacteria in the community: Trends and lessons learned. Infectious Disease Clinics of North America. 2016;**30**(2):377-390

[3] Tong SYC, Davis JS, Eichenberger E, Holland TL, Fowler VG Jr. *Staphylococcus aureus* infections: Epidemiology, pathophysiology, clinical manifestations, and management. Clinical Microbiology Reviews. 2015;**28**(3):603-623

[4] Foster TJ. Antibiotic resistance in *Staphylococcus aureus*. Current status and future prospects. FEMS Microbiology Reviews. 2017;**41**(3):43-49

[5] Zapotoczna M, O'Neill E, O'Gara JP. Untangling the diverse and redundant mechanisms of *Staphylococcus aureus* biofilm formation. PLoS Pathogens. 2016;**12**(7):1-6

[6] Ribeiro M, Monteiro FJ, Ferraz MP. Infection of orthopedic implants with emphasis on bacterial adhesion process and techniques used in studying bacterial-material interactions. Biomatter. 2012;**2**(4):176-194

[7] Moormeier DE, Bayles KW. *Staphylococcus aureus* biofilm: A complex developmental organism. Molecular Microbiology. 2017;**104**(3):365-376

[8] Krifa M, El Mekdad H, Bentouati N, Pizzi A, Ghedira K, Hammami M, et al. Immunomodulatory and anticancer effects of *Pituranthos tortuosus* essential oil. Tumor Biology. 2015;**36**(7):5165-5170

[9] Puškárová A, Bučková M, Kraková L, Pangallo D, Kozics K. The antibacterial and antifungal activity of six essential oils and their cyto/genotoxicity to human HEL 12469 cells. Scientific Reports. 2017;**7**(1):1-11

[10] Swamy MK, Akhtar MS, Sinniah UR. Antimicrobial properties of plant essential oils against human pathogens and their mode of action: An updated review. Evidence-based Complementary and Alternative Medicine. 2016;**2016**(1):1-21

[11] Myles IA, Datta SK. *Staphylococcus aureus*: An introduction. Seminars in Immunopathology. 2012;**34**(2):181-184

[12] Lindsay JA. Genomic variation and evolution of *Staphylococcus aureus*. International Journal of Medical Microbiology. 2010;**300**(2-3):98-103

[13] Kuroda M, Ohta T, Uchiyama I, Baba T, Yuzawa H, Kobayashi I, et al. Whole genome sequencing of meticillin-resistant *Staphylococcus aureus*. Lancet. 2001;**357**(9264):1225-1240

[14] Alibayov B, Baba-Moussa L, Sina H, Zdeňková K, Demnerová K. *Staphylococcus aureus* mobile genetic elements. Molecular Biology Reports. 2014;**41**(8):5005-5018

[15] Hodille E, Rose W, Diep BA, Goutelle S, Lina G, Dumitrescu O. Virulence in *Staphylococcus aureus*. Clinical Microbiology and Infection. 2017;**30**(4):887-917

[16] Kim HK, Missiakas D, Schneewind O. Mouse models for infectious diseases caused by *Staphylococcus aureus*. Journal of Immunological Methods. 2014;**410**:88-99

[17] Sakr A, Brégeon F, Mège JL, Rolain JM, Blin O. *Staphylococcus aureus* nasal colonization: An update on mechanisms, epidemiology, risk factors, and subsequent infections. Frontiers in Microbiology. 2018;**9**:1-15

[18] Ondusko DS, Nolt D. *Staphylococcus aureus*. Pediatrics in Review. 2018;**39**(6):287-298

[19] Todd JK. Staphylococcal infections. Pediatrics in Review. 2005;**26**(12):444-450

[20] Tanwar J, Das S, Fatima Z, Hameed S. Multidrug resistance: An emerging crisis. Interdisciplinary Perspectives on Infectious Diseases. 2014;**2014**:1-7

[21] Agyeman PKA, Schlapbach LJ, Giannoni E, Stocker M, Posfay-Barbe KM, Heininger U, et al. Epidemiology of blood culture-proven bacterial sepsis in children in Switzerland: A population-based cohort study. Lancet Child & Adolescent Health. 2017;**1**(2):124-133

[22] Otto M. Community-associated MRSA: What makes them special? International Journal of Medical Microbiology. 2013;**303**(6-7):324-330

[23] Lakhundi S, Zhang K. Methicillin-resistant *Staphylococcus aureus*: Molecular characterization, evolution, and epidemiology. Clinical Microbiology Reviews. 2018;**31**(4):1-103

[24] Chen CJ, Su LH, Chiu CH, Lin TY, Wong KS, Chen YY, et al. Clinical features and molecular characteristics of invasive community-acquired methicillin-resistant *Staphylococcus aureus* infections in Taiwanese children. Diagnostic Microbiology and Infectious Disease. 2007;**59**(3):287-293

[25] Karakonstantis S, Kalemaki D. Evaluation and management of *Staphylococcus aureus* bacteriuria: An updated review. Infection. 2018;**46**(3):293-301

[26] Marcó DPJ. Community-acquired *Staphylococcus aureus*, a recent problem. Archivos Argentinos de Pediatría. 2016;**114**(6):500-502

[27] Chen M, Yu Q, Sun H. Novel strategies for the prevention and treatment of biofilm related infections. International Journal of Molecular Sciences. 2013;**14**(9):18488-18501

[28] Fisher RA, Gollan B, Helaine S. Persistent bacterial infections and persister cells. Nature Reviews. Microbiology. 2017;**15**(8):453-464

[29] Karauzum H, Datta SK. Adaptive immunity against *Staphylococcus aureus*. Current Topics in Microbiology and Immunology. 2017;**409**:419-439

[30] Liu G. Molecular pathogenesis of *Staphylococcus aureus* infection George. Pediatric Research. 2010;**65**(5):71-77

[31] Mirzaee M, Najar-Peerayeh S, Behmanesh M, Moghadam MF. Relationship between adhesin genes and biofilm formation in vancomycin-intermediate *Staphylococcus aureus* clinical isolates. Current Microbiology. 2015;**70**(5):665-670

[32] Herman-Bausier P, Labate C, Towell AM, Derclaye S, Geoghegan JA, Dufrêne YF. *Staphylococcus aureus* clumping factor A is a force-sensitive molecular switch that activates bacterial adhesion. Proceedings of the National Academy of Sciences of the United States of America. 2018;**115**(21):5564-5569

[33] Lacey KA, Geoghegan JA, McLoughlin RM. The role of *Staphylococcus aureus* virulence factors in skin infection and their potential as vaccine antigens. Pathogens. 2016;**5**(1):1-17

[34] Campoccia D, Speziale P, Ravaioli S, Cangini I, Rindi S, Pirini V, et al. The presence of both bone sialoprotein-binding protein gene and collagen adhesin gene as a typical virulence trait of the major epidemic cluster in isolates from orthopedic implant infections. Biomaterials. 2009;**30**(34):6621-6628

[35] Tristan A, Ying L, Bes M, Etienne J, Vandenesch F, Lina G. Use of multiplex PCR to identify *Staphylococcus aureus* adhesins involved in human hematogenous infections. Journal of Clinical Microbiology. 2003;**41**(9):4465-4467

[36] Ghasemian A, Peerayeh SN, Bakhshi B, Mirzaee M. Detection of accessory gene regulator groups genes and cassette chromosome mec types among *Staphylococcus aureus* isolated from intensive care unit patients. Asian Pacific Journal of Tropical Disease. 2015;**5**(2):153-157

[37] Pantosti A, Sanchini A, Monaco M. Mechanisms of antibiotic resistance in *Staphylococcus aureus*. Future Microbiology. 2007;**2**(3):323-334

[38] Chambers HF, Deleo FR. Waves of resistance: *Staphylococcus aureus* in the antibiotic era. Nature Reviews. Microbiology. 2009;**7**(9):629-641

[39] Hanssen AM, Ericson Sollid JU. SCCmec in staphylococci: Genes on the move. FEMS Immunology and Medical Microbiology. 2006;**46**(1):8-20

[40] Tiwari HK, Sen MR. Emergence of vancomycin resistant *Staphylococcus aureus* (VRSA) from a tertiary care hospital from northern part of India. BMC Infectious Diseases. 2006;**6**(1):1-6

[41] Weigel LM, Clewell DB, Gill SR, Clark NC, McDougal LK, Flannagan SE, et al. Genetic analysis of a high-level vancomycin-resistant isolate of *Staphylococcus aureus*. Science. 2003;**302**(5650):1569-1571

[42] Santos AL, Santos DO, Freitas CC, Ferreira BLA, Afonso IF, Rodrigues CR, et al. *Staphylococcus aureus*: Visitando uma cepa de importância hospitalar *Staphylococcus aureus*: Visiting a strain of clinical importance. Jornal Brasileiro de Patologia e Medicina Laboratorial. 2007;**43**(6):413-423

[43] Cázares-Domínguez V, Cruz-Córdova A, Ochoa SA, Escalona G, Arellano-Galindo J, Rodríguez-Leviz A, et al. Vancomycin tolerant, methicillin-resistant *Staphylococcus aureus* reveals the effects of vancomycin on cell wall thickening. PLoS ONE. 2015;**10**(3):1-16

[44] Hall-Stoodley L, Stoodley P, Kathju S, Høiby N, Moser C, William Costerton J, et al. Towards diagnostic guidelines for biofilm-associated infections. FEMS Immunology and Medical Microbiology. 2012;**65**(2):127-145

[45] Yin W, Wang Y, Liu L, He J. Biofilms: The microbial "protective clothing" in extreme environments. International Journal of Molecular Sciences. 2019;**20**(14):1-18

[46] Flemming HC, Wingender J. The biofilm matrix. Nature Reviews. Microbiology. 2010;**8**(9):623-633

[47] Lebeaux D, Ghigo JM, Beloin C. Biofilm-related infections: Bridging the gap between clinical management and fundamental aspects of recalcitrance toward antibiotics. Microbiology and Molecular Biology Reviews. 2014;**78**(3):510-543

[48] Archer NK, Mark JM, Costerton JW, Leid JG, Powers ME, Shirtliff ME. *Staphylococcus aureus* biofilms properties, regulation and roles in human disease. Virulence. 2011;**2**(5):445-459

[49] Rani SA, Pitts B, Beyenal H, Veluchamy RA, Lewandowski Z, Davison WM, et al. Spatial patterns of DNA replication, protein synthesis, and oxygen concentration within bacterial biofilms reveal diverse physiological states. Journal of Bacteriology. 2007;**189**(11):4223-4233

[50] Khatoon Z, McTiernan CD, Suuronen EJ, Mah TF, Alarcon EI. Bacterial biofilm formation on

implantable devices and approaches to its treatment and prevention. Heliyon. 2018;**4**(12):e01067

[51] Toole GO, Kaplan HB, Kolter R. Biofilm formation as microbial development. Annual Reviews in Microbiology. 2000;**54**(1):49-79

[52] Heilmann C. Chapter 7 Adhesion mechanisms of staphylococci. Europe. Vol. 715(1). 2011. pp. 143-158

[53] Rohde H, Burandt EC, Siemssen N, Frommelt L, Burdelski C, Wurster S, et al. Polysaccharide intercellular adhesin or protein factors in biofilm accumulation of *Staphylococcus epidermidis* and *Staphylococcus aureus* isolated from prosthetic hip and knee joint infections. Biomaterials. 2007;**28**(9):1711-1720

[54] O'Gara JP. ica and beyond: Biofilm mechanisms and regulation in *Staphylococcus epidermidis* and *Staphylococcus aureus*. FEMS Microbiology Letters. 2007;**270**(2):179-188

[55] Marques VF, Motta CC, Soares BD, Melo DA, Coelho SM, Coelho ID, et al. Biofilm production and beta-lactamic resistance in Brazilian *Staphylococcus aureus* isolates from bovine mastitis. Brazilian Journal of Microbiology. 2017;**48**(1):118-124

[56] Yap PSX, Yiap BC, Ping HC, Lim SHE. Essential oils, a new horizon in combating bacterial antibiotic resistance. Open Microbiology Journal. 2014;**8**(1):6-14

[57] Elshafie HS, Camele I. An overview of the biological effects of some mediterranean essential oils on human health. BioMed Research International. 2017;**2017**(1):1-14

[58] Sharifi-Rad J, Sureda A, Tenore GC, Daglia M, Sharifi-Rad M, Valussi M, et al. Biological activities of essential oils: From plant chemoecology to traditional healing systems. Molecules. 2017;**22**(1):1-55

[59] O'bryan CA et al. Potential of plant essential oils and their components in animal agriculture—In vitro studies on antibacterial mode of action. Frontiers in Veterinary Science. 2015;**2**(35):1-8

[60] Barbosa L, Filomeno C, Teixeira R. Chemical variability and biological activities of *Eucalyptus* spp. essential oils. Molecules. 2016;**21**(1671):1-33

[61] Raymond C, George N. Natural products chemistry: Sources, separations, and structures. Natural Products Chemistry. 2014;**1**:24-29

[62] Yang L, Wen KS, Ruan X, Zhao YX, Wei F, Wang Q. Response of plant secondary metabolites to environmental factors. Molecules. 2018;**23**(4):1-26

[63] Pazouli L, Niinenets Ü. Multi-substrate terpene synthases: Their occurrence and physiological significance. Frontiers in Plant Science. 2016;**7**(1019):1-16

[64] Dhifi W, Bellili S, Jazi S, Bahloul N, Mnif W. Essential oils' chemical characterization and investigation of some biological activities: A critical review. Medicines. 2016;**3**(4):25

[65] Bakkali F, Averbeck S, Averbeck D, Idaomar M. Biological effects of essential oils—A review. Food and Chemical Toxicology. 2008;**46**(2):446-475

[66] Jean-Michel M, Céline R. Natural Microbiol Agents. Vol. 8(1). Springer. 2018. p. 337

[67] Başer KHC, Demirci F. Essential oils. Kirk-Othmer Encyclopedia of Chemical Technology. 2000:1-37

[68] Chamorro ER, Zambón SN, Morales WG, Sequeira AF,

Velasco GA. Study of the Chemical Composition of Essential Oils by Gas Chromatography. Vol. 1(1). Intech; 2012. pp. 307-324

[69] Wińska K, Mączka W, Łyczko J, Grabarczyk M, Czubaszek A, Szumny A. Essential oils as antimicrobial agents—Myth or real alternative? Molecules. 2019;**24**(11):1-21

[70] Tiwari BK, Valdramidis VP, O'Donnell CP, Muthukumarappan K, Bourke P, Cullen PJ. Application of natural antimicrobials for food preservation. Journal of Agricultural and Food Chemistry. 2009;**57**(14):5987-6000

[71] Nazzaro F, Fratianni F, De Martino L, Coppola R, De Feo V. Effect of essential oils on pathogenic bacteria. Pharmaceuticals. 2013;**6**(12):1451-1474

[72] Chimnoi N, Reuk-ngam N, Chuysinuan P, Khlaychan P, Khunnawutmanotham N, Chokchaichamnankit D, et al. Characterization of essential oil from *Ocimum gratissimum* leaves: Antibacterial and mode of action against selected gastroenteritis pathogens. Microbial Pathogenesis. 2018;**118**:290-300

[73] Burt S. Essential oils: Their antibacterial properties and potential applications in foods—A review. International Journal of Food Microbiology. 2004;**94**(3):223-253

[74] Yap PSX, Yang SK, Lai KS, Lim SHE. Essential oils: The ultimate solution to antimicrobial resistance in *Escherichia coli*? Pathogenesis and Biotechnological Applications. 2017;**1**(2017):299-313

[75] Rai M, Paralikar P, Jogee P, Agarkar G, Ingle AP, Derita M, et al. Synergistic antimicrobial potential of essential oils in combination with nanoparticles: Emerging trends and future perspectives. International Journal of Pharmaceutics. 2017;**519**(1-2):67-78

[76] Lee K, Lee JH, Kim SI, Cho MH, Lee J. Anti-biofilm, anti-hemolysis, and anti-virulence activities of black pepper, cananga, myrrh oils, and nerolidol against *Staphylococcus aureus*. Applied Microbiology and Biotechnology. 2014;**98**(22):9447-9457

[77] Rabin N, Zheng Y, Opoku-Temeng C, Du Y, Bonsu E, Sintim HO. Agents that inhibit bacterial biofilm formation. Future Medicinal Chemistry. 2015;**7**(5):647-671

[78] Vázquez-Sánchez D, Cabo ML, Rodríguez-Herrera JJ. Antimicrobial activity of essential oils against *Staphylococcus aureus* biofilms. Food Science and Technology International. 2015;**21**(8):559-570

[79] Lopes-Lutz D, Alviano DS, Alviano CS, Kolodziejczyk PP. Screening of chemical composition, antimicrobial and antioxidant activities of *Artemisia* essential oils. Phytochemistry. 2008;**69**(8):1732-1738

[80] Jafri H, Husain FM, Ahmad I. Antibacterial and antibiofilm activity of some essential oils and compounds against clinical strains of *Staphylococcus aureus*. Journal of Biomedical and Therapeutic Sciences. 2014;**1**(1):65-71

[81] Pontes EKU, Melo HM, Nogueira JWA, Firmino NCS, Carvalho MG, Catunda Júnior FEA, et al. Antibiofilm activity of the essential oil of citronella (*Cymbopogon nardus*) and its major component, geraniol, on the bacterial biofilms of *Staphylococcus aureus*. Food Science and Biotechnology. 2019;**28**(3):633-639

[82] Tohidpour A, Sattari M, Omidbaigi R, Yadegar A, Nazemi J. Antibacterial effect of essential oils from two medicinal plants against Methicillin-resistant *Staphylococcus*

aureus (MRSA). Phytomedicine. 2010;**17**(2):142-145

[83] Porfírio EM, Melo HM, Pereira AMG, Cavalcante TTA, Gomes GA, De Carvalho MG, et al. In vitro antibacterial and antibiofilm activity of *Lippia alba* essential oil, citral, and carvone against *Staphylococcus aureus*. Scientific World Journal. 2017;**2017**(1):1-7

[84] Saeide S, Mahboubeh M, Javad RF, Fereshteh JMH. Antimicrobial and anti-biofilm effects of *Mentha piperita* and *Zataria multiflora* on pathogenic bacteria. Journal of Medical Bacteriology. 2019;**8**(5):1-7

[85] Kwieciński J, Eick S, Wójcik K. Effects of tea tree (*Melaleuca alternifolia*) oil on *Staphylococcus aureus* in biofilms and stationary growth phase. International Journal of Antimicrobial Agents. 2009; **33**(4):343-347

[86] Aleksic V, Knezevic P. Antimicrobial and antioxidative activity of extracts and essential oils of *Myrtus communis* L. Microbiological Research. 2014;**169**(4):240-254

[87] Melo RS, Azevedo AMA, Helena T, Rodrigues S, Ponte IL, et al. Effectiveness of *Ocimum gratissimum* L. essential oil against multidrug-resistant isolates of *Staphylococcus aureus* and *Escherichia coli*. Molecules. 2019;**24**(21):3864-3879

[88] Scandorieiro S, de Camargo LC, Lancheros CAC, Yamada-Ogatta SF, Nakamura CV, de Oliveira AG, et al. Synergistic and additive effect of oregano essential oil and biological silver nanoparticles against multidrug-resistant bacterial strains. Frontiers in Microbiology. 2016;**7**:1-14

[89] Lu M, Dai T, Murray CK, Wu MX. Bactericidal property of oregano oil against multidrug-resistant clinical isolates. Frontiers in Microbiology. 2018;**9**:1-14

[90] Walmiki MR, Ravishankar RV. Cell attachment inhibition and anti-biofilm activity of *Syzygium aromaticum*, *Cuminum cyminum* and *Piper nigrum* essential oils against pathogenic bacteria. Journal of Essential Oil-Bearing Plants. 2017;**20**(1):59-68

[91] Jardak M, Elloumi-Mseddi J, Aifa S, Mnif S. Chemical composition, anti-biofilm activity and potential cytotoxic effect on cancer cells of *Rosmarinus officinalis* L. essential oil from Tunisia. Lipids in Health and Disease. 2017;**14**(1):1-10

[92] Sharifi A, Mohammadzadeh A, Zahraei Salehi T, Mahmoodi P. Antibacterial, antibiofilm and antiquorum sensing effects of *Thymus daenensis* and *Satureja hortensis* essential oils against *Staphylococcus aureus* isolates. Journal of Applied Microbiology. 2018;**124**(2):379-388

[93] Vitanza L, Maccelli A, Marazzato M, Scazzocchio F, Comanducci A, Fornarini S, et al. *Satureja montana* L. essential oil and its antimicrobial activity alone or in combination with gentamicin. Microbial Pathogenesis. 2019;**126**:323-331

[94] Perez AP, Perez N, Lozano CMS, Altube MJ, de Farias MA, Portugal RV, et al. The anti MRSA biofilm activity of *Thymus vulgaris* essential oil in nanovesicles. Phytomedicine. 2019;**57**:339-351

[95] Shahabi N, Tajik H, Moradi M, Forough M, Ezati P. Physical, antimicrobial and antibiofilm properties of *Zataria multiflora* Boiss essential oil nanoemulsion. International Journal of Food Science and Technology. 2017;**52**(7):1645-1652

[96] Parrino B, Diana P, Cirrincione G, Cascioferro S. Bacterial biofilm inhibition in the development of effective anti-virulence strategy. Open Medicinal Chemistry Journal. 2018;**12**(1):84-87

Chapter 12

Methods for Searching of Potential Beneficial Bacteria and Their Products in Dental Biofilm

Marián Maďar, Jana Kačírová, Eva Styková, Michaela Maďarová and Radomíra Nemcová

Abstract

Dental microbiota is associated with different types of organisms with dentition including humans and is responsible for many oral diseases all over the world. Bacteria in a dental biofilm are important also in other diseases, i.e., endocarditis, pulmonary fibrosis, and arthritis, and some findings predict the connection of dental microbiota with cancerogenesis. Not all oral bacterial representatives are pathogenic or potentially pathogenic. Dental biofilm consists of numerous different bacteria that may have beneficial characteristics for good condition of dental and oral health. Searching for bacteria or their products with the beneficial effect is important in the development of new biologically based strategies for the prevention or treatment of oral and dental diseases. For searching of potential probiotic candidates are useful methods that could map phenotypic or genotypic characteristics of studied bacteria. This chapter is focused on the spectrum of these basic methods searching for beneficial bacteria and their products.

Keywords: dental, biofilms, probiotic, bacteria, methods

1. Introduction

Each form of life on earth needs to obtain water and some substances from the external environment for its growth. From viruses to whales, every form of life needs some substances. Differences are only in the mechanism of obtaining. Many types of organisms on earth for this purpose developed the digestive tract with the oral cavity during the evolution. The same mechanism is still on earth millions and millions of years. For example, dinosaurs had the same mechanism and during evolution developed dentition for good mechanical preparing of eaten food like humans today with some differences of course. We can deduct, that the dental problems in Jurassic age had the same cause as today if we are thinking about mechanical destruction. In the case of special dental diseases, like periodontitis or dental caries, the comparing is debatable. Maybe in Jurassic age were also some pathogens something like *Streptococcus mutans* nowadays, which were responsible for the destruction of dental enamel in Tyrannosaurus rex. Nowadays problems in dental diseases have the same causative mechanism. Many of them are caused by the effect of dental biofilm bacteria.

1.1 Biofilm, dental biofilm

A biofilm comprises any syntrophic consortium of microorganisms in which cells stick to each other and often also to a surface. Biofilms are highly organized bacterial agglomeration, which diversity is depending on the external and internal conditions of together growing bacteria.

Bacterial biofilms are also characteristic of the growth of one type of bacteria, i.e. a biofilm of *Staphylococcus aureus* [1]. Biofilms may form on living or non-living surfaces and can be founded in natural, industrial, and hospital conditions. In humans, a typical exam for biofilm is dental plaque. This microcosm was deeply characterized with the help of numerous basic or sophisticated methods of research. Microbiology procedures, microscopic techniques, genomic and proteomic methods bring new light on new findings in dental plaque (biofilm) research.

It is interesting, that the knowledge about dental biofilm from the discoveries of Anton van Leeuwenhoek (1632–1723) to today age is still not perfect because we are not able to decrease the number of dental diseases in the world [2].

Dental caries and periodontal diseases are the most common diseases in the world especially in areas with bad quality of dental medicine and in poor regions of the world. On the other side, it is also a disease, which is a wide range presented in all countries and all social communities.

Bacterial pathogens founded in dental enamel lesions are many times highly pathogenic and cause also systematic diseases like endocarditis, meningitis, pulmonary fibrosis, arthritis, and some findings predict the connection of dental microbiota with cancerogenesis [3, 4].

1.2 Dental biofilm bacterial composition

Opinions on the number of bacteria living in the oral cavity vary. It has been estimated that about 500 species of bacteria inhabit the oral cavity in humans [5]. Molecular-based studies have shown that bacterial communities found in the oral cavity are highly complex with about 1000 species and have been shown to be the second most complex microbial community in the body after the colon [6]. Although the animal microbiocenosis of animals and humans has similar properties, there are also significant differences in relation to the microbial species and the relative proportions of these species in the oral cavity [7]. For example, rodents lack gender representatives *Peptostreptococcus, Bacteroides* (currently *Prevotella* and *Porphyromonas*), *Treponema*, *Vibrio* and *Leptotrichia* [8]. Oral microbiocenosis of dogs is believed to be more diverse than oral microbiocenosis in humans [9]. However, bacteria in dental biofilm that are responsible for periodontal infectious diseases in humans and animals have been shown to be similar [7].

The microbiota of the dental biofilm differs from the microbiota on the mucosal surfaces and the composition of the microbiota of the dental biofilm varies in different anatomical sites. Gingival crevice supplies nutrients to bacteria and has low redox potential; therefore, it is colonized predominantly by anaerobic species such as *Prevotella* spp., *Veillonella* spp. and *Fusobacterium* spp. In contrast, supragingival plaque consists mainly of Gram-positive facultatively anaerobic bacteria, especially *Streptococcus* spp. and *Actinomyces* spp. The composition of the oral microbiota is highly dependent on the clinical condition of the teeth and gingivae. Healthy oral plaque contains predominantly facultatively anaerobic Gram-positive species, while in periodontal diseases microbiota turns into obligate anaerobic Gram-negative species [10]. In the formation of dental biofilm, primarily Gram-positive cocci, especially

Streptococcus sanguis and *Streptococcus mitis*, are involved in primary colonization, which colonizes the teeth for the first 4 hours after professional cleansing [11]. Other early colonizers include *Actinomyces* spp., *Capnocytophaga* spp., *Eikenella* spp., *Haemophilus* spp., *Prevotella* spp., *Propionibacterium* spp., and *Veillonella* spp. [12]. One of the major bacteria that serve as a bridge between early and late oral biofilm colonizers is *Fusobacterium nucleatum* [13]. Although it is an anaerobic bacterium, it could tolerate oxygen in the biofilm. This ability allows *F. nucleatum* to promote the growth of other strictly anaerobic bacteria such as *Porphyromonas gingivalis* [14]. Later colonizers are *Lactobacillus* spp., *Porphyromonas* spp., *Actinobacillus* spp., *Prevotella* spp., *Eubacterium* spp., *Selenomonas* spp., *Tannerella* spp., *Aggregatibacter* and *Treponema* spp. [15, 16]. Which type of bacteria, pathogens or potential beneficial members of dental microbiota will be chosen for research depends on the researcher. Currently, preparations containing probiotic strains such as: *Lactobacillus reuteri* (BioGaia Prodentis), *Bacillus coagulans* (Life Extension Advanced Oral Hygiene) and *Streptococcus salivarius* K12 (Bactoral) are available on the market [17–20]. *Bacillus subtilis* in form of tablet [VITALREXTM (VL)] is also used in the treatment of periodontal diseases [21]. Depending on the findings of the beneficial effect of living bacteria, there is also the possibility to use only their metabolic products for research aimed at preventing and treating dental diseases.

2. Recommended methods

2.1 Selection criteria useful for studying of dental biofilm and sample obtaining

In the oral cavity area, it is possible to study apart from dental biofilm also other biofilms, i.e. buccal, lingual, prosthesis, filled live or death teeth, soft tissue biofilms, etc. Our preferred place for obtaining of dental biofilm samples are sites of tooth surfaces close to the salivary duct orifices, because proteins produced in saliva could help to form biofilm and calculus. In humans, it is the lingual surface of the lower front teeth and decreases towards the third molar teeth. On the upper jaw, the supragingival calculus is often formed on the buccal surfaces of the first molars [22]. Also, in veterinary patients, supragingival calculus usually accumulates more rapidly and in larger amounts on the buccal surfaces of the upper jaws [23]. Places for sampling are variable depending on the anatomical proportion of hosts that are used for research as volunteers. Except for humans, it is possible to study dental biofilm also on domesticated or wild animals. Important criteria in the case of human biofilm are smoke, veganism, celiac disease, age, health condition, therapy with medication and so on. Each external and internal factor could change the composition of biofilm and each human has individual microbiota in the mouth. It is better when the group of volunteers has similar dental care (a type of toothpaste used) and similar food consumption habits. The selection of volunteers should be based on the targeted microbiota from the dental biofilm e.g. autochthonous or allochthonous or obtaining of pathogenic bacteria from target pathological lesions in the oral cavity, e.g. caries, etc. Autochthonous microbiota is isolated from volunteers who starve overnight after carefully brushing their teeth. The dental biofilm sample has to be obtained immediately after waking up. Volunteers could not eat, drink or brush their teeth before sampling. The composition of autochthonous or allochthonous microbiota depends on sampling time. If sampling takes place during the day, samples also contain allochthonous microbiota. Better condition for obtaining samples of autochthonous microbiota is from volunteers, which several days do not brush the their teeth.

2.2 Taking of dental biofilm samples

Samples of dental biofilms are easy to obtain, sampling is very simple, painless and noninvasive. Each human volunteer should confirm it with the signed agreement with taking samples, their next processing and provide the data in the anamnestic questionnaire concerning GDPR. In the case of domestic animals, dog or cat, owners have to agree with the possibility of sample taking and processing.

All things that are needed for the researcher are a sterile syringe needle and a sterile Eppendorf tube filled with sterile filtrated PBS commercial produced or according https://www.protocolsonline.com/recipes/phosphate-buffered-saline-pbs/. Cultivation liquid medium can be use for this purpose too.

We provide Brain hearth infusion broth (Merck K GaA Darmstadt, Germany). In the case of lactic acid bacteria isolation, we use deMan, Rogosa and Sharpe MRS (CONDA S.A, Madrid Spain) broth. The blood agar (Tryptic soy agar (TSA)) with 5% ram's blood (BBL, Microbiology Systems, Cockeysville, USA) is often chosen as the first medium for the cultivation of bacteria in bacteriology. In case of selection of major streptococcal species it is good to use Mitis Salivarius Agar (Merck K GaA Darmstadt, Germany). The classical cultivation method is at 37.5°C during 24–48 hours under anaerobic or aerobic conditions, depending on target bacterial members of dental biofilm. We provide BD GasPak™ systems (Becton, Dickinson and Company) for anaerobic cultivation. The further selection of strains is according to the cultivation characteristics of selected colonies. Selected strains could be stored in the glycerol stock or Microbank system (Pro Lab Diagnostic). Each isolated strain has to be identified for further analysis. We provide MALDI-TOF mass spectrometry or Blast n analysis of 16S rRNA sequence for identification. The biochemical tests could help with the identification and reveal the characteristics of the tested strain.

2.3 Methods useful for identification of bacterial composition of dental biofilm

For the study of the bacterial community and its composition, it is possible to use numerous methods. At first, it needs to be mentioned the classical microbiology. By classical bacteriology cultivation methods, we could select different types of cultivable bacteria in samples of dental biofilm. For this purpose, we could use different types of media, from liquid to solid, from basic to highly specific and selective media. Different conditions are also used in aerobic and anaerobic cultivation. The most numerous bacterial resident in the dental biofilm has better start line as low representative bacteria. On the other hand, the conditions in a cultivation medium could bring sometimes better conditions for the growth of former less presented bacteria in a tested sample. Due to this problem, it is hard to declare the ratio of different types of cultivable bacteria. Colonies forming units (CFU) method could reveal the approximate ratio of bacteria, but only the cultivable ones. Quantitative real-time PCR is a cultivation-independent perfect toll for declaring of the bacterial composition of cultivable, hard cultivable or uncultivable bacteria in tested sample, but it is limited due to numbers of selected bacterial groups. Amplicon sequencing is a sensitive method that is cultivation independent and good for declaring the composition of all bacterial members in the tested sample and it could quantify the ratio between bacterial groups [24, 25]. This method is cultivation free and principle is based on the amplification of total DNA isolated from the sample and next-generation sequencing (NGS) analysis. Big data obtained after sequencing are analyzed *in silico*.

If we combine the amplicon sequencing method with 16S rRNA identification of selected and isolated bacteria, we obtain perfect strategy and tools for confirmation of identified cultivable and uncultivable bacteria and also their semiquantitative ratio in our sample of dental biofilm.

It is necessary to know the numbers and ratios of bacteria in the sample because it can bring light to physiological or pathological parameters. On the other hand, it is hard to study this topic, due to the different bacterial composition of individual dental biofilms. Many isolated bacteria are autochthonous and host specific, and still found in a dental biofilm of the individuals. Based on these findings we can predict approximately similar conditions.

Cultivation, isolation, identification, and storage of the strains are necessary steps for deep research of pathogens, potential pathogens, and potentially probiotic strains and research of their interaction.

2.4 Classical cultivation necessary step in research

This method is still necessary for valid research of potentially beneficial bacteria and their products in dental biofilm. For testing of potential candidates as probiotic bacteria from dental biofilm at first, we need to isolate and store it by microbiological cultivation techniques. The same goes for pathogenic bacteria. A very important step in bacteriology research is the identification of bacteria. Form of growth, Gram staining, catalase activity, biochemical parameters are helpful in the analysis of solitary bacterial colonies. These methods are in some cases imperfect for the exact identification of bacteria. In comparison with the methods mentioned above, the sequencing of genes coding 16S RNA or other PCR products and next Blast n analysis or MALDI-TOF mass spectrometry identification are more sensitive.

Other growth characteristics as the possibility of growth inhibition of other bacteria are helpful in the selection of candidates with the production of bioactive substances, especially in the case of biosurfactants, bacteriocins, or bacteriocin-like inhibitory substances (BLIS) [26, 27].

The presence of genes coding bioactive substances could be easily detected by PCR, but better is to check the possibility of their production at first. For example, *Streptococcus salivarius* inhibition potential against *Micrococcus luteus* could help unfold the production of salivaricin [28]. *Streptococcus thermophilus* could induce cell lysis of *Pediococcus acidilactici* and reveal the potential of Thermophilin 110 production [29]. Generally, if the presence of bacteriocins is detected in tested potential probiotics by growth inhibition during cultivation with the target organism, the next step with the help of cultivation and proteomic methods is the overlaying gel test. It can detect the mass size of bacteriocins, i.e. in the case of *Lactobacillus plantarum* or *Streptococcus thermophilus* or other tested bacteria [29, 30]. Cultivation is also necessary to obtain a large volume of bacteriocin for further testing with the help of large volume fermentation. If bacteriocins are soluble and are produced in cultivation media, flow centrifuge is needed for their isolation. If bacteriocins are insoluble it is necessary to use ultracentrifuge in this step. The extraction of bacteriocins is an important step in isolation [31]. If we have pure bacteriocins for testing, cultivation is still needed. For example, salivaricin isolated from *Streptococcus salivarius* K12 is active against bacterial species involved in halitosis, by inhibition of *Micrococcus luteus* Il, *Streptococcus anginosus* T29, *Eubacterium saburreum* ATCC 33271 and *Micromonas micros* ATCC 33270 [28].

Cultivation procedures are needed in case of studying of the capability of the other bioactive substances like biosurfactants or exopolysaccharides. These two products have antagonist effects. Exopolysaccharides enhance adherence and biosurfactants promote disruption of adherence.

It was detected that biosurfactant produced by *Lactobacillus reuteri* could very significantly down-regulate expression of *Streptococcus mutans* glucosyltransferase genes (*gtfB, gtfC*) and fructosyltransferase gene *ftf* [32]. These genes are very important for the production of exopolysaccharides which are responsible for adherence of oral streptococci [33, 34]. In the case of *Streptococcus mutans* glucosyltransferase genes are responsible for cariogenic activity [35]. These genes are also useful for the differentiation of streptococcal candidates which are often difficult to differentiate because they have high homologous sequences in the 16S rRNA gene.

3. Possibility to produce bioactive substances detected in tested potential bacterial probiotic candidates by PCR (bacteriocins, biosurfactants, and exopolysaccharides)

3.1 Recommended isolation of DNA

The isolation of DNA from bacteria that are difficult to isolate, i.e. lactobacilli strains, is performed by the NucleoSpin® Tissue kit (Macherey-Nagel GmbH and Co. KG, Düren, Germany) using a lysis solution during overnight incubation at 95°C. The next steps of DNA isolation are according to the manufacturer's procedure. It is possible to use other kits for DNA isolation. It depends on researcher choice and routine practice in PCR laboratory. After isolation of DNA it is better to verify DNA quality and quantity. We use Nanodrop spectrophotometric (Wilmington, Delaware USA) analysis for this purpose.

For quick isolation of DNA it is also possible to use one bacterial colony and 100 μl DNAzol direct (Molecular research centre Inc. Cincinnati. USA), and heat it to 95°C during 15 min for isolation of DNA without measuring of DNA quantity, but storage of DNA samples for next analysis is time limited. For storage of DNA isolated by both methods we recommended −20°C. The isolation steps are according to the manufacturer and specific sample.

For PCR we could use Mastermix: One Taq 2X Master Mix (England Biolabs, Ipswich, Massachusetts, USA) and specific primers (**Tables 1–3**) in concentration of 33 μmol at volume 0.6 and 1–2 μl of DNA isolated with help of DNAzol direct.

3.2 Bacteriocins and methods for their detection

A large number of lactic acid bacteria produce bacteriocins that kill other microorganisms. Lactobacilli bacteriocins have potential utility as pathogen inhibitors in humans [36]. Also, oral streptococci have their bacteriocins for example *Streptococcus mutans* have mutacin, and *Streptococcus salivarius* has salivaricin [37, 38]. There are a number of factors influencing the efficacy of bacteriocins *in vivo* and *in situ*, including the survival of the production strain, specific activity, and animal model and targeted pathogen. However, bacteriocins have a great deal of promise to manage various infections and may become an alternative to existing antibiotics. Bacteriocins will need to undergo the same rigorous, costly research and validation process as all other previously approved therapies used in therapy [26]. Recommended conditions for detection of genes coding bacteriocins of some oral potential beneficial bacteria by PCR are described in (**Table 1**).

The researcher could study probiotic or pathogenic bacteria depending of the particular relationship to diseases. For example, PCR condition for bacteriocin detection from *Lactobacillus* spp. is mentioned in the publication [46]. Detection of genes coding production of bacteriocins is only the start of the research. By this method, we could select potential candidates for further research. Inhibition potential can be detected by preferred sensitive bacterial strain for example like in case of *Streptococcus salivarius* salivaricin the sensitive strain is *Micrococcus luteus* [28]. After confirmation of bacteriocin gene presence in tested isolates, there is still much work to be done with purification, fractionation, and isolation of bacteriocins. Not

Target gene	Primers	PCR protocol	Product size	Source
Streptococcus salivarius Salivaricin *sal*A	SalAUS 5′GTAGAAAATATTTACTACATACT3′ SalADS 5′GTTAAAGTATTCGTAAAACTGATG3′	95°C, 13 min, 30× (95°C, 30 sec, 55°C, 1 min, 72°C, 1 min) 72°C, 5 min	338 bp	[38–40]
Lactobacillus reuteri glycerol dehydrogenase *gldC* (reuterin)	GD1f 5′GTTCAGTCCGCCGCATATC3′ GD1r 5′GCCGCTCTTCGTGGATTTC3′	94°C, 5 min, 34× (94°C, 1 min, 58°C, 30 sec, 72°C, 50 sec) 72°C, 7 min	562 bp	[41]
Lactobacillus plantarum *Plantaricin*	plaF 5′-GGCATAGTTAAAATTCCCCCC-3′ plaR 5′-CAGGTTGCCGCAAAAAAAG-3′	94°C, 5 min, 30× (94°C 45 sec, 53.2°C, 45 sec, 72°C, 45 sec) 72°C, 5 min	428 bp	[42]
Lactobacillus plantarum *Plantaricin S*	plnF 5′-GCCTTACCAGCGTAATGCCC-3′ plnR 5′-CTGGTGATGCAATCGTTAGTTT-3′	94°C, 5 min, 30× (94°C, 45 sec, 62.3°C, 30 sec, 68°C, 2 min sec) 68°C, 5 min	475 bp	[43]
Streptococcus mutans *Mutacin*	F 5′-AGTTTCAATAGTTACTGTTGC-3′ R 5′-GCCAAACGGAGTTGATCTCGT-3′	94°C, 5 min, 34× (94°C, 1 min, 58°C, 30 sec, 72°C, 50 sec) 72°C, 7 min	750/450 bp	[44]
Bacillus subtilis *Subtilisin*	spaSFwd 5′CAAAGTTCGATGATTTCGATTTGGATGT3′ spaSRev 5′GCAGTTACAAGTTAGTGTTTGAAGGAA3′	94°C, 5 min, 34× (94°C, 30 sec, 55°C, 30 sec, 65°C, 60 sec) 65°C, 7 min	722 bp	[45]
Bacillus subtilis *Subtilosin*	sboAFwd 5′CGCGCAAGTAGTCGATTTCTAACA3′ sboARev R 5′CGCGCAAGTAGTCGATTTCTAACA3′	94°C, 5 min, 34× (94°C, 30 sec, 50°C, 30 sec, 65°C, 60 sec) 65°C, 7 min	565 bp	[45]

Table 1.
PCR conditions used for the detection of gene coding production of bactericcins.

Target gene	Primers	PCR protocol	Product size	Source
Bacillus subtilis surfactin sfp	sfp F 5′ATGAAGATTTACGGAATTTA3′ sfp R 5′TTATAAAAGCTCTTCGTACG3′	95°C, 3 min, 30× (95°C, 30 sec, 50°C, 30 sec, 72°C, 45 sec) 72°C, 10 min	675	[55]
Bacillus subtilis surfactin *srfAA*	*srfAA* F 5′TCGGGACAGGAAGACATCAT3′ *srfAA* R 5′CCACTCAAACGGATAATCCTGA3′	95°C, 3 min, 30× (95°C, 30 sec, 60°C, 30 sec, 72°C, 30 sec) 72°C, 10 min	201	[55]
Bacillus subtilis fengycin*fenB*	*fenB* F 5′CCTGGAGAAAGAATATACCGTACCY3′ *fenB* R 5′GCTGGTTCAGTT KGATCACAT3′	95°C, 3 min, 30× (95°C, 30 sec, 57°C, 30 sec, 72°C, 45 sec) 72°C, 10 min	201	[55]
Bacillus subtilis fengycin*fenD*	*fenD* R 5′GCTGGTTCAGTT KGATCACAT3′ *fenD* F 5′GGCCCGTTCTCTAAATCCAT3′	95°C, 3 min, 30× (95°C, 30 sec, 60°C, 30 sec, 72°C, 1 min) 72°C, 5 min	670	[55]
Bacillus subtilis iturín *ituD*	*ituD* F 5′TTGAAYGTCAGYGCSCCTTT3′ *ituD* R 5′TGCGMAAATAATGGSGTCGT3′	95°C, 3 min, 30× (95°C, 30 sec, 57°C, 30 sec, 72°C, 32 sec) 72°C, 10 min	482	[55]
Bacillus subtilis iturín *ituC*	*ituC* F 5′GGCTGCTGCAGATGCTTTAT3′ *ituC* R 5′TCGCAGATAATCGCAGTGAG3′	95°C, 3 min, 30× (95°C, 30 sec, 58°C, 30 sec, 72°C, 30 sec) 72°C, 10 min	423	[55]

Table 2.
PCR conditions used for the detection of gene coding production of biosurfactants.

all bacteria which present genes for bacteriocins are also capable inhibit pathogens. Some inhibition effects are caused by bacteriocins like inhibitory substances or by others active molecules which are waiting to discovered.

3.3 Biosurfactants and methods for their detection

Biosurfactants are naturally produced molecules that demonstrate potentially useful properties such as the ability to reduce surface tensions between different phases [47]. The release of biosurfactants by adhering microorganisms as a defense mechanism against other colonizing strains on the same substratum surface has been described previously for probiotic bacteria in the urogenital tract, the intestines, and the oropharynx, but not for microorganisms in the oral cavity [48]. The antimicrobial properties observed in dialyzed biosurfactants produced by the tested lactobacilli open possibilities for their use against microorganisms responsible for oral diseases [49]. Biosurfactants (BS) obtained from *Lactobacillus* spp. exhibit antibiofilm and antiadhesive activity against a broad spectrum of microbes [50]. For example, they are active against biofilm formation of *Candida albicans* [51] or *Staphylococcus aureus* [52]. Biosurfactants produced by the *Bacillus subtilis* SPB1 strain (HQ392822) revealed a wide spectrum of actions including antimicrobial activity towards multidrug-resistant microorganisms [53, 54]. For the detection of biosurfactants production, i.e. in the case of *Bacillus subtilis*, it is recommended to use PCR with the help of specific primers listed in **Table 2**.

Target gene	Primers	PCR protocol	Product size	Source
Str. mutans. Glucosyltransferase gene (*gtf*)	MKD-F 5′GGCACCACAACATTGGGAAGCTCAGTT3′ MKD-R 5′GGAATGGCCGCTAAGTCAACAGGAT3′	95°C, 13 min, 30× (95°C, 30 sec, 67°C, 1 min, 72°C, 1 min) 72°C, 5 min	433 bp	[60, 61]
Str. salivarius. Glucosyltransferase gene (*gtf*)	MKK-F 5′GTGTTGCCACATCTTCACTCGCTTCG3′ MKK-R 5′CGTTGATGTGCTTGAAAGGGCACCATT3′	95°C, 13 min, 30× (95°C, 30 sec, 66°C, 1 min, 72°C, 1 min) 72°C, 5 min	544 bp	[60]
Str. oralis. Glucosyltransferase gene (*gtf*)	*gtfR* MKR-F 5′TCCCGGTCAGCAAACTCCAGCC3′ *gtfR* MKR-R 5′GCAACCTTTGGATTTGCAAC3′	95°C, 13 min 30× (95°C, 30 sec, 66°C, 1 min, 72°C, 1 min) 72°C, 5 min	374 bp	[60]
Lactobacillus spp. Glucosyltransferase gene (*gtf*)	DexreuV 5′GTGAAGGTAACTATGTTG3′ DexreuR 5′ATCCGCATTAAAGAATGG3′	94°C, 5 min, 31× (94°C, 1 min, 47°C, 1 min, 72°C, 1 min) 72°C, 10 min	600 bp	[62]

Table 3.
PCR conditions used for the detection of gene coding production of exopolysaccharides.

Other species producing biosurfactants and condition for their detection are able in research papers for example: *Lactobacillus paracasei* produced biosurfactants with anti-adhesive properties [56]. *Streptococcus mitis* biosurfactants plays a protective role in the oral cavity and protects against colonization of saliva-coated surfaces by cariogenic *Streptococcus mutans* [48]. Based on *Bacillus subtilis* SPB1 lipopeptides production researcher predict their possibility used in toothpaste formulation [53]. Biosurfactants are promising bioactive molecules for oral-related health applications [47].

3.4 Exopolysaccharides and methods for their detection

Lactic acid bacteria are the most frequently mentioned in studies of exopolysaccharides (EPS) in oral microbiota [57]. Except for lactobacilli, which are participated in the later stages of dental biofilm formation, streptococci are one of the first bacteria capable of producing EPS. Streptococci are able to assert themselves and adhere to the hard tissues of the oral cavity immediately after washing the teeth. This property of adherence is predetermined and is encoded in genes that are also responsible for the production of glucosyltransferases. Glucosyltransferases (Gtfs) are produced by several types of lactic acid bacteria [58]. Gtfs are generally characterized as Gtf-S (glucosyltransferase-soluble) or Gtf-I (glucosyltransferase-insoluble) enzymes, depending on whether the glucan they produce is water soluble or insoluble [59]. For detection of exopolysaccharides production in oral lactic acid bacterial members is useful PCR with help of specific primers see in **Table 3**.

4. Testing of growth inhibition activity against pathogens

Testing of bacterial isolates as potential beneficial candidates or their products is necessary step in new discoveries. We are able declarate production of bioactive substance by very easy PCR reactions, as mentioned above in part 3. Activity of these substances is easy to declare by simply *in vitro* tests. At first for activity it is possible to use spot on or disc diffusion test. Same mechanism of declaration is for live bacteria isolates as for isolated bioactive substances.

If we found bacteria with interesting effect in spot or disc diffusion test it predict selection criteria of former characterized bacteria for next research.

4.1 The disc diffusion method for *Lactobacillus reuteri* for testing of growth inhibition activity against pathogens

We recommend the disc diffusion test for the detection of the inhibitory properties of beneficial microorganisms. Selected lactobacilli strains were grown on MRS agar (CONDA S.A, Madrid Spain) for 48 hours. anaerobically (Gas Pak Plus, BBL, Microbiology Systems, Cockeysville, USA) at 37°C. Then, a standardized suspension with an optical density of 1 McFarland by dissolving several solitary colonies in 5 ml of physiological saline was prepared. Sterile clean discs (6 mm diameter, BBL, Cockeysville, USA) were placed on Petri dishes (Ø 90 mm) with 20 ml of PYG agar (HiMedia Laboratories GmbH Einhausen, Germany). The sterile paper discs were inoculated with 5 μl of standardized suspensions of lactobacilli.

As a negative control, one Petri dish with PYG agar is served with a clean paper discs soaked with sterile MRS broth.

The plates with discs were incubated for 48 hours. anaerobically (Gas Pak Plus, BBL, Microbiology Systems, Cockeysville, USA) at 37°C. The discs were removed with a sterile syringe needle or tweezer after incubation. Subsequently, 3 ml of 0.7% PYG agar was inoculated with 0.3 ml of the indicator pathogenic strain and put into plates with lactobacilli. Pathogenic strains were incubated for 18 hours in PYG broth at 37°C. The plates with YPG medium inoculated with pathogen were incubated for 24 hours aerobically at 37°C. After incubation, the diameter of the inhibition zones was measured. The results were recorded in the table as the arithmetic means of the three measurements ± standard deviation.

4.2 The disc diffusion method for *Streptococcus salivarius* for testing of growth inhibition activity against pathogens

The disc diffusion test with *Micrococcus luteus* was used for the preliminary testing of *Streptococcus salivarius* inhibition [39]. This test analyses the activity of the BLIS produced in agar and determines the activity spectrum of Sal9 producers. Briefly, the tested strain was inoculated across the surface of the Blood agar medium (BBL, Microbiology Systems, Cockeysville, USA) in a glass Petri dish (Ø 90 mm) as a 1 cm-wide streak. After incubation, the strain growth was stopped by its exposure to chloroform vapor for 30 min. The plate was then aired for 15 min before 24 hours inoculating cultures as the indicator strains across the original tested strain. The plate was incubated for 24 hours and examined for the zones of the indicator strain growth inhibition. The inhibition activity against the selected standard indicators was recorded in code form by inoculating the indicators in three triplets. The inhibition of the first member of a triplet was given a score of 4, the second a score of 2, and the third a score of 1. The absence of the inhibitory action against an indicator was scored as 0. The code was recorded as a sequence of three numbers representing the sum of each triplet. All tests were performed in duplicate, and further testing was undertaken until the consistency of the inhibition patterns was obtained [63].

5. Conclusion

It is necessary to know the composition of the dental biofilm of healthy individuals and the bacterial composition in pathological conditions to identify species responsible for disease initiation and progression. Identification of species and their characterization is essential for the selection of pathogenic, potentially pathogenic and potentially probiotic species. Blast n analysis of 16S RNA or MALDI-TOF mass spectrometry identification is perfect tools for identification of bacterial species. The ability to modulate the microbiocenosis of the dental biofilm by bacteria living together in the biofilm should be studied. The some bacteria are capable of producing bioactive substances whose presence we can quickly and easily declare with help of PCR. Sequencing and comparing of genes coding bioactive substances can uncover differences between tested bacteria isolates. Presence of these genes and prove the ability to inhibit the growth of other bacterial species are important steps in selection of potentially probiotic candidates. These bacteria are of great interest for further study and may be useful in the development of new antibacterial agents. Bioactive substances can be extracted by physical methods (centrifugation, separation and fractionation), by chemical methods (purification) and detected by modern analytical method (HPLC) or proteomic methods (MALDI-TOF MS). Next important step is declaration of activity pure extracted substance. Bioactive substances of bacterial origin can be used in dental preparations and serve as prevention or supplementary therapy of periodontal diseases. During recent years there has occurred a shift towards ecological and microbial community based approach to the therapy of oral cavity diseases. With the increasing resistance to antibiotics, the use of probiotics appears as a prospective alternative treatment or preventative measure in the control of periodontal diseases. From the clinical point of view, it is not yet possible to give direct recommendations for the use of probiotics. However, the available scientific evidence indicates that probiotic therapy is a promising approach also in the field of stomatology. The potential beneficial strains of *Streptococcus salivarius* or *Lactobacillus reuteri* and others bacterial strains isolated from many oral biofilms can be selected for next research based on their production of bioactive substances and on growth inhibition level against oral pathogenic bacteria not only in human but also in social animals like dogs and cats.

Acknowledgements

This publication was supported by the project of the Ministry of education science Research and sport of the Slovak Republic VEGA 1/0788/19: "*Study of changes in microbiota of dental biofilms in humans and dogs in order to harmonize the oral microbiocenosis by using the selected oral probiotics*" and by the Slovak Research and Development Agency under the contract no. APVV-16-0203.

Author details

Marián Maďar[1*], Jana Kačírová[1], Eva Styková[1], Michaela Maďarová[2] and Radomíra Nemcová[1]

1 University of Veterinary Medicine and Pharmacy, Košice, Slovakia

2 Institute of Animal Physiology, Center of Biosciences, Slovak Academy of Sciences, Košice, Slovakia

*Address all correspondence to: madarmarian@gmail.com

References

[1] Pontes EKU, Melo HM, Nogueira JWA, et al. Antibiofilm activity of the essential oil of citronella (*Cymbopogon nardus*) and its major component, geraniol, on the bacterial biofilms of *Staphylococcus aureus*. Food Science and Biotechnology. 2019;**28**(3):633-639. DOI: 10.1007/s10068-018-0502-2502

[2] Hicks J, Garcia-Godoy F, Flaitz C. Biological factors in dental caries: Role of remineralization and fluoride in the dynamic process of demineralization and remineralization (part 3). The Journal of Clinical Pediatric Dentistry. 2004;**28**(3):203-214

[3] Debelian GJ, Olsen I, Tronstad L. Systemic diseases caused by oral microorganisms. Endodontics & Dental Traumatology. 1994;**10**(2):57-65

[4] Karpinski TM. Role of Oral microbiota in cancer development. Microorganisms. 2019;**7**(1):20. DOI: 10.3390/microorganisms7010020

[5] Paster BJ, Boches SK, Galvin JL, et al. Bacterial diversity in human subgingival plaque. Journal of Bacteriology. 2001;**183**(12):3770-3783. DOI: 10.1128/JB.183.12.3770-3783.2001

[6] Wade WG. The oral microbiome in health and disease. Pharmacological Research. 2012;**69**(1):137-143. DOI: 10.1016/j.phrs.2012.11.006

[7] Percival S, Knottenbelt D, Cochrane C. Biofilms and Veterinary Medicine. Vol. 6. Berlin Heidelberg: Springer-Verlag; 2011

[8] Elliott DR, Wilson M, Buckley CM, et al. Cultivable oral microbiota of domestic dogs. Journal of Clinical Microbiology. 2005;**43**(11):5470-5476. DOI: 10.1128/jcm.43.11.5470-5476.2005

[9] Oh C, Lee K, Cheong Y, et al. Comparison of the oral microbiomes of canines and their owners using next-generation sequencing. PLoS One. 2015;**10**(7):e0131468. DOI: 10.1371/journal.pone.0131468

[10] Kleessen B, Bezirtzoglou E, Mättö J. Culture-based knowledge on biodiversity, development and stability of human gastrointestinal microflora. Microbial Ecology in Health and Disease. 2000;**12**(Suppl. 2):53-63

[11] Elmar Hellwig JK, Attin T. Záchovná stomatologie a parodontologie. Praha: Grada Publishing; 2003

[12] Kolenbrander PE, Andersen RN, Blehert DS, et al. Communication among oral bacteria. Microbiology and Molecular Biology Reviews. 2002;**66**(3):486-505

[13] Benitez-Paez A, Belda-Ferre P, Simon-Soro A, et al. Microbiota diversity and gene expression dynamics in human oral biofilms. BMC Genomics. 2014;**15**:311. DOI: 10.1186/1471-2164-15-311

[14] Diaz PI, Rogers AH, Zilm PS. *Fusobacterium nucleatum* supports the growth of *Porphyromonas gingivalis* in oxygenated and carbon-dioxide-depleted environments. Microbiology. 2002;**148**(Pt 2):467-472. DOI: 10.1099/00221287-148-2-467

[15] Badet C, Thebaud NB. Ecology of lactobacilli in the oral cavity: A review of literature. The Open Microbiology Journal. 2008;**2**:38-48. DOI: 10.2174/1874285800802010038

[16] Struzycka I. The oral microbiome in dental caries. Polish Journal of Microbiology. 2014;**63**(2):127-135

[17] Stensson M, Koch G, Coric S, et al. Oral administration of *Lactobacillus reuteri* during the first year of life reduces caries prevalence in the primary

dentition at 9 years of age. Caries Research. 2013;**48**(2):111-117. DOI: 10.1159/000354412

[18] Konuray G, Erginkaya Z. Potential use of *Bacillus coagulans* in the food industry. Food. 2018;7(6):92. DOI: 10.3390/foods7060092

[19] Ahola AJ, Yli-Knuuttila H, Suomalainen T, et al. Short-term consumption of probiotic-containing cheese and its effect on dental caries risk factors. Archives of Oral Biology. 2002;**47**(11):799-804. DOI: 10.1016/s0003-9969(02)00112-7

[20] Caglar E, Cildir SK, Ergeneli S, et al. Salivary mutans streptococci and lactobacilli levels after ingestion of the probiotic bacterium *Lactobacillus reuteri* ATCC 55730 by straws or tablets. Acta Odontologica Scandinavica. 2006;**64**(5):314-318. DOI: 10.1080/00016350600801709

[21] Tsubura S, Mizunuma H, Ishikawa S, et al. The effect of *Bacillus subtilis* mouth rinsing in patients with periodontitis. European Journal of Clinical Microbiology & Infectious Diseases. 2009;**28**(11):1353-1356. DOI: 10.1007/s10096-009-0790-9

[22] Jin Y, Yip HK. *Supragingival calculus*: Formation and control. Critical Reviews in Oral Biology and Medicine. 2002;**13**(5):426-441

[23] Borah BM, Halter TJ, Xie B, et al. Kinetics of canine dental calculus crystallization: An in vitro study on the influence of inorganic components of canine saliva. Journal of Colloid and Interface Science. 2014;**425**:20-26. DOI: 10.1016/j.jcis.2014.03.029

[24] Rasmussen K, Nikrad J, Reilly C, et al. N-Acetyl-l-cysteine effects on multi-species oral biofilm formation and bacterial ecology. Letters in Applied Microbiology. 2015;**62**(1):30-38. DOI: 10.1111/lam.12513

[25] Anderson AC, Rothballer M, Altenburger MJ, et al. In-vivo shift of the microbiota in oral biofilm in response to frequent sucrose consumption. Scientific Reports. 2018;**8**(1):1-13. DOI: 10.1038/s41598-018-32544-6

[26] Hammami R, Fernandez B, Lacroix C, et al. Anti-infective properties of bacteriocins: An update. Cellular and Molecular Life Sciences. 2012;**70**(16):2947-2967. DOI: 10.1007/s00018-012-1202-3

[27] Plaza G, Chojniak J, Rudnicka K, et al. Detection of biosurfactants in *Bacillus* species: Genes and products identification. Journal of Applied Microbiology. 2015;**119**(4):1023-1034. DOI: 10.1111/jam.12893

[28] Burton JP, Chilcott CN, Wescombe PA, et al. Extended safety data for the oral cavity probiotic *Streptococcus salivarius* K12. Probiotics and Antimicrobial Proteins. 2010;**2**(3):135-144. DOI: 10.1007/s12602-010-9045-4

[29] Gilbreth SE, Somkuti GA. Thermophilin 110: A bacteriocin of *Streptococcus thermophilus* ST110. Current Microbiology. 2005;**51**(3):175-182. DOI: 10.1007/s00284-005-4540-7

[30] Rattanachaikunsopon P, Phumkhachorn P. Isolation and preliminary characterization of a bacteriocin produced by *Lactobacillus plantarum* N014 isolated from nham, a traditional Thai fermented pork. Journal of Food Protection. 2006;**69**(8):1937-1943

[31] Elayaraja S, Annamalai N, Mayavu P, et al. Production, purification and characterization of bacteriocin from *Lactobacillus murinus* AU06 and its broad antibacterial spectrum. Asian Pacific Journal of Tropical Biomedicine. 2014;**4**(1):305-311. DOI: 10.12980/APJTB.4.2014C537apjtb-04-s1-s305

[32] Salehi R, Savabi O, Kazemi M, et al. Effects of *Lactobacillus reuteri*-derived biosurfactant on the gene expression profile of essential adhesion genes (gtfB, gtfC and ftf) of *Streptococcus mutans*. Advanced Biomedical Research. 2014;**3**:169. DOI: 10.4103/2277-9175.139134

[33] Lairson LL, Henrissat B, Davies GJ, et al. Glycosyltransferases: Structures, functions, and mechanisms. Annual Review of Biochemistry. 2008;**77**:521-555. DOI: 10.1146/annurev.biochem.76.061005.092322

[34] Tieking M, Korakli M, Ehrmann MA, et al. In situ production of exopolysaccharides during sourdough fermentation by cereal and intestinal isolates of lactic acid bacteria. Applied and Environmental Microbiology. 2003;**69**(2):945-952. DOI: 10.1128/aem.69.2.945-952.2003

[35] Bowen WH, Koo H. Biology of *Streptococcus mutans*-derived glucosyltransferases: Role in extracellular matrix formation of cariogenic biofilms. Caries Research. 2011;**45**(1):69-86. DOI: 10.1159/000324598

[36] Ruiz FO, Gerbaldo G, Garcia MJ, et al. Synergistic effect between two bacteriocin-like inhibitory substances produced by lactobacilli strains with inhibitory activity for *Streptococcus agalactiae*. Current Microbiology. 2012;**64**(4):349-356. DOI: 10.1007/s00284-011-0077-0

[37] Woodruff WA, Novak J, Caufield PW. Sequence analysis of mutA and mutM genes involved in the biosynthesis of the lantibiotic mutacin II in *Streptococcus mutans*. Gene. 1998;**206**(1):37-43

[38] Wescombe PA, Upton M, Dierksen KP, et al. Production of the lantibiotic salivaricin A and its variants by oral streptococci and use of a specific induction assay to detect their presence in human saliva. Applied and Environmental Microbiology. 2006;**72**(2):1459-1466. DOI: 10.1128/aem.72.2.1459-1466.2006

[39] Barbour A, Philip K, Muniandy S. Enhanced production, purification, characterization and mechanism of action of salivaricin 9 lantibiotic produced by *Streptococcus salivarius* NU10. PLoS One. 2013;**8**(10):e77751. DOI: 10.1371/journal.pone.0077751

[40] O'Shea EF, Gardiner GE, O'Connor PM, et al. Characterization of enterocin- and salivaricin-producing lactic acid bacteria from the mammalian gastrointestinal tract. FEMS Microbiology Letters. 2009;**291**(1):24-34. DOI: 10.1111/j.1574-6968.2008.01427.x

[41] Kinova Sepova H, Bilkova A. Isolation and identification of new lactobacilli from goatling stomach and investigation of reuterin production in *Lactobacillus reuteri* strains. Folia Microbiologia. 2012;**58**(1):33-38. DOI: 10.1007/s12223-012-0166-x

[42] Suwanjinda D, Eames C, Panbangred W. Screening of lactic acid bacteria for bacteriocins by microbiological and PCR methods. Biochemistry and Molecular Biology Education. 2007;**35**(5):364-369. DOI: 10.1002/bmb.84

[43] Stephens SK, Floriano B, Cathcart DP, et al. Molecular analysis of the locus responsible for production of plantaricin S, a two-peptide bacteriocin produced by *Lactobacillus plantarum* LPCO10. Applied and Environmental Microbiology. 1998;**64**(5):1871-1877

[44] Kamiya RU, Napimoga MH, Hofling JF, et al. Frequency of four different mutacin genes in *Streptococcus mutans* genotypes isolated from caries-free and caries-active individuals. Journal of Medical Microbiology. 2005;**54**(6):599-604. DOI: 10.1099/jmm.0.45870-0

[45] Sutyak KE, Wirawan RE, Aroutcheva AA, et al. Isolation of the *Bacillus subtilis* antimicrobial peptide subtilosin from the dairy product-derived *Bacillus amyloliquefaciens*. Journal of Applied Microbiology. 2008;**104**(4):1067-1074. DOI: 10.1111/j.1365-2672.2007.03626.x

[46] Chervinets Y, Chervinets V, Shenderov B, et al. Adaptation and probiotic potential of lactobacilli, isolated from the oral cavity and intestines of healthy people. Probiotics and Antimicrobial Proteins. 2017;**10**(1):22-33. DOI: 10.1007/s12602-017-9348-9

[47] Elshikh M, Marchant R, Banat IM. Biosurfactants: promising bioactive molecules for oral-related health applications. FEMS Microbiology Letters. 2016;**363**(18):1-7, fnw213

[48] van Hoogmoed CG, van Der Kuijl-Booij M, van Der Mei HC, et al. Inhibition of *Streptococcus mutans* NS adhesion to glass with and without a salivary conditioning film by biosurfactant-releasing *Streptococcus mitis* strains. Applied and Environmental Microbiology. 2000;**66**(2):659-663. DOI: 10.1128/aem.66.2.659-663.2000

[49] Ciandrini E, Campana R, Casettari L, et al. Characterization of biosurfactants produced by *Lactobacillus* spp. and their activity against oral streptococci biofilm. Applied Microbiology and Biotechnology. 2016;**100**(15):6767-6777. DOI: 10.1007/s00253-016-7531-7

[50] Satpute SK, Mone NS, Das P, et al. Inhibition of pathogenic bacterial biofilms on PDMS based implants by *L. acidophilus* derived biosurfactant. BMC Microbiology. 2019;**19**(1):39. DOI: 10.1186/s12866-019-1412-z

[51] Cochis A, Fracchia L, Martinotti MG, et al. Biosurfactants prevent in vitro *Candida albicans* biofilm formation on resins and silicon materials for prosthetic devices. Oral Surgery, Oral Medicine, Oral Pathology, Oral Radiology. 2012;**113**(6):755-761. DOI: 10.1016/j.oooo.2011.11.004

[52] Merghni A, Dallel I, Noumi E, et al. Antioxidant and antiproliferative potential of biosurfactants isolated from *Lactobacillus casei* and their anti-biofilm effect in oral *Staphylococcus aureus* strains. Microbial Pathogenesis. 2017;**104**:84-89. DOI: 10.1016/j.micpath.2017.01.017

[53] Bouassida M, Fourati N, Krichen F, et al. Potential application of *Bacillus subtilis* SPB1 lipopeptides in toothpaste formulation. Journal of Advanced Research. 2017;**8**(4):425-433. DOI: 10.1016/j.jare.2017.04.002

[54] Ghribi D, Abdelkefi-Mesrati L, Mnif I, et al. Investigation of antimicrobial activity and statistical optimization of *Bacillus subtilis* SPB1 biosurfactant production in solid-state fermentation. Journal of Biomedicine & Biotechnology. 2012;**2012**:373682. DOI: 10.1155/2012/373682

[55] Chung S, Kong H, Buyer JS, et al. Isolation and partial characterization of *Bacillus subtilis* ME488 for suppression of soilborne pathogens of cucumber and pepper. Applied Microbiology and Biotechnology. 2008;**80**(1):115-123. DOI: 10.1007/s00253-008-1520-4

[56] Gudina EJ, Teixeira JA, Rodrigues LR. Isolation and functional characterization of a biosurfactant produced by *Lactobacillus paracasei*. Colloids and Surfaces. B, Biointerfaces. 2009;**76**(1):298-304. DOI: 10.1016/j.colsurfb.2009.11.008

[57] Schwab C, Walter J, Tannock GW, et al. Sucrose utilization and impact of sucrose on glycosyltransferase expression in *Lactobacillus reuteri*. Systematic and Applied Microbiology.

2007;**30**(6):433-443. DOI: 10.1016/j.syapm.2007.03.007

[58] Argimon S, Alekseyenko AV, DeSalle R, et al. Phylogenetic analysis of glucosyltransferases and implications for the coevolution of mutans streptococci with their mammalian hosts. PLoS One. 2013;**8**(2):e56305. DOI: 10.1371/journal.pone.0056305

[59] Kingston KB, Allen DM, Jacques NA. Role of the C-terminal YG repeats of the primer-dependent streptococcal glucosyltransferase, GtfJ, in binding to dextran and mutan. Microbiology. 2002;**148**(2):549-558. DOI: 10.1099/00221287-148-2-549

[60] Hoshino T, Kawaguchi M, Shimizu N, et al. PCR detection and identification of oral streptococci in saliva samples using gtf genes. Diagnostic Microbiology and Infectious Disease. 2004;**48**(3):195-199. DOI: 10.1016/j.diagmicrobio.2003.10.002

[61] Al-Ahmad A, Auschill TM, Braun G, et al. Overestimation of *Streptococcus mutans* prevalence by nested PCR detection of the 16S rRNA gene. Journal of Medical Microbiology. 2006;**55**(1):109-113. DOI: 10.1111/j.1365-2672.2005.02638.x

[62] Tieking M, Kaditzky S, Valcheva R, et al. Extracellular homopolysaccharides and oligosaccharides from intestinal lactobacilli. Journal of Applied Microbiology. 2005;**99**(3):692-702. DOI: 10.1111/j.1365-2672.2005.02638.x

[63] Wescombe PA, Upton M, Renault P, et al. Salivaricin 9, a new lantibiotic produced by *Streptococcus salivarius*. Microbiology. 2011;**157**(5):1290-1299. DOI: 10.1099/mic.0.044719-0

Chapter 13

Composition, Structure, and Formation of Biofilms Constituted by Periodontopathogenic Microorganisms

Juliana Cabrini Carmello, Sarah Raquel de Annunzio and Carla Raquel Fontana

Abstract

Microorganisms that compose the oral microbiota maintain complex interactions with each other, especially pathogens related to periodontal disease. It is possible to characterize the etiology of this multifactorial and polymicrobial disease by the accumulation of biofilms formed in the supra- and subgingival environments associated to the immunological response and the susceptibility of the host, being responsible for a large part of the dental loss especially in the adult phase. Periodontal treatment has been carried out mainly by scaling and root planing. This therapy is limited due to the difficult access in some areas of the teeth, impairing the removal of biofilms. So, this chapter will focus on the composition and formation of the biofilm as well as the host's immune response to periodontopathogenic microorganisms. Additionally, the therapeutic challenges and the treatments that are currently being studied in order to eliminate this biofilm, such as antimicrobial phototherapy, will be discussed.

Keywords: bacterial biofilm, periodontal diseases, oral infections, phototherapy, photodynamic therapy

1. Introduction

The human oral microbiota are composed of a wide variety of microorganisms, among the various species of bacteria, fungi, viruses, and protozoa, which live in commensalism, without cause damage to the host [1, 2]. Alterations in the microbial composition due to changes in the environmental conditions or decrease of the host immunity may lead some commensal microorganisms, for instance, *Streptococcus* sp., *Fusobacterium* sp., *Porphyromonas* sp., *Eimeria* sp., *Haemophilus* sp., *Lactobacillus* sp., and *Staphylococcus* sp., to act as opportunists causing infections such as periodontal diseases [2].

Periodontal diseases affect a large part of the population, being one of the main causes of tooth loss in humans [3]. This infection is dependent on the result of the interaction of bacteria with different virulence, present in the dental biofilm, with factors that modify the host immunoinflammatory response [3]. The dental biofilm is a highly organized structure of microorganisms, in which the microbial species are connected to each other, embedded into an extracellular polymeric matrix forming a highly protective system for the resident species [2, 4, 5].

This infection is not just a local phenomenon, since the microorganisms can penetrate the bloodstream and colonize other niches of the human body, causing bacteremia. Bacteremia is common in individuals who have oral infections, especially in patients with deficient immune systems [2]. Additionally, it has been suggested that there is a relationship between periodontal pathogens and the onset of pulmonary, cardiovascular, diabetes, rheumatoid arthritis, and gestational complications [6–9].

For the treatment of periodontal disease, mechanical removal of the biofilm has been performed as well as the use of antibiotics and antiseptics for bacterial decontamination or as adjuvants to the mechanical removal of the subgingival and supragingival plaque [10]. However, the reinfection occurs very often, and the control of the inflammatory response is difficult. In some individuals, the inflammatory response may reflect a systemic dysregulation, and thus, the resolution of inflammation is impaired using conventional treatment [11]. In this context, phototherapy has been considered as an alternative to antimicrobial agents, such as antibiotics, to suppress subgingival bacterial species and to act as an adjuvant to the conventional treatments to combat periodontal disease.

It is believed that the future in healthcare is to search more efficient treatment alternatives that reduce operating time by improving the final result, eliminating the side effects of the treatment. Thus, the expectancy regarding the application of phototherapy for the treatment of bacterial infections is high, since this therapy has been effective in eliminating the microorganisms present in biofilms without causing systemic side effects to the host tissues.

2. Periodontal disease: classification, epidemiology, and etiology

Recently, a new classification for periodontal diseases has been suggested [12]. In general, the gingivitis can be defined as gingival inflammation caused by bacterial biofilm. Periodontitis includes gingival inflammation accompanied by bone loss and is classified into three different forms: necrotizing periodontitis, periodontitis as a manifestation of systemic disease, and periodontitis. In the last one, periodontitis is classified as "chronic" and "aggressive" [12].

Periodontal disease has been considered multifactorial, episodic, and site-dependent in nature [13–16]. Despite being a multifactorial infection, over the years several studies have demonstrated the importance of microorganisms in the installation and progression of the disease [17–21]. It has been estimated that the presence of plaque and gingivitis is very prevalent in humans, affecting more than 90% of the adult individuals. However, the same cannot be said for periodontitis where, despite the abundance of plaque in most people, the prevalence of periodontitis is relatively low, affecting about 20% of the individuals [22].

In periodontal pockets, the location or distribution of pathogens may be related to periodontal destruction. Noiri et al. [23] reported the presence of *Prevotella nigrescens* in the middle portion of periodontal pockets (epithelial tissue) and the presence of *Fusobacterium nucleatum* and *Treponema denticola* (in areas of non-adherent plaque), related to areas of adhered plaque and *Aggregatibacter actinomycetemcomitans*, in the apical region of the pockets. According to Slots [17] regarding the presence of bacteria in the periodontal pockets, 89.5% were obligatory anaerobic, and 74.9% were Gram-negative. Of all Gram-positive bacilli, 78.4% (deep pockets) and 19.9% (healthy groove) were anaerobic. It can be hypothesized that gingival inflammation initiated by the supragingival plaque may produce favorable environmental conditions for the colonization of Gram-negative bacteria [17].

In 1988, Socransky and Haffagee [18] reported that destructive periodontal disease depends on the compatible nature of the host or beneficial species colonizing the gingival margin that favors the colonization of other species. Combination of *F. nucleatum*, *Tannerella forsythia* and *Wolinella recta* or *Bacteroides gingivalis*, *Bacteroides intermedius*, and *Staphylococcus intermedius* were associated with sites with greater insertion loss and deep pockets. Clusters of *Veillonella parvula* and *Actinomyces* sp. or combinations of *Streptococcus sanguis* II, *Streptococcus mitis*, *V. parvula*, and *S. intermedius* were associated with sites of lower disease activity and responded more favorably to therapy. Kamma et al. [24] reported that 93.6% of the collected sites presented probing bleeding, and 23.5% were positive for suppuration. *Prevotella intermedia*/*P. nigrescens*, *Porphyromonas gingivalis*, and *Campylobacter rectus* were detected in 77.3–85.9% of the samples using culture methods and in 85.6–91.3% using immunofluorescence. *Peptostreptococcus micros* and *A. actinomycetemcomitans* were found respectively in 63.3 and 25.0% of all sites using culture method and in 58.7 and 27.7% of sites using immunofluorescence. *P. gingivalis*, *T. forsythia*, *P. intermedia*/*P. nigrescens*, and *C. rectus* were observed in 62.1% of the tested sites and 89.4% of the studied patients. The sensitivity found for immunofluorescence of *T. forsythia*, *C. rectus*, *P. intermedia*/*P. nigrescens*, and *P. gingivalis* was high (0.99–0.94) using culture as a reference detection method. The agreement between culture and immunofluorescence in detecting the presence or absence of the investigated species was 85.2–88.1% *for P. gingivalis*, *P. intermedia*/*P. nigrescens*, *C. rectus*, and *T. forsythia*, 75.9% for *A. actinomycetemcomitans*, and 70.4% for *P. micros*.

Comparing the subgingival microbiota of healthy individuals with gingivitis and early periodontitis, using the culture method and DNA probes for hybridization diagnosis, it was initially observed by the culture method that *Bacteroides forsythus*, *Campylobacter rectus*, and *Selenomonas noxia* were predominant species associated with active interproximal lesions. *Actinomyces naeslundii* and *Streptococcus oralis* were dominant in the colonization of active vestibular sites. *Actinomyces naeslundii*, *Campylobacter gracilis,* and *T. forsythia* (at lower levels than periodontitis) were predominant in gingivitis. Health-associated species were *Streptococcus oralis*, *Actinomyces naeslundii*, and *Actinomyces gerencseriae*. By DNA probe diagnosis, higher averages of *Bacteroides forsythus* and *Campylobacter rectus* were identified in periodontitis. *Porphyromonas gingivalis* and *A. actinomycetemcomitans* were detected less frequently in the studied subjects [25].

It has been reported that the microbiota may also vary depending on the teeth involved [26]. Evaluating the microbiota in primary teeth, Kamma et al. [26] found that *Gemella morbillorum* and *Peptostreptococcus magnus* were more frequent in incisive teeth, while *P. micros*, *Streptococcus intermedius*, *Bacteroides forsythus* (*T. forsythia*), *Fusobacterium nucleatum*, *Prevotella loeschei*, *Prevotella melaninogenica*, and *Selenomonas sputigena* were more frequent. The bacterial species *Streptococcus constellatus*, *P. micros*, *Pseudoramibacter alactolyticus*, *Eikenella corrodens*, and *F. nucleatum* were associated with non-blooded sites, while *S. intermedius*, *Campylobacter concisus*, *P. intermedia*, and *Prevotella loescheii* were more frequently found at sites with bleeding [26].

Some authors define that the pathogenesis of periodontitis involves anaerobic bacteria in the oral cavity and that tissue damage occurs as a result of complex bacterial pathogenic interaction and the host's immunoinflammatory response to infection [27–30]. Additionally, although each microorganism has an important role, it is believed that Gram-negative anaerobic rods (*A. actinomycetemcomitans*,

P. gingivalis, *P. intermedia*, *Bacteroides forsythus*, *C. rectus*, *Eubacterium nodatum*, *P. micros*, *S. intermedius*, and *Treponema* sp.), mobile rods, and spirochetes are mainly responsible for causing periodontal disease [31].

As the periodontal diseases are mixed with synergistic infections, it is difficult to determine the role played by a particular species. Studies have shown the relationship of *A. actinomycetemcomitans* with localized aggressive periodontitis and its association with *F. nucleatum*, *P. gingivalis*, *T. forsythia*, and *T. denticola* in chronic periodontitis. *Tannerella forsythia* also shows a remarkable ability to stay in periodontal sites undergoing mechanical or antimicrobial treatment and, because of this feature, is associated with refractory periodontitis [30, 32, 33]. Colombo et al. [34] reported that individuals with refractory periodontitis had a significantly higher frequency of periodontopathogens, such as *Parvimonas micra* (previously *Peptostreptococcus micros* or *Micromonas micros*), *Campylobacter gracilis*, *Eubacterium nodatum*, *Selenomonas noxia*, *Tannerella forsythia*, *P. gingivalis*, *Prevotella* sp., and *Eikenella corrodens*. In addition to these species, some unusual were also identified: *Pseudoramibacter alactolyticus*, *TM7* sp. *[OT] 346/356*, *Bacteroidetes* sp. *OT 272/274*, *Solobacterium moorei*, *Desulfobulbus* sp. *OT 041*, *Brevundimonas diminuta*, *Sphaerocytophaga* sp. *OT 337*, *Shuttleworthia satelles*, *Filifactor alocis*, *Dialister invisus/pneumosintes*, *Granulicatella adiacens*, *Mogibacterium timidum*, *Veillonella atypica*, and *Mycoplasma salivarium*. Accordingly, increased proportions of *P. gingivalis*, *Bacteroides forsythus*, *Prevotella*, *Fusobacterium*, *Campylobacter*, and *Treponema* species were more prevalent in supra- and subgingival samples from individuals with periodontitis [20].

3. Bacterial plaque: biofilm structure, composition, and formation

The positive association of bacterial plaque (biofilm) accumulation and periodontal tissue inflammation was evidenced in 1965 by Loe et al. [35] establishing the theory of the "nonspecific plaque hypothesis." This theory related gingival inflammation and periodontal destruction from an accumulation of nonspecific microorganisms on the gingival margin. However, later Loe et al. [36] observed that some individuals did not have periodontal disease despite having a large accumulation of gingival plaque, contradicting the "nonspecific plaque hypothesis." Thus, the "hypothesis of specific plaque" emerged, which associates the progression of the disease with the microbial composition. However, this hypothesis did not justify cases in which periodontopathogens were found in places where the disease was not detected or cases in which periodontal disease was diagnosed but microorganisms were not found [37].

In the early 1990s, a new hypothesis called the "ecological plate hypothesis" was described [38]. This hypothesis proposes that the development of gingivitis occurs due to nonspecific plaque accumulation that causes inflammation in the gingival tissues, causing changes in the gingival sulcus environment that make it an environment conducive to the development of Gram-negative bacteria. These environmental changes lead to immunomodulated tissue and inflammatory changes and tissue destruction and result in a greater predominance of periodontopathogens in this microenvironment [22]. This hypothesis corroborates the current concept that the cause of periodontal disease may depend on the host's environmental and immunological factors and not on a particular microorganism or plaque buildup [39]. This concept led researchers to gain a greater understanding of the pathogenesis of periodontal disease [22].

Biofilms that are formed on tooth surfaces and epithelial cells lining the periodontal/gingival sulcus are among the most complex and diverse biofilms formed by up to 800 different species described so far [40]. It has been reported in the literature that Gram-negative anaerobic bacteria are generally related to periodontal disease. However, facultative anaerobic Gram-positive bacteria are considered beneficial for periodontal health, such as *Streptococcus sanguinis*, which has the ability to produce hydrogen peroxide, which is cytotoxic to *A. actinomycetemcomitans*, a periodontopathogen that is already established in the literature [41, 42].

Bacteria organized in biofilms form microcolonies surrounded by a matrix consisting of extracellular polysaccharides and glycoproteins. This matrix gives protection to bacterial cells and can make these microorganisms up to 1500 times more resistant to antimicrobial treatments in the oral cavity compared to planktonic bacteria [43]. In addition, biofilms are permeated by circulatory channels (which allow the entry and exit of nutrients, metabolites, and residues) and have a mechanism of communication between bacteria called quorum sensing [44]. From this mechanism it is possible to coordinate the bacterial behavior in relation to the environment, being able to regulate the expression of specialized genes according to the population density and to intervene in physiological processes such as the induction of virulence factors [45, 46].

The diversity among the bacterial population in biofilms is due to the existence of microenvironments that present variations in chemical and metabolite concentrations and pH values, so that species with varied metabolic needs can survive [47, 48]. This variety of bacteria present in biofilm ensures that polymicrobial infections caused by dental plaque formed are more difficult to control and makes identifying one or more specific organisms that may be responsible for the infection more difficult [49].

The periodontal biofilm is constantly formed in the supragingival region, and if not removed within 2–4 days, the volume formed will cause this plaque to extend below the gingival margin and into the groove. In a healthy furrow, the number of bacteria found is approximately 10^3; however, in a deep pocket this number can range from 10^8 to 10^{10} [37, 50].

In the process of biofilm formation, subsequent layers of microorganisms bind to existing bacteria through coaggregation. This coaggregation will only occur if these microorganisms share characteristics and/or symbiotic relationships as with the bacteria *T. denticola* and *P. gingivalis*. From the fermentation of amino acids present in the *T. denticola*, gingival plaque produces succinate which is used by *P. gingivalis*, which produces fatty acids which can contribute to *T. denticola* growth [42].

As the bacterial population increases in the biofilm due to the addition of more layers, oxygen runs out making it an environment conducive to anaerobic bacterial colonization [39, 48, 51].

Until the late 1980s, the diagnostic methods used up to now, such as bacterial culture, have not been able to detect and quantify periodontopathogens of subgingival biofilms, given that in this biofilm there are anaerobic bacteria that need adequate growth conditions, besides the difficulty in cultivating the microorganisms that were smaller in the periodontal biofilm samples, preventing the identification and characterization of this biofilm [51]. In this context, in 1998 Dr. Sigmund Socransky described the technique called checkerboard DNA–DNA hybridization for microbiological diagnosis using deoxyribonucleic acid (DNA) probes. From this technique it was possible to develop researches that would improve the knowledge of the periodontal disease microbiota, making it possible to evaluate a large number of samples and microorganisms present in the oral cavity [52].

In this study, Socransky and Haffagee [52] grouped the bacteria in the samples into six complexes named by different colors: red, orange, yellow, green, purple, and blue complex. **Table 1** describes the bacterial species that are part of each

Complex	Bacteria
Red	*Porphyromonas gingivalis*
	Tenarella forsythia
	Treponema denticola
Orange	*Fusobacterium nucleatum*
	Fusobacterium periodonticum
	Prevotella intermedia
	Prevotella nigrescens
	Parvimonas micra
	Campylobacter rectus
	Eubacterium nodatum
	Campylobacter gracilis
	Canpylobacter showae
	Fusobacterium nucleatum ssp. *vicentii*
	Fusobacterium nucleatum ssp. *polimorphum*
	Streptococcus constellatus
Green	*Capnocytophaga sputigena*
	Capnocytophaga gingivalis
	Capnocytophaga ochracea
	Eikenella corrodens
	Aggregatibacter actinomycetemcomitans
Yellow	*Streptococcus gordoni*
	Streptococcus mitis
	Streptococcus sanguinis
	Streptococcus oralis
	Streptococcus intermedius
Purple	*Actinomyces odontolyticus*
	Veillonella parvula
Blue	*Actinomyces gerencseriae*
	Actinomyces naeslundi
	Actinomyces israelli

Table 1.
Representation of bacteria divided into complexes established in the study by Socransky and Haffagee [52].

complex. Bacteria that were grouped in the red complex are considered as etiological agents of chronic periodontitis and related to gingival bleeding and increased pocket depth. The bacteria under the complex named orange, which proceeds the installation of the red complex and its constituents, are considered possible periodontal pathogens. The complexes named green, yellow, purple, and blue are integrated by bacteria that colonize the dental surface in the early stages of biofilm formation and are compatible with periodontal health. However, these complexes provide receptors and provide an ecosystem conducive to the emergence of bacteria present in the orange complex and in turn the red complex, which are in fact related to the pathogenesis of periodontal disease.

4. Host immune response to pathogenic microorganisms

The main periodontopathogens, such as *A. actinomycetemcomitans*, *P. gingivalis*, *T. denticola*, and *T. forsythia*, have important proteolytic and exopeptidase activity, which have trypsin-like activity. In *T. denticola* these proteases behave like chymotrypsin-like serine proteases and are responsible for the invasion of this microorganism into tissues. Moreover, they play an important role in the development of necrosis in periodontal disease and amino acid fermentation by releasing ammonia, hydrogen sulfide, methyl mercaptans, and highly toxic fatty acids, which exert direct cytotoxic activity and reduce the speed of tissue repair [53–55]. *A. actinomycetemcomitans* is capable of producing an active thermolabile leukotoxin on neutrophils, monocytes, macrophages, and T lymphocytes, producing degranulation of these cells, with subsequent tissue disorganization and local immunosuppression [56, 57].

A specific bacterial etiology for the development of periodontitis from longitudinal studies with individuals infected with *A. actinomycetemcomitans* has been suggested [58]. A cohort study of 96 students included a test group of 38 students positive for *A. actinomycetemcomitans* and 58 healthy controls for this bacterium. The patients were studied longitudinally for 2–3 years. During the study period, 7 of the 37 individuals that are actinomycete-positive (i.e., 18%) developed bone loss compared to none of the *A. actinomycetemcomitans*-negative subjects. The authors suggested that *A. actinomycetemcomitans* is a significant risk marker for the development of aggressive periodontitis [58].

The interaction between the host and the microorganisms is clearly responsible for the development of gingivitis injury. With regard to periodontitis, it can be argued that the specific bacteria observed so far are present as a result of the disease, but not necessarily caused the disease. This argument is no different from the most mucosal bacterial biofilm infections in which the relationship between disease and inflammation is not clear. What comes first: host response or change in biofilm microorganisms? [59].

Although many studies evaluate the subgingival microbiota of healthy and diseased periodontal sites, further investigations are needed to fully understand these infections and host-pathogen interaction and to study new treatment options for this disease. One such approach is the phototherapy or photodynamic therapy described below.

5. Conventional treatments and therapeutic challenges

The treatment of periodontal disease is focused on the elimination of biofilm and calculus and the prevention of its formation. As a conventional treatment, scaling and root planing (SRP) is performed by removing plaque accumulation and calculating below the gingival margin, preventing disease progression and bacterial recolonization on the tooth surface [60].

This treatment has caused a decrease in pathogens, considering that after this procedure, it was reported in the literature that the bacterial load of *T. denticola* and *P. gingivalis* was reduced after 1 year of SRP. In addition, this treatment has other benefits, such as the gain in clinical insertion level and reduction of periodontal pocket depth [61, 62].

However, this procedure is limited due to the technical difficulty in removing biofilms located in hard to reach areas, such as very deep periodontal pockets, root concavities, bifurcations, and large invaginations. Additionally, a possible relapse may occur as some periodontopathogens such as *A. actinomycetemcomitans* and

P. gingivalis can invade the tissue, so the persistence of these bacteria on the root surface can cause recolonization in sites that have already been treated [39].

In order to optimize the effects of SRP treatment, protocols have been proposed to associate systemic or local antibiotics to eliminate persistent bacteria after the SRP procedure. Studies have shown that this association provides improvement in the patient's clinical condition [63]. The main antibiotics commonly used in the treatment of periodontal disease are amoxicillin, metronidazole, clindamycin, azithromycin, ciprofloxacin, doxycycline, and minocycline [64]. However, the use of these drugs as an adjuvant to this disease has limitations, such as the emergence of bacteria resistant to these antibiotics as well as the side effects caused by these antimicrobial agents, such as diarrhea and vaginal candidiasis, which result from the commensal microbiota imbalance. In addition, drug interaction may occur between antibiotics and other drugs being used by patients, resulting in ineffectiveness or other adverse effects [49, 65].

Thus, in recent years, in the area of dentistry, promising antimicrobial adjuvant therapies have been studied, such as phototherapy and photodynamic therapy [66, 67].

5.1 Phototherapy and photodynamic therapy

Studies have shown that some bacteria related to periodontal disease have the ability to produce a photosensitive substance intrinsically, such as protoporphyrin IX. Even without the addition of a photosensitizing drug, pigmented bacteria have been more susceptible when applied to phototherapy [66, 68]. Photosensitizers are molecules that when irradiated by a light source at a suitable wavelength undergo photochemical reactions to emit fluorescence. This process is used by photodynamic therapy to produce reactive oxygen species [69, 70]. Most bacteria do not have endogenous photosensitive compounds. Thus, cells lacking these compounds may become susceptible to light when an exogenous photosensitizing molecule is added [71, 72].

The mechanism of action of photodynamic therapy happens when the photosensitive substance (intrinsic or extrinsic) is activated when irradiation is applied by a light source compatible with the length of the substance. This process will form reactive oxygen species, such as hydrogen peroxide, hydroxyl radical, and singlet oxygen, causing the death of the bacteria. This interaction of light and photosensitizer can occur through two types of reactions, called type I and II. In the type I reaction, charge transfer occurs between the photosensitizer and biomolecules, resulting in radicals and radical ions that react with molecular oxygen, forming reactive oxygen species. During the type II reaction, the excited triplet state photosensitizer transfers energy directly to the fundamental triplet state oxygen, forming singlet oxygen [69, 73, 74].

Studies involving photodynamic therapy and periodontal disease have investigated different light sources such as light-emitting diodes, low-power lasers, and conventional light [75–79]. As for photosensitizers, there are several molecules studied aiming at inactivation of periodontopathogens such as poly-L-lysine-chlorin-6 conjugate and phenothiazine dyes (toluidine blue and methylene blue) [80, 81].

Photodynamic action is being increasingly studied to complement the microbial reduction achieved by conventional mechanical periodontal therapy. In vitro studies have shown that periodontopathogens have been suppressed in planktonic phase and biofilm, and after the application of photodynamic therapy, it has been verified that virulence factors of these bacteria have been decreased, such as lipopolysaccharides and proteases [82–84].Clinical trials have also shown that this therapy is effective as an adjuvant in the treatment of periodontal disease [85–87]. However, several parameters must be considered for this therapy to be successful, such as the photosensitizer used, its concentration, and the irradiation parameters. Thus, further studies should be conducted to develop clinically applied protocols.

6. Conclusions

The etiology of periodontal disease is multifactorial and directly associated with biofilm accumulation in the supra- and subgingival region, immune response and host susceptibility. In recent decades, several studies have sought to investigate the complex interactions of periodontopathogens in biofilm as well as adjuvant antimicrobial therapies that do not cause adverse effects in patients nor bacterial resistance. Phototherapy and photodynamic therapy are examples of treatments that have shown promising results in vitro and in clinical trials. Further investigations need to be done in order to establish parameters which allow the safe and efficient application of the therapy.

Conflict of interest

The authors declare no conflict of interest.

Author details

Juliana Cabrini Carmello[1]*, Sarah Raquel de Annunzio[2] and Carla Raquel Fontana[2]

1 São Paulo State University (Unesp), School of Dentistry, Department of Dental Materials and Prosthodontics, Araraquara, Brazil

2 São Paulo State University (Unesp), School of Pharmaceutical Sciences, Department of Clinical Analysis, Araraquara, Brazil

*Address all correspondence to: cabrini.juliana@gmail.com

References

[1] Flemming HC, Wingender J, Szewzyk U, Steinberg P, Rice SA, Kjelleberg S. Biofilms: An emergent form of bacterial life. Nature Reviews. Microbiology. 2016;**14**:563-575. DOI: 10.1038/nrmicro.2016.94

[2] Bowen WH, Burne RA, Wu H, Koo H. Oral biofilms: Pathogens, matrix, and polymicrobial interactions in microenvironments. Trends in Microbiology. 2018;**26**:229-242. DOI: 10.1016/j.tim.2017.09.008

[3] Socransjy SS, Haffagee AD. Periodontal microbial ecology. Periodontology 2000. 2000;**38**:135-187

[4] Watnick P, Kolter R. Biofilm, city of microbes. MINIREVIEW, Journal of Bacteriology. 2000;**182**:2675-2679

[5] Rickard AH, Gilbert P, High NJ, Kolenbrander PE, Handley PS. Bacterial coaggregation: An integral process in the development of multi-species biofilms. Trends in Microbiology. 2003;**11**:94-100

[6] Wang JT, Liu ZQ, Zhang TY, Chen Y, Zhou X, Li GX, et al. Screening of periodontal and salivary parameters in patients with frequent acute exacerbation of chronic obstructive pulmonary disease. Zhonghua Kou Qiang Yi Xue Za Zhi. 2019;**54**:410-415. DOI: 10.3760/cma.j.issn.1002-0098.2019.06.013

[7] Khumaedi AI, Purnamasari D, Wijaya IP, Soeroso Y. The relationship of diabetes, periodontitis and cardiovascular disease. Diabetes and Metabolic Syndrome: Clinical Research and Reviews. 2019;**13**:1675-1678. DOI: 10.1016/j.dsx.2019.03.023

[8] Lee YH, Lew PH, Cheah CW, Rahman MT, Baharuddin NA, Vaithilingam RD. Potential mechanisms linking periodontitis to rheumatoid arthritis. Journal of the International Academy of Periodontology. 2019;**21**:99-110

[9] Iheozor-Ejiofor Z, Middleton P, Esposito M, Glenny AM. Treating periodontal disease for preventing adverse birth outcomes in pregnant women. Cochrane Database of Systematic Reviews. 2017;**12**(6):CD005297

[10] Sanz M, Teughels W. Group a of European workshop on periodontology. Innovations in nonsurgical periodontal therapy: Consensus report of the sixth European workshop on periodontology. Journal of Clinical Periodontology. 2008;**35**(8 Suppl):3-7

[11] Müller Campanile V, Megally A, Campanile G, Gayet-Ageron A, Giannopoulou C, Mombelli A. Risk factors for recurrence of periodontal disease in patients in maintenance care in a private practice. Journal of Clinical Periodontology. 2019;**46**:918-926

[12] Caton JG, Armitage G, Berglundh T, Chapple ILC, Jepsen S, Kornman KS. A new classification scheme for periodontal and peri-implant diseases and conditions - introduction and key changes from the 1999 classification. Journal of Periodontology. 2018;**89**:S1-S8. DOI: 10.1111/jcpe.12935

[13] Caffesse R, Motta L, Morrison E. The rationale for periodontal therapy. Periodontology 2000. 1995;**2000**(9):7-13

[14] Socransky SS, Haffagee AD. Dental biofilms: Difficult therapeutic targets. Periodontology 2000. 2000;**28**:12-55

[15] Socransky SS, Haffagee AD, Smith C, Duff GW. Microbiological parameters associated with IL-1 gene polymorphisms in periodontitis patients. Journal of Clinical Periodontology. 2000;**27**:810-818

[16] Socransky SS, Haffagee AD, Smith C, Martin L, Haffagee JA, Uzel NG, et al. Use of checkerboard DNA-DNA hybridization to study complex microbial ecosystems. Oral Microbiology and Immunology. 2004;**19**:352-362. DOI: 10.1111/j.1399-302x.2004.00168.x

[17] Slots J. Microflora in the healthy gingival sulcus in man. Scandinavian Journal of Dental Research. 1977;**85**:247-254

[18] Socransky SS, Haffagee AD, Dzink JL, Hillman JD. Associations between microbial species in subgingival plaque samples. Oral Microbiology and Immunology. 1988;**3**:1-7

[19] Dahlen G. Putative periodontopathogens in "diseased" and "non-diseased" persons exhibiting poor bucal hygiene. Journal of Clinical Periodontology. 1992;**19**:35-42

[20] Ximénes-Fyvie LA, Hafajee AD, Socransk SS. Comparison of the microbiota of supra- and subgingival plaque in health and periodontitis. Journal of Clinical Periodontology. 2000;**27**:648-657

[21] Moreira AN, Caniggia LF, Ferreira RC, Verónica C, Alonso C, Piovano S. Effect of supragingival plaque control on subgingival microflora and periodontal tissues. Pesquisa Odontológica Brasileira. 2001;**15**:119-126

[22] Bartold PM, Van Dyke TE. Periodontitis: A host-mediated disruption of microbial homeostasis. Unlearning learned concepts. Periodontology 2000. 2013;**62**:203-217. DOI: 10.1111/j.1600-0757.2012.00450.x

[23] Noiri Y, Li L, Ebisu S. The localization of periodontal-disease-associated bacteria in human periodontal pockets. Journal of Dental Research. 2001;**80**:1930-1934

[24] Kamma JJ, Nakou M, Gmür R, Baehni PC. Microbiological profile of early onset/aggressive periodontitis patients. Oral Microbiology and Immunology. 2004;**19**:314-321

[25] Tanner A, Maiden MF, Macuch PJ, Murray LL, Kent RL Jr. Microbiota of health, gingivitis, and initial periodontitis. Journal of Clinical Periodontology. 1998;**25**:85-98

[26] Kamma JJ, Diamanti-Kipioti A, Nakou M, Mitsis FJ. Profile of subgingival microbiota in children with primary dentition. Journal of Periodontal Research. 2000;**35**:33-41

[27] Nishida M, Grossi SG, Dunford RG, Ho AW, Trevisan M, Genco RJ. Calcium and the risk for periodontal disease. Journal of Periodontology. 2000;**71**:1057-1066

[28] Hayashi C, Gudino CV, Gibson FC 3rd, Genco CA. Review: Pathogen-induced inflammation at sites distant from oral infection: Bacterial persistence and induction of cell-specific innate immune inflammatory pathways. Molecular Oral Microbiology. 2010;**25**:305-316. DOI: 10.1111/j.2041-1014.2010.00582.x

[29] Bakthavatchalu V, Meka A, Sathishkumar S, Lopez MC, Bhattacharyya I, Boyce BF, et al. *Tannerella forsythia* infection-induced calvarial bone and soft tissue transcriptional profiles. Molecular Oral Microbiology. 2010;**25**:317-330. DOI: 10.1111/j.2041-1014.2010.00583.x

[30] Holla LI, Hrdlickova B, Linhartova P, Fassmann A. Interferon-γ *+874A/T* polymorphism in relation to generalized chronic periodontitis and the presence of periodontopathic bacteria. Archives of Oral Biology. 2011;**56**:153-158. DOI: 10.1016/j.archoralbio.2010.09.005

[31] Lovegrove JM. Dental plaque revisited: Bacteria associated with

periodontal disease. Journal of the New Zealand Society of Periodontology. 2004;**87**:7-21

[32] Haubek D. The highly leukotoxic JP2 clone of *Aggregatibacter actinomycetemcomitans*: Evolutionary aspects, epidemiology and etiological role in aggressive periodontitis. APMIS. Supplementum. 2010;**130**:1-53. DOI: 10.1111/j.1600-0463.2010.02665.x

[33] Saygun I, Nizam N, Keskiner I, Bal V, Kubar A, Açikel C, et al. Salivary infectious agents and periodontal disease status. Journal of Periodontal Research. 2011;**46**:235-239. DOI: 10.1111/j.1600-0765.2010.01335.x

[34] Colombo APV, Teles RP, Torres MC, Rosalém JRW, Mendes MCS, Souto R, et al. Effects of non-surgical mechanical therapy on the subgingival microbiota of brazilians with untreated chronic periodontitis: 9-month results. Journal of Periodontology. 2005;**76**:778-784

[35] Loe H, Theilade E, Jensen SB. Experimental gingivitis in man. Journal of Periodontology. 1965;**36**:177. DOI: 10.1902/jop.1965.36.3.177

[36] Loe H, Aerud A, Boysen H, Morrison E. Natural history of periodontal disease in man. Rapid, moderate and no loss of attachment in Sri Lankan laborers 14 to 46 years of age. Journal of Clinical Periodontology. 1986;**13**:431-440. DOI: 10.1111/j.1600-051x.1986.tb01487.x

[37] Teughels W, Quirynen M, Jakubovics N. Periodontal microbiology. In: Newman MG, Takei HH, Klokkevold PR, et al., editors. Carranza's Clinical Periodontology. 11th ed. St Louis (MO): Elsevier; 2012. pp. 232-270

[38] Marsh PD. Microbial ecology of dental plaque and its significance in health and disease. Advances in Dental Research. 1994;**8**:263-271. DOI: 10.1177/08959374940080022001

[39] Papapanou P. Periodontal diseases: General concepts. In: Lamont RJ, Hajishengallis GN, Jenkinson HF, editors. Oral Microbiology and Immunology. 2nd ed. Washington, DC: ASM Press; 2014. pp. 251-259, 261-271

[40] Lourenço TG, Heller D, Silva-Boghossian CM, Cotton SL, Paster BJ, Colombo AP. Microbial signature profiles of periodontally healthy and diseased patients. Journal of Clinical Periodontology. 2014;**41**:1027-1036. DOI: 10.1111/jcpe.12302

[41] Hillman JD, Socransky SS, Shivers M. The relationships between *Streptococcal* species and periodontopathic bacteria in human dental plaque. Archives of Oral Biology. 1985;**30**:791. DOI: 10.1016/0003-9969(85)90133-5

[42] Lamont RJ, Hajishengallis GN, Jenkinson HF, editors. Oral Microbiology and Immunology. 2nd ed. Washington, DC: ASM Press; 2014. pp. 21-22

[43] Taraszkiewicz A, Fila G, Grinholc M, Nakonieczna J. Innovative strategies to overcome biofilm resistance. BioMed Research International. 2013:1-13. DOI: 10.1155/2013/150653

[44] Egland PG, Marquis RE. Oral microbial physiology. In: Lamont RJ, Hajishengallis GN, Jenkinson HF, editors. Oral Microbiology and Immunology. 2nd ed. Washington, DC: ASM Press; 2014. p. 113, 130, 134-138

[45] Ammor MS, Michaelidis C, Nychas GJ. Insights into the role of quorum sensing in food spoilage. Journal of food protection, Des Moines. 2008;**71**:1510-1525. DOI: 10.4315/0362-028x-71.7.1510

[46] Bai AJ, Rai VR. Bacterial quorum sensing and food industry.

Comprehensive Reviews in Food Science and Food Safety, Amsterdam. 2011;**10**:183-193. DOI: 10.1111/j.1541-4337.2011.00150.x

[47] Newman MG, Takei HH, Klokkevold PR, Carranza F, editors. Carranza's Clinical Periodontology. 11th ed. St Louis (MO): Elsevier; 2012. pp. 232-270

[48] Scannapieco FA. The oral environment. In: Lamont RJ, Hajishengallis GN, Jenkinson HF, editors. Oral Microbiology and Immunology. 2nd ed. Washington, DC: ASM Press; 2014. pp. 57-62, 66, 72

[49] Harvey JD. Periodontal microbiology. Dental Clinics of North America. 2017;**61**:253-269. DOI: 10.1016/j.cden.2016.11.005

[50] Dibart S, Skobe Z, Snapp KR, Socransky SS, Smith CM, Kent R. Identification of bacterial species on or in crevicular epithelial cells from healthy and periodontally diseased patients using DNA-DNA hybridization. Oral Microbiology and Immunology. 1998;**13**:30

[51] Gomes SC, Piccinin FB, Oppermann RV, Susin C, Nonnenmacher CI, Mutters R, et al. Periodontal status in smokers and never smokers: Clinical findings and real time polymerase chain reaction quantification of putative periodontal pathogens. Journal of Periodontology. 2006;**77**:1483-1490. DOI: 10.1902/jop.2006.060026

[52] Socransky SS, Haffajee AD, Cugini MA, Smith C, Kent RL Jr. Microbial complexes in subgingival plaque. Journal of Clinical Periodontology. 1998;**25**:134. DOI: 10.1111/j.1600-051x.1998.tb02419.x

[53] Holt S, Kesavalu L, Walker S, Genco CA. Virulence of *Porphyromonas gingivalis*. Periodontology. 2000. 1999;**20**: 168-238

[54] Holt SC, Ebersole JL. *Porphyromonas gingivalis, Treponema denticola*, and *Tannerella forsythia*: The "red complex", a prototype polybacterial pathogenic consortium in periodontitis. Periodontology 2000. 2005;**38**:72-122

[55] Brook I. The role of anaerobic bacteria in mediastinitis. Therapy in practice. Drugs. 2006;**66**:315-320. DOI: 10.2165/00003495-200666030-00004

[56] Tsai CC, Shenker BJ, Dirienzo JM, Malamud D, Taichman NS. Extraction and isolation of a leukotoxin from *Actinobacillus actinomycetemcomitans* with polymyxin B. Infection and Immunity. 1984;**43**:700-705

[57] Haubek D, Westergaard J. Detection of a highly toxic clone of *Actinobacillus actinomycetemcomitans* (JP2) in a Moroccan immigrant family with multiple cases of localized aggressive periodontitis. International Journal of Paediatric Dentistry. 2004;**14**:41-48

[58] Fine DH, Markowitz K, Furgang D, Fairlie K, Ferrandiz J, Nasri C, et al. *Aggregatibacter actinomycetemcomitans* and its relationship to initiation of localized aggressive periodontitis: Longitudinal cohort study of initially healthy adolescents. Journal of Clinical Microbiology. 2007;**45**:3859-3869. DOI: 10.1128/JCM.00653-07

[59] Dongari-Bagtzoglou A. Pathogenesis of mucosal biofilm infections: Challenges and progress. Expert Review of Anti-Infective Therapy. 2008;**6**:201-208. DOI: 10.1586/14787210.6.2.201

[60] Gomes BC, Esteves CT, Palazzo ICV, Darini ALC, Felis GE, Sechi LA, et al. Prevalence and characterization of *Enterococcus* sp. isolated from Brazilian foods. Food Microbiology. 2008;**25**:668-675. DOI: 10.1016/j.fm.2008.03.008

[61] Cugini MA, Haffajee AD, Smith C, Kent RL, Socransky SS. The effect of scaling and root planing

on the clinical and microbiological parameters of periodontal diseases: 12-month results. Journal of Clinical Periodontology. 2000;**27**:30-36. DOI: 10.1034/j.1600-051x.2000.027001030.x

[62] Simonson LG, Mcmahon KT, Childers DW, Morton HE. Bacterial synergy of *Treponema denticola* and *Porphyromonas gingivalis* in a multinational population. Oral Microbiology and Immunology. 1992;**7**:111-112

[63] Guerrero A, Echeverria JJ, Tonetti MS. Incomplete adherence to an adjunctive systemic antibiotic regimen decreases clinical outcomes in generalized aggressive periodontitis patients: A pilot retrospective study. Journal of Clinical Periodontology. 2005;**34**:897-902. DOI: 10.1111/j.1600-051X.2007.01130.x

[64] Newman MG, Takei HH, Carranza F, editors. Carranza's Clinical Periodontology. 11th ed. St Louis (MO): Elsevier; 2012. pp. 482-491

[65] Herrera D, Alonso B, Leon R, Roldan S, Sanz M. Antimicrobial therapy in periodontitis: The use of systemic antimicrobials against the subgingival biofilm. Journal of Clinical Periodontology. 2008;**35**:45-66. DOI: 10.1111/j.1600-051X.2008.01260.x

[66] Fontana CR, Song X, Polymeri A, Goodson JM, Wang X, Soukos NS. The effect of blue light on periodontal biofilm growth in vitro. Lasers in Medical Science. 2015;**30**:2077-2086. DOI: 10.1007/s10103-015-1724-7

[67] Meimandi M, Talebi Ardakani MR, Esmaeil Nejad A, Yousefnejad P, Saebi K, Tayeed MH. The effect of photodynamic therapy in the treatment of chronic periodontitis: A review of literature. Journal of Lasers in Medical Sciences. 2017;**8**(Suppl 1):S7-S11. DOI: 10.15171/jlms.2017.s2

[68] Allaker RP, Douglas CW. Novel antimicrobial therapies for dental plaque-related diseases. International Journal of Antimicrobial Agents. 2009;**33**:8-13. DOI: 10.1016/j.ijantimicag.2008.07.014

[69] Dolmans DE, Fukumura D, Jain RK. Photodynamic therapy for cancer. Nature Reviews. Cancer. 2003;**3**:380-387. DOI: 10.1038/nrc1071

[70] Nishie H, Kataoka H, Yano S, Kikuchi JI, Hayashi N, Narumi A, et al. A next-generation bifunctional photosensitizer with improved water-solubility for photodynamic therapy and diagnosis. Oncotarget. 2016;**7**:74259-74268. DOI: 10.18632/oncotarget.12366

[71] Wilson M, Dobson J, Harvey W. Sensitization of oral bacteria to killing by low power laser radiation. Current Microbiology. 1992;**25**:77-81

[72] Wilson M. Photolysis of oral bacteria and its potential use in the treatment of caries and periodontal disease. The Journal of Applied Bacteriology. 1993;**75**:299-306. DOI: 10.1111/j.1365-2672.1993.tb02780.x

[73] Kharkwal GB, Sharma SK, Huang YY, Dai T, Hamblin MR. Photodynamic therapy for infections: Clinical applications. Lasers in Surgery and Medicine. 2011;**43**:755-767. DOI: 10.1002/lsm.21080

[74] Hamblin MR. Antimicrobial photodynamic inactivation: A bright new technique to kill resistant microbes. Current Opinion in Microbiology. 2016;**33**:67-73. DOI: 10.1016/j.mib.2016.06.008

[75] Bevilacqua IM, Nicolau RA, Khouri S, Brugnera A Jr, Teodoro GR, Zangaro RA, et al. The impact of photodynamic therapy on the viability of *Streptococcus mutans* in a planktonic culture. Photomedicine and Laser Surgery. 2007;**25**:513-518. DOI: 10.1089/pho.2007.2109

[76] Soukos NS, Ximenez-Fyvie LA, Hamblin MR, Socransky SS, Hasan T. Targeted antimicrobial photochemotherapy. Antimicrobial Agents and Chemotherapy. 1998;**42**:2595-2601

[77] Wood S, Nattress B, Kirkham J, Shore R, Brookes S, Griffiths J, et al. An in vitro study of the use of photodynamic therapy for the treatment of natural oral plaque biofilms formed in vivo. Journal of Photochemistry and Photobiology. B. 1999;**50**:1-7. DOI: 10.1016/S1011-1344(99)00056-1

[78] Matevski D, Weersink R, Tenenbaum HC, Wilson B, Ellen RP, Lepine G. Lethal photosensitization of periodontal pathogens by a red-filtered xenon lamp in vitro. Journal of Periodontal Research. 2003;**38**:428-435. DOI: 10.1034/j.1600-0765.2003.00673.x

[79] Prates RA, Yamada AM Jr, Suzuki LC, Eiko Hashimoto MC, Cai S, Gouw-Soares S, et al. Bactericidal effect of malachite green and red laser on *Actinobacillus actinomycetemcomitans*. Journal of Photochemistry and Photobiology. B. 2007;**86**:70-76. DOI: 10.1016/j.jphotobiol.2006.07.010

[80] Soukos NS, Hamblin MR, Hasan T. The effect of charge on cellular uptake and phototoxicity of polylysine chlorin(e6) conjugates. Photochemistry and Photobiology. 1997;**65**:723-729. DOI: 10.1111/j.1751-1097.1997.tb01916.x

[81] Jori G, Fabris C, Soncin M, Ferro S, Coppellotti O, Dei D, et al. Photodynamic therapy in the treatment of microbial infections: Basic principles and perspective applications. Lasers in Surgery and Medicine. 2006;**38**:468-481. DOI: 10.1002/lsm.20361

[82] Qin Y, Luan X, Bi L, He G, Bai X, Zhou C, et al. Toluidine blue-mediated photoinactivation of periodontal pathogens from supragingival plaques. Lasers in Medical Science. 2008;**23**:49-54. DOI: 10.1007/s10103-007-0454-x

[83] Pfitzner A, Sigusch BW, Albrecht V, Glockmann E. Killing of periodontopathogenic bacteria by photodynamic therapy. Journal of Periodontology. 2004;**75**:1343-1349. DOI: 10.1902/jop.2004.75.10.1343

[84] Komerik N, Wilson M, Poole S. The effect of photodynamic action on two virulence factors of gram-negative bacteria. Photochemistry and Photobiology. 2000;**72**:676-680. DOI: 10.1562/0031-8655(2000)072<0676:teopao>2.0.co;2

[85] Braun A, Dehn C, Krause F, Jepsen S. Short-term clinical effects of adjunctive antimicrobial photodynamic therapy in periodontal treatment: A randomized clinical trial. Journal of Clinical Periodontology. 2008;**35**:877-884. DOI: 10.1111/j.1600-051X.2008.01303.x

[86] Christodoulides N, Nikolidakis D, Chondros P, Becker J, Schwarz F, Rossler R, et al. Photodynamic therapy as an adjunct to non-surgical periodontal treatment: A randomized, controlled clinical trial. Journal of Periodontology. 2008;**79**:1638-1644. DOI: 10.1902/jop.2008.070652

[87] Andersen R, Loebel N, Hammond D, Wilson M. Treatment of periodontal disease by photodisinfection compared to scaling and root planing. The Journal of Clinical Dentistry. 2007;**18**:34-38

Chapter 14

Biofilms Formed by Pathogens in Food and Food Processing Environments

Leontina Grigore-Gurgu, Florentina Ionela Bucur, Daniela Borda, Elena-Alexandra Alexa, Corina Neagu and Anca Ioana Nicolau

Abstract

This chapter presents the ability of some pathogenic (*Listeria monocytogenes*, *Escherichia coli*, *Salmonella enterica*, *Campylobacter jejuni*, *Pseudomonas aeruginosa*) and toxigenic bacteria (*Bacillus cereus*, *Staphylococcus aureus*) to form biofilms and contribute to the persistence of these microorganisms in the food industry. Particularities regarding attachment and composition of biofilms formed in food and food processing environments are presented and genes involved in biofilm production are mentioned. To give a perspective on how to fight against biofilms with new means, nonconventional methods based on bacteriocins, bacteriophages, disruptive enzymes, essential oils, nanoemulsions and nanoparticles, and use of alternative technologies (cold plasma, ultrasounds, light-assisted technologies, pulsed electric field, and high pressure processing) are shortly described.

Keywords: bacteriocin, essential oils, bacteriophages, nanoemulsion, alternative technologies

1. Introduction

Food matrices having water activities above 0.9 and wet food processing environments are wonderlands for microorganism multiplication and biofilm development. Biofilms are considered of great concern in regard to functioning of mechanical parts that may be blocked, to energy consumption, which becomes higher when heat transfer decreases, and to corrosion as corrosion rate of surfaces increases underneath biofilms (corrosion grows 10–1000 times faster causing loss of material and increasing porosity) but their presence in food and food processing environments is also a serious public health risk due to problems associated with foodborne illnesses and food spoilage [1].

The biofilms that are threatening the safety of food products are produced by some pathogenic bacteria such as *Listeria monocytogenes*, *Escherichia coli*, *Salmonella enterica*, *Campylobacter jejuni*, and *Pseudomonas aeruginosa* and toxigenic bacteria such as *Staphylococcus aureus* and *Bacillus cereus* [2]. Biofilms are responsible for persistence of such bacteria in food processing environments and (re)contamination of processed foods [3]. When contamination of food products happens, recalls are necessary. These actions present large economic burden to industry and are also associated with brand damage.

2. Biofilm formation

Biofilms are formed on all types of surfaces existing in food plants ranging from plastic, glass, metal, cement, to wood and food products [4]. Usually, biofilms form a monolayer or more often multilayers, in which bacteria may undergo a significant change in physiology with an increased tolerance to environmental stresses [5].

L. monocytogenes, the pathogen that proliferates at low temperatures, is able either to form pure culture biofilms or to grow in multispecies biofilms [6]. Prevalent strains in food processing environments have good adhesion ability due to the presence of flagella, pili, and membrane proteins [7]. Composition of biofilms produced by *L. monocytogenes* is different in comparison with that produced by other bacteria. For example, exopolysaccharides like alginate in Pseudomonas or poly-N-acetylglucosamine in *Staphylo-coccus* have not been put into evidence [8].

Salmonella spp. express proteinaceous extracellular fibers called curli that are involved in surface and cell-cell contacts and promotion of community behavior and host colonization [9]. Besides curli, different fimbrial adhesins have been identified to have implications in biofilm formation, dependent of serotype. The presence of cellulose in the biofilm matrix contributes to cells' resistance to mechanical forces and improved adhesion to abiotic surfaces [10]. Significant differences between serovars were put into evidence regarding biofilm formation the most persistent in food processing environments being the ones that are capable to form biofilms [11].

Flagella, pili, and membrane proteins are also used by *E. coli* to initiate attachment on inanimate surfaces. Flagella are lost after attachment and bacteria start producing an extracellular polymeric substance (EPS) that provides a better resistance of bacteria to disinfectants as hypochlorite [12]. Similarities in biofilm structure and composition as well as regulatory mechanisms with *Salmonella* spp. have been demonstrated for *E. coli*, mostly in terms of expression of small RNAs leading to a change in bacterial physiology regarding the cell motility and production of curli or EPS [13].

In general terms, different *E. coli* serotypes have been reported to enhance flexibility and adaptability in forming biofilms when exposed to different stresses. For example, *E. coli* seropathotype A isolates associated with human infection, O157:H7 and O157:NM, showed greater ability to form biofilms than those belonging to seropathotype B or C associated with outbreaks and hemolytic-uremic syndrome (HUS) or sporadic HUS cases but no epidemics, respectively [14]. In addition, synergistic interactions are taking place in a fresh-cut produce processing plant in which *E. coli* is interacting with *Burkholderia caryophylli* and *Ralstonia insidiosa* with the formation of mixed biofilms [15].

C. jejuni, which is known as an anaerobic bacterium, is able to develop biofilms both in microaerophilic conditions (5% O_2 and 10% CO_2) and in aerobic conditions (20% O_2) [1]. The cells embedded in the biofilm matrix are better protected from oxygen and survive for days in food processing environments [1].

Pseudomonas spp. produce high amounts of EPS and have been shown to attach and form biofilms on stainless steel surfaces. They coexist within biofilms with *Listeria, Salmonella*, and other pathogens forming multispecies biofilms, more stable and resistant [6].

B. cereus is a cause of biofilm formation on many food contact surfaces such as conveyor belts, stainless steel pipes, and storage tanks [16], but it is also able to form immersed or floating biofilms, and to secrete within the biofilm a vast array of metabolites, surfactants, bacteriocins, enzymes as lipases and proteases affecting the sensorial qualities of foods, and toxins. For floating biofilms, the production of kurstakin, a lipopeptide biosurfactant, that is regulated via quorum sensing (QS) signaling is important [17].

Within the biofilm, *B. cereus* exists either in vegetative or in sporal form, the spores being highly resistant and adhesive, properties that increase the resistance of the bacterium to antimicrobials and cleaning procedures.

Four mechanisms based on the flagellar motility of *B. cereus* are described as being involved in biofilm formation. The first mechanism is used in static conditions when the bacterium must reach on its own suitable places for biofilm formation [18], at the air-liquid interface. The second one is represented by the creation of channels in the biofilm matrix to facilitate nutrients' access on one hand and penetration of toxic substances on the other hand [19]. The third mechanism refers to motile planktonic bacteria that penetrate the biofilm and increase its biomass [18, 19], while the fourth represents the extension of the biofilm based on the ability of motile bacteria located at the edge of the biofilm to colonize the surroundings [18].

It has been showed that, in its planktonic form, *S. aureus* does not appear resistant to disinfectants, compared to other bacteria, but it may be among the most resistant ones when attached to a surface [20]. It seems that different stress-adaptive responses may enhance biofilm formation, with certain differences in terms of their composition and architecture, especially for the wild-type biofilms colonizing the food and related processing environments. Examples include protein-based sources responsible for the structure of biofilms formed by *S. aureus* of food origin [21] similar to those put into evidence for the coagulase-negative ones. However, other studies demonstrated that simple carbohydrates, such as milk lactose, can modulate the biofilm formation especially by inducing the production of polysaccharide intercellular adhesins [22].

3. Genes involved in the biofilm formation

Over time, beside the conditions that favor the biofilm formation in food processing plants, the genetic background of biofilm forming microorganisms was also intensively studied. At each step of biofilm development and dispersal, there is a specific genetic signal control.

The *L. monocytogenes* pattern of the microarray gene expression was analyzed at different time intervals (4, 12, and 24 h) in order to depict genes' expression at different stages of biofilm formation. The results showed that more than 150 genes were upregulated after 4 h of biofilm formation and a total of 836 genes highlighted a slow increase in expression with time [23]. Although for many bacterial species the genome sequencing allowed the identification of genes that were involved in biofilm synthesis, for *L. monocytogenes*, these genes could not be identified using just the bioinformatics analysis.

In the biofilm formation, the attachment step is a prerequisite in which flagella and type I pili-mediated motilities are critical for the initial interaction between the cells and surface.

In order to find out the roles of the genes and regulatory pathway controlling the biofilm formation, researchers applied one or two genome-wide approaches, like transposon insertion mutagenesis or/and transcriptome analyses. With a transposon mutagenesis library, it was possible to identify 70 *L. monocytogenes* mutants, with Himar1 mariner transposon insertion, which produced less biofilms [24]. From a total of 38 genetic loci identified, 4 of them (**Table 1**) were found to be involved in bacterial motility (*fli*D, *fli*Q, *fla*A, and *mot*A), a required property for initial surface

Gene/KEGG/protein encoded	Gene function	Role	Bacterium	Ref.
Initial attachment				
*fli*Q/LMON_0682/Flagellar biosynthesis protein	Motility	Cell adhesion and bacterial attachment	*L. monocytogenes*	[23–25]
*fla*A/lmo0690/Flagellin	Flagella bio-synthesis			
*fli*D/Flagellar hook-associated protein 2	Enable the polymerization of the flagellin monomers; flagellar capping protein			[148]
*mot*A/BN418_0793/Flagellar motor protein	Flagellar motor rotation			
*prf*A/IJ09_09365/Listeriolysin positive regulatory factor A	DNA-binding transcription factor activity	Positive regulation of single species biofilm formation	*L. monocytogenes*	
fimA/JW4277/Type-1 fimbrial protein, A chain	Enable bacteria to colonize the host epithelium	Cell adhesion	*E. coli*	[30, 31]
fhiA/ECUMN_0250/Flagellar biosynthesis protein	Motility bacterial-type flagellum assembly			
yadL/ECs0141/*yadM*/*yad*K/*yad*C/Fimbrial protein	Fimbrial bio-synthesis			
*tab*A/yjgK/b4252/toxin-antitoxin biofilm protein	Represses fimbria genes	Single-species biofilm		
icaA/Poly-beta-1,6-N-acetyl-D-glucosamine synthase from *ica*ADBCR operon	Acetylglucosaminyl transferase activity, cell adhesion	Involved in the polymerization of a biofilm adhesin polysaccharide	*S. aureus*	[149]
*tpi*A/SAR0830/Triosephosphate isomerase	Involved in gluconeogenesis pathway	Role in adherence		[150]
*sra*P/SAOUHSC_02990/ Serine-rich adhesin for platelets	Mediates binding to human platelets	Plays a positive role in biofilm formation		[151]
*Spo*0A/BSU24220/Stage 0 sporulation protein A	Regulatory role in sporulation	Single-species surface biofilm formation	*B. cereus, B. subtilis*	[152] [153]
*deg*S/BSU35500/Signal transduction histidine-protein kinase/phosphatase	Transition to growth phase; flagellum formation	Biofilm formation		[154, 155]
*fli*L/STM1975/Flagellar protein	Controls the rotational direction of flagella	Motility, cell adhesion	*S. enterica*	[156]
*ycf*R/Outer membrane protein	Promotes the attachment to the surface			[157]

Gene/KEGG/protein encoded	Gene function	Role	Bacterium	Ref.
Microcolonies development				
*dlt*A/LMOf2365_099/D-alanine-D-alanyl carrier protein ligase	Catalyzes the first step in the D-alanylation of lipoteichoic acid (LTA)	Cell wall biogenesis	*L. monocytogenes S. aureus*	[24]
*dlt*C/LMOf2365_099/D-alanyl carrier protein	Carrier protein involved in the D-alanylation of LTA			
*dlt*B/lmo0973/DltB	Involved in the transport of activated D-alanine through the membrane		*S. aureus, B. subtilis*	
sdrC/NWMN_0523/Serine-aspartate repeat-containing protein C *sdrH*/SAUSA300_1985 Serine-aspartate repeat family protein	Cell adhesion	Mediates interactions with components of the extracellular matrix to promote bacterial adhesion	*S. aureus*	[158]
*bhs*A/STY1254/Multiple stress resistance protein	Stress response, response to copper ion	Regulation of biofilm formation. May repress cell–cell interaction and cell surface interaction	*E. coli*	[159]
*bsm*A/*yjf*O/Lipoprotein	Stress response to hydrogen peroxide and to DNA damage	Single-species biofilm formation; enhanced flagellar motility	*E. coli, S. enterica*	[160]
*csg*D/b1040/CsgBAC operon transcriptional regulatory protein	DNA-binding transcription activator activity	The master regulator for adhesive curli fimbriae expression		[161]
*mlr*A/b2127/HTH-type transcriptional regulator	DNA-binding transcription factor activity	Activates transcription of *csg*D		[162]
*sin*R/BSU24610/HTH-type transcriptional regulator	Negatively regulates transcription of the *eps* operon	DNA-binding protein master regulator of biofilm formation	*B. subtilis, B. cereus*	[163, 164]
*eps*G (*yve*Q)/BSU34310/Transmembrane protein	Production of exopolysaccharide	Biofilm maintenance		[165]
*eps*H (*yve*R)/BSU34300/Putative glycosyl-transferase				[166]
*ymd*B/BSU16970/2′,3′-cyclic-nucleotide 2′-phospho-diesterase	Regulatory role. Induces genes involved in biofilm formation	Directing the early stages of colony development		
*pgc*A/Phosphoglucomutase	Catalyzes the interconversion between glucose-6-phosphate and alpha-glucose-1-phosphate	Exopolysaccharide synthesis		[167]

Gene/KEGG/protein encoded	Gene function	Role	Bacterium	Ref.
*gcp*A/SL1344_191/Biofilm formation in nutrient-deficient medium	Biofilm production under low-nutrient concentrations		*S. enterica*	[156]
Biofilm maturation				
*tas*A/ BSU24620/major biofilm matrix component	Identical protein binding	Major component of the biofilm extracellular matrix	*B. cereus*	[168]
*tap*A/ BSU24640/TasA anchoring/assembly protein	Important for proper anchoring and polymerization of TasA fibers at the cell surface	Essential for biofilm formation	*B. subtilis* No paralog in *B. cereus* genome	[169]
*sip*W/BSU24630/Signal peptidase IW	Cleavage of the signal sequence of TasA and TapA		*B. cereus*	
*bsl*A (*yua*B)/BSU31080/Biofilm-surface layer protein A	Confers a specific microstructure to the biofilm surface	Confers hydrophobicity to the biofilm	*B. subtilis,* No paralog in *B. cereus* genome	[170–171]
wcaF/b2054/Putative colanic acid biosynthesis acetyl-transferase	Synthesis of colanic acid	Involved in the pathway slime polysaccharide biosynthesis	*E. coli*	[172]
*wca*L/STM2100/Putative colanic acid biosynthesis glycosyl-transferase			*S. enterica*	[173]
*bss*R (*yli*H)/JW0820/Biofilm regulator	Regulation of biofilm formation	In the glucose presences, cells showed increased biofilm formation	*E. coli*	[33]
*mqs*R/b3022/mRNA interferase toxin	Motility-quorum sensing cell proliferation	Biofilm architecture		[172]
*tqs*A/b1601/AI-2 transport protein	Efflux transmembrane transporter activity	Represses biofilm formation and motility		[31]
*bdc*A/b4249/Cyclic-di-GMP-binding biofilm dispersal mediator protein	Controls cell motility, size, aggregation, and production of extracellular DNA and extracellular polysaccharides	Biofilm dispersal	*E. coli, S. enterica*	[174]
*ihf*AB/Integration host factor	Specific DNA-binding protein	Matrix density Cellulose production	*S. enterica, S. aureus*	[175–177]
*bap*A/biofilm-associated protein	Large surface proteins family	Bacterial adhesion Biofilm maturation		

Gene/KEGG/protein encoded	Gene function	Role	Bacterium	Ref.
clfA/ / Clumping factor A; *clfB*/ NWMN_2529/ Clumping factor B	Cell surface-associated protein implicated in bacterial attachment	Aggregation of unicellular organisms; cell adhesion	*S. aureus*	[178]
*ica*C/SAOUHSC_03005/poly-beta-1,6-N-acetyl-D-glucosamine export protein (PNAG)	Export of PNAG across the cell membrane		*E. coli, S. aureus*	[149]
*pfl*A/SAOUHSC_00188/Pyruvate formate lyase-activating enzyme *pfl*B/SACOL020/Formate acetyltransferase	Enzymes that catalyze the first step in the acetogenesis from pyruvate	Organic free radical synthesis		[29]
*sar*A/Transcriptional regulator	Global regulator of a few genes with important roles in biofilm development	Biofilm formation process in a cell density-dependent manner	*S. aureus*	[179]
*agr*D/LMM7_0043/Putative autoinducing peptide	Involved in proteolytic processing	Quorum Sensing	*L. monocytogenes*	[180]
lmo0048/Putative AgrB-like protein	Involved in proteolytic processing		*L. monocytogenes* *B. cereus*	
*agr*C/Accessory gene regulator	Histidine kinase activity		*S. aureus*	[181]
*agr*A/CQ02_00305/BN389_00610/ Accessory gene regulator	A response regulator			[182] [183]
*agr*B/MF_00784/Accessory gene regulator	Proteolytic processing of *Agr*D		*S. aureus*	[184]
luxS/lmo1288/S-ribosyl-homo-cysteine lyase	Catalysis of precursor molecules of AI-2		*L. monocytogenes* *E. coli, B. cereus, S. enterica*	[48] [49]
*lux*Q/Autoinducer 2 sensor kinase/phosphatase	Phospho-relay sensor kinase activity		*E. coli, B. cereus, S. enterica*	

Table 1.
List of genes with significant role in biofilm formation within pathogenic microorganisms (UniprotKB database).

attachment. Another gene with increased expression at 4 h and decreased expression after 12 h from biofilm initiation was *prf*A, the listeriolysin positive regulatory factor A. It seems that this regulatory factor is necessary just in the initial stages of biofilm formation and aggregation but not in the colonization stage [23, 25, 26].

Extracellular and surface proteins such as internalin A and BapL, respectively, have been found to be involved in the initial bacterial adhesion in *L. monocytogenes* EGD-e [27]. Moreover, its mobility is ensured by flagella and is temperature-dependent affecting the biofilm formation. As such, above 30°C, the transcription of *flaA* is stopped.

S. aureus genes responsible for cell adhesion to the surface are included in the icaADBC operon with functions in biosynthesis of the glucosamine polymer and polysaccharide intercellular adhesins [28]. Therefore, other genes encoding a number of transporter proteins (*pro*P, *opu*D, *aap*A, and *dlt*A) were upregulated after 8 hours from the biofilm initiation [29]. For *E. coli*, the genes involved in the cell adhesion, like *fim*A, *yad*K, *yad*N, *yad*M, and *yad*C-encoding fimbriae-like proteins-are coexpressed with the integral cell membrane genes, with outer membrane proteins (*htr*E), with transcriptional regulators (*mng*R and *nha*R), or other genes, but this network appears to be strain specific [30, 31].

In the case of *S. enterica*, differential expression analysis revealed that *ycf*R is highly conserved as in many Gram-negative bacteria, being upregulated under chlorine stress and responsible for the virulence and attachment of bacterium to the glass or polystyrene [32, 33].

Moreover, *Salmonella* spp.-related biofilms are driven by a transcriptional regulatory CsgD protein that activates the expression of curli and cellulose. The transcription of *csg*BAC operon, which encodes the structural subunits for curli, indirectly activates the transcription of the second mechanism, *adr*A, associated with cellulose production [10]. Important factors in the activation of *Salmonella* spp. biofilms are the c-di-GMP that is behaving like a secondary messenger molecule when the CsgD content is elevated [34].

Microcolonies are formed by cell proliferation, and many genes involved in cell division, cell wall biogenesis, virulence and motility, stress response, and transcriptional regulation factors are expressed.

Table 1 shows a selection of the genes that are expressed in all the steps of biofilm formation or are upregulated under influence of different biotic or abiotic factors. It was reported that the Δ*dlt*ABC *L. monocytogenes* strains are defective in biofilm formation, validating by transposon mutagenesis, the critical role of D-alanylation of teichoic acids, for biofilm synthesis [24]. So, the mutants without D-alanine on the surface of teichoic acids have a higher negative charge and develop a biofilm-negative phenotype.

The mature biofilm evolves from microcolonies and this development is associated with EPS production. The biofilm matrix of *B. cereus* is similar to other *Bacillus* sp., but the *eps* genes, responsible for the EPS synthesis, are not mandatory for *B. cereus* compared to *B. subtilis* [35]. Little is known about the regulatory networks in *B. cereus*, but studies have shown that CodY and SpoOA may as well play a crucial role in biofilm formation [36].

Furthermore, the structural proteins encoded by *tap*A and *bsl*A from *B. subtilis* genome are absent in the matrix of *B. cereus* because these genes have no paralog in *B. cereus* genome. Instead the *tas*A gene is essential for *B. cereus* biofilm development, being responsible for the matrix fiber synthesis [37].

An important polysaccharide identified in the matrix biofilm of many pathogenic bacteria is the colanic acid, which plays an important physiological role for bacteria living in biofilm. This EPS is synthesized by specific enzymes encoded by *wca*L gene (*S. enterica*) or *wca*F (*E. coli*). It has been also shown that *rpo*S gene, the main regulator of the general stress response, may be seen as a key factor in the development of mature biofilms in *E. coli* [38].

Consequently, the transition from the planktonic state to the biofilm state is critical and it is subjected to a strict gene regulation, essential for matrix synthesis, cell aggregation, and cell signaling.

Nevertheless, bacteria of multiple genetic backgrounds communicate by regulating their relationship of cooperativeness through a mechanism called quorum sensing (QS) in which the bacterial cells are having social interactions with each other through small diffusible signal molecules called autoinducers, thus contributing to the biofilm development [10].

Quorum sensing process described in the 1970s is involved in the control of various gene expressions through chemical signaling molecules that are synthesized in response to cell population density [39]. When bacteria start to sense their critical biomass, they answer by activating or repressing genes from 10% of bacteria genome [40]. The system has been described for both Gram-negative and Gram-positive bacteria.

Among QS, other two important regulators are known to control biofilm shape and structure: cyclic diguanosine-5′-monophosphate (c-di-GMP) and small RNAs. For example, *S. aureus* biofilm development is regulated by many environmental conditions and genetic signals. A significant constituent in biofilm formation is mediated by the polysaccharide intercellular adhesin composed mainly of polymeric N-acetyl-glucosamine (PNAG) and eDNA, encoded by the ica operon [41]. In certain cases, such as *S. aureus*, biofilm-associated protein (Bap) is involved in biofilm maturation rather than polysaccharide intercellular adhesion (polysaccharide intercellular adhesins) expression [42].

The c-di-GMP involvement in *S. aureus* is an important biofilm regulator that allosterically switches on enzymes of exopolysaccharide biosynthesis [43], while the function of small RNA genes involved is still not yet studied in detail [44]. Although it has been noticed to show an increased susceptibility to disinfectants in planktonic state, however, in biofilm state, it may be among the most resistant ones equally important for food as well as for the medical sectors.

Gram-positive bacteria such as *S. aureus*, *B. subtilis*, and *L. monocytogenes* are communicating through inducers encoded by accessory gene regulator (*Agr*) system (**Table 1**). It seems like the *Agr* complex regulates more than 100 genes in the *S. aureus* genome [45], and its deletion from *L. monocytogenes* genome affects more than 600 genes [46].

The accessory gene regulator of *S. aureus* modulates the expression of virulence factors and toxins in response to autoinducing peptides (AIPs) while luxS synthesizes AI-2, which inhibits exopolysaccharide synthesis through an unknown QS cascade [47].

For *S. enterica* and *E. coli*, the QS system is mediated by two genes, *lux*S and *lux*R, homolog to *Sdi*A in order to reach intercellular signaling [48, 49].

The *L. monocytogenes* QS signaling triggers the transcriptional activation of one of the virulence PrfA-regulated genes a*ctA*, resulting in the bacterial aggregation and biofilm formation [10]. Another gene involved in the cell-to-cell interactions is *secA2* gene. Its deletion may inactivate the SecA2 pathway with an increased cell aggregation and sedimentation [50].

4. Fighting against biofilms with nonconventional methods

Since biofilms act as a barrier that protects the embedded cells against cleaning and disinfecting agents [51], the control of biofilm is an issue that is currently addressed to find effective solutions that can prevent biofilm formation or eliminate the already formed one. Biocontrol of biofilms by using bacteriocins, disruptive enzymes, essential oils, or bacteriophages is gaining importance, as well as using nanoemulsions and nanoparticles. These new methods are promising strategies with remarkable results in the fight against biofilms.

4.1 Bacteriocins used to control biofilms

Bacteriocins are antimicrobial peptides ribosomally produced by an extensive range of bacteria to inhibit or kill competing microorganisms in a micro-ecological system [52, 53]. The most studied bacteriocin and the only one allowed presently as food-grade additive is nisin, a lantibiotic with proven effects against many Gram-positive bacteria including foodborne pathogens [54]. This bacteriocin was shown to penetrate the biofilm formed by *S. aureus* and permeate the sessile bacterial cells by real-time monitoring [55]. Moreover, nisin and its bioengineered derivatives were able to enhance the capability of conventional antibiotics such as chloramphenicol of decreasing *S. aureus* biofilm viability [56]. Nevertheless, a study assessing the effect of neutral electrolyzed water and nisin and their combination against listerial biofilm on glass and stainless steel surfaces indicated the potency of this bacteriocin to improve the efficacy of sanitizers used in food industry [57]. Nisin was also indicated to be effective against biofilms formed by Gram-negative bacteria such as *Salmonella* typhimurium when combined with P22 phage and EDTA, a synergistic combination that reduced 70% of the mature biofilm [58].

Another way to prevent biofilms development is represented by the adsorption of these bioactive compounds on the surfaces that come into contact with foods [59]. In this case, Nisaplin adsorbed to three types of food-contact surfaces commonly encountered in food processing plants, namely stainless steel, polyethylene terephthalate (PET), and rubber, reduced the adhesion ability of food-isolated *L. monocytogenes* strains [60]. Other studies showing the efficacy of nisin in preventing surface colonization by *L. monocytogenes* were conducted by Daeschel et al. [61] and Bower et al. [62].

A bacteriocin found to markedly inhibit the biofilm formed by *S. aureus* is sonorensis, a member of the heterocycloanthracin subfamily produced by *Bacillus sonorensis* MT93 [63].

4.2 Disruptive enzymes for fighting against biofilms

Disruptive enzymes, such as proteases, glycosidases, amylases, cellulases, or DNAses, are considered a green alternative to chemical treatments often used in the fight against biofilms' formation in food-related environments [2]. Such enzymes do not have toxic effects and are used both alone and as part of the industrial detergents' composition to improve their cleaning efficacy [64–66].

Proteases are a class of enzymes that catalyzes the cleavage of proteins' peptide bonds. Although they are produced by all living organisms, microbial proteolytic enzymes are preferred over animal or plant origin proteases. The most commonly used source of bacterial proteases is represented by those produced by the genus *Bacillus* since they have remarkable properties such as tolerance to extreme temperatures, large pH domain, organic solvents, detergents, and oxidizing compounds [67]. Given their low substrate specificity, extracellularly produced proteases were shown to be more effective in degrading organic-based aging biofilms compared to amylases [68]. Combinations of a buffer that contained surfactants and dispersing and chelating agents with serine proteases and polysaccharidases were shown to be efficient in removing the biofilms formed by *B. cereus* and *P. fluorescens*, respectively, on stainless steel slides by the cleaning-in-place procedure [69]. Purified alkaline proteases from *B. subtilis* were reported to degrade biofilms produced by both *P. mendocina* and *E. coli* within 10 minutes [70]. Mold-origin proteases, such as proteinase K, were proved to be effective agents against biofilms formed by *L. monocytogenes* when used either alone or in combination with other biofilms' inhibitors. In a study, proteinase K was capable of complete dispersion of *L. monocytogenes* biofilms grown for 72 h on both plastic and stainless steel surfaces at concentrations above 25 μg/mL. The same study also emphasized the synergistic effect between DNases and proteinase K regarding *L. monocytogenes*-established biofilm dispersion [71].

Polysaccharide-hydrolyzing enzymes were indicated to remove the biofilms formed by *Staphylococcus* spp. and *Pseudomonas* spp. on steel and polypropylene substrata. However, these enzymes did not exhibit a significant bactericidal effect, so they were combined with oxidoreductases for an improved performance [72]. Experimental studies showed that cellulase in conjunction with cetyltrimethylammonium bromide had the capacity of removing 100% of the *S. enterica* mature biofilm at the phase of irreversible attachment. This finding suggests an alternative strategy for removing *Salmonella* biofilms in meat processing facilities [73].

4.3 Using essential oils against biofilms

Plant essential oils (EOs) are rich in phytochemical compounds, which are secondary metabolites produced by plants as defense mechanism against pathogens [74]. Regarding microbial inactivation, EOs have been reported to mainly affect the cellular membrane by permeabilization [75]. This leads to the disruption of vital cellular processes, including energy production, membrane transport, and metabolic regulatory functions [76].

Studies evaluating the potential of EOs as disinfectants were conducted. Leonard et al. [77] assessed the bioactivity of *Syzygium aromaticum* (clove), *Mentha spicata* (spearmint), *Lippia rehmannii*, *Cymbopogon citratus* (lemongrass) EOs, and their major components on the listerial biofilm. The assessment revealed that *M. spicata* and *S. aromaticum* EOs inhibited the growth of listerial biofilm, while, surprisingly, in the presence of their main compounds alone, namely R-(−) carvone and eugenol, respectively, the biofilm biomass increased. Similar phenomenon was previously noticed by [78] in the case of α-pinene, 1,8-cineole, (+)-limonene, linalool, and geranyl acetate, with researchers arguing that bacterial cells in biofilms have a reduced metabolic activity, which make them more resistant to deleterious agents. These results suggest that antimicrobial activity of EOs is rather due to the synergism among the chemical substances that compose them, than due to an individual component's activity. On the other hand, a disinfectant solution based on *Cymbopogon citratus* and *Cymbopogon nardus* EOs was reported to completely reduce the number of *L. monocytogenes* stainless steel surface-adhered cells residing in a 240 h biofilm after 60 min of interaction [79].

Thyme EO has proven antimicrobial properties [80]. In terms of biofilm inhibition capacity, this EO was shown to inhibit significantly the biofilm formed by *B. cereus* [81] and biofilms formed by other food-related pathogens, including *S. aureus* and *E. coli* [82, 83]. Thymol and carvacrol are principal constituents of thyme oil [84], and their potential regarding biofilm inhibition is intensively studied. Surfactant-encapsulated carvacrol was effective against biofilms produced by *E. coli* O157:H7 and *L. monocytogenes* on stainless steel coupons [85]. This natural biocide was also shown to control a dual-species biofilm formed by *S. aureus* and *S. enterica* at quasi-steady state [86]. However, scientists emphasized that carvacrol concentration should be seriously considered when used to combat strong biofilm producers, such as *S. aureus* strains isolated from food-contact surfaces, since low concentrations may exhibit an inductive effect. In the case of the biofilm formed by *Salmonella* typhimurium on stainless steel surfaces, exposure to thymol resulted in a more pronounced decrease in the biofilm mass compared to exposure to carvacrol or eugenol [87]. Moreover, these compounds enhanced the susceptibility of this pathogen to the treatments with antibiotics such as nalidixic acid [88].

Eugenol is a phytochemical compound preponderantly found in aromatic plants [89]. Interestingly, a study showed that this substance was able to inhibit the intracellular signaling pathway called quorum sensing in the case of biofilms formed by methicillin-resistant *S. aureus* strains isolated from food handlers. This mechanism has an important role in the host colonization, biofilm development, and defense strategies against harmful agents, allowing bacterial cells to act as social communities [90]. EOs of bay, clove, pimento berry, and their major constituent, eugenol, were proved to inhibit significantly the biofilm formed by *E. coli* O157:H7. The antibiofilm activity was assigned to the benzene ring of eugenol. Moreover, eugenol led to the downregulation of genes associated with the biofilm formation, attachment, and effacement phenotype, such as curli, fimbriae, and toxin genes [91].

4.4 Fighting against biofilms with bacteriophages

Bacteriophages are viruses that infect bacterial cells. They use the genetic machinery of their host cells to replicate, killing bacteria when reaching a sufficiently high number to produce lysis [92]. They are abundantly encountered anywhere host bacteria live [93] and, therefore, their potential is presently harnessed as natural antimicrobial agents to control pathogenic bacteria in food products and food-related environments [94]. One of the bacteriophages' applications that is intensively explored targets biofilm-forming bacteria that are relevant for food industry, including *L. monocytogenes, S. aureus, E. coli, B. cereus*, and *S. enterica*. However, the success of this approach in fighting biofilms depends on a series of factors such as composition and structure of biofilms, biofilms' maturity, and physiological state of bacterial host residing within biofilms, concentration of bacterial host, or extracellular matrix [95].

Although it is generally thought that biofilms confer resistance to bacteriophages, these bacterial predators developed several mechanisms to destroy bacteria communities. Once they reach the EPS (extracellular polymeric substances) producing host, they start to replicate, resulting in an increased number and, implicitly, in a progressive degradation of the biofilms and prevention of their regeneration. Bacteriophages can also express or induce the expression from within host genome of depolymerizing enzymes that degrade EPS. Nevertheless, they can also infect persister cells, which are dormant variants of regular bacterial cells that are highly resistant to antibiotics. In this case, the lysis process is triggered once persister bacteria are reactivated [96].

Scientists [97] reported the ability of a bred phage to reduce L-form biofilms formed by *L. monocytogenes* on stainless steel surfaces. This bacteriophage was as effective as lactic acid (130 ppm) in the eradication of preformed L-form biofilms. P100 phage treatment was also shown to reduce the number of *L. monocytogenes* cells under biofilm conditions on stainless steel coupon surface regardless of serotype [98]. The potency of three bacteriophages, namely LiMN4L, LiMN4p, and LiMN17, used as a cocktail or individually at ~9 log10 PFU/mL was evaluated to inactivate *L. monocytogenes* cells residing within 7-day biofilms strongly adhered to clean or fish broth-coated stainless steel coupons and dislodged biofilm cells [99]. These phages exhibited a higher efficiency in the case of dislodged cells compared to intact biofilms when applied for short periods of time. Therefore, for high efficiency, short-term phage treatments in fish processing environments may require prior processes aiming at disrupting the biofilms [99]. The ability of *Salmonella* spp. to develop biofilms was shown to depend on the attachment surface types that may be encountered in chicken slaughterhouses. With regard to this, surfaces such as glass and stainless steel favored the formation of *Salmonella* biofilms, while polyvinyl chloride surface sustained less the development of them. The antibiofilm activity of a pool of bacteriophages isolated from hospital and poultry wastewater was concentrated at 3 h of action for all types of surfaces. Curiously, biofilms attached to the glass surface were resistant to a 6-h treatment. Bacteriophages were able to degrade the glass-attached biofilms after 9 h of interaction [100]. A bacteriophage BPECO 19 was evaluated as possible inhibitor of a three *E. coli* O157:H7 strain biofilm grown on both abiotic (stainless steel, rubber, and minimum biofilm eradication concentration device) and biotic (lettuce leaves) surfaces. This bacteriophage showed great biofilm inhibition activity on all the tested surfaces, being suggested as effective antibiofilm agent in food industry [101].

4.5 Nanotechnology-based antimicrobials used to control biofilms

Currently, controlling biofilm formation by nanotechnology-based antimicrobials is of industrial interest, nanoemulsions and nanoparticles (NPs) with antibiofilm activity being an alternative to conventional methods.

Recently, some studies made on model system (polystyrene well plates) and real systems (fresh pineapple, tofu, and lettuce) indicated that nanoemulsions of EOs have significantly higher antibiofilm activity compared to pure EOs (**Table 2**). Antimicrobial efficacy of nanoemulsions is dependent on the droplet size and electrical properties of nanoemulsions [102, 103], nature of bacteria [75, 104], and food matrix [105–107].

Nanoparticles (NPs) can be used for both inhibition of biofilm formation and eradication of already formed ones [108].

In the last period, NPs with natural compounds gained increased interest because it was demonstrated that the inorganic capsules can protect the natural products with antimicrobial activity [109]. In this respect, cinnamaldehyde-encapsulated chitosan nanoparticles, garlic-silver NPs, and "tree of tee" oil NPs were used to combat biofilm formation by *P. aeruginosa* on polystyrene well plates and glass pieces [110–112]. Meanwhile, the biofilm formed by *S. aureus* on glass slide was inhibited by applying gold NPs with EO of *Nigella sativa* [113] and garlic-silver NPs [111].

Nanoemulsion	Particle size, nm	Biofilm-forming bacteria	Mode of action	Ref.
EO of *Citrus medica* L. var. *sarcodactylis*	73	*S. aureus*	Inhibit the ability of bacteria to attach to surfaces	[185]
EO of *Cymbopogon flexuosus* (lemongrass)	78.46 ± 0.51	*P. aeruginosa* (PA01) and *S. aureus* (ATCC 29213)	Reduce the adhesion of pathogenic bacteria to surfaces	[186]
Trans-CA	>100 <100	*P. aeruginosa* (CMCC 10104), *S.* typhimurium and *S. aureus*	Membrane disruption by destabilization of lipids	[187]
Linalool	10.9 ± 0.1	*S.* typhimurium (ATCC 1331)	Cell membrane integrity disruption	[107]

Table 2.
Antibiofilm activity of essential oil (EO) nanoemulsions.

Metal-based NPs (silver, gold, and metal oxides) with antimicrobial activity can be used to create different nanocomposite materials able to prevent bacterial adhesiveness to food-contact surfaces and equipment. Wu and coworkers [114] showed that cysteine dithiothreitol and beta-mercaptoethanol were able to reduce *S. aureus* biofilm formation on polystyrene polymer. Liang and coworkers [115] revealed that silver salt of 12-tungstophosphoric acid NPs (AgWPA-NPs) can be used to develop new materials for preserving foods, since they were able to inhibit *S. aureus* biofilm formation by damaging bacterial cells' membrane. Moreover, genes related to biofilm formation, such as *icaA*, *sarA*, and *cidA* were shown to be downregulated as a consequence of AgWPA-NPs' application. Naskar and coworkers [116] tested the antibiofilm activity of polyethylene glycol-coupled Ag-ZnO-rGO (AZGP) nanocomposite on both Gram-positive bacteria (*S. aureus* ATCC 25923) and Gram-negative bacteria (*P. aeruginosa* MTCC 2453). These NPs, at a concentration of 31.25 μg/mL, reduced the biofilm formed by *S. aureus* with ~95% and that formed by *P. aeruginosa* with ~93%. Zinc oxide NPs were used for the destruction of the biofilm formed on glass slide by *S. aureus* and *P. aeruginosa* [117]. Titania nanoparticles can be used to prevent the formation of *P. fluorescens* biofilm on the surfaces of TiO_2/polystyrene nanocomposite film [118]. It has been shown that nanostructured TiO_2 combined with UVA irradiation can be used to destroy *L. monocytogenes* biofilm, while silver NPs at a concentration of 15 μg/mL had the capacity to inhibit *S. aureus* and *E. coli* biofilms [119, 120].

The ability of two types of superparamagnetic iron oxide (IONs and IONs coated with 3-aminopropyltriethoxysilane) to inhibit biofilm formation by *B. subtilis* was successfully tested by [121].

5. Food technologies to control the biofilm formation

Some food technologies belonging to alternative technologies seem to be successful for preventing the biofilm formation and/or for targeting resistant microorganisms and making them more susceptible to molecular interventions in order to hinder their biofilm formation ability. Among these technologies are included plasma treatments, ultrasound treatments, light-based technologies, pulsed electric fields (PEF), and high hydrostatic pressures. With the exception of ultrasound treatments that can be used to fight against biofilms formed on mechanical parts or pipes, the others are mostly applied for food matrix decontamination.

5.1 Plasma treatments

Plasma is generated when the added energy ionizes a gas, which is composed of ions, neutrals, and electrons. Plasma treatment is a surface treatment that has a low penetration depth and was reported to be effective against biofilms, depending on the type of surface biofilms are formed on, the distance between plasma and surface, and the thickness or the microbial load.

Plasma sources for producing nonthermal plasma at atmospheric pressure are plasma jets, dielectric barrier discharges (DBD), corona discharges, and microwave discharges. Different other characteristics of the plasma have been reported to influence the biofilms' inactivation such as the setup (electrode configuration), the exposure mode, the operating gas, the frequency, the plasma intensity (voltage), and the time of exposure [122].

Researches [123] showed that the efficacy of DBD in-package atmospheric cold plasma (ACP) against S. typhimurium, *L. monocytogenes*, and *E. coli* could reach up to 5 log CFU/g after 300 s of treatment at 80 kV. Other researchers [124] studied the effect of ACP on monoculture biofilms (*E. coli*, *S. enterica*, *L. monocytogenes*, and *P. fluorescens*) and mixed culture biofilms (*L. monocytogenes* and *P. fluorescens*) and demonstrated that the latest are more difficult to inactivate than the former ones. *L. monocytogenes* and *P. fluorescens* inoculated as mixed cultures on lettuce were reduced by 2.2 and 4 log CFU/g, respectively, and the biofilms formed at 4°C were more resistant than the ones formed at 15°C.

Govaert et al. [122] studied the influence of different plasma characteristics on the inactivation of *L. monocytogenes* and *S.* typhimurium biofilms and showed that inactivation can vary from 1 log to approximately 3.5 log (CFU/cm^2), but the highest reduction was obtained for a DBD electrode with He and no O_2 in the gas mixture and an input voltage of 21.88 V. A high efficiency of the inactivation of bacterial biofilm was achieved by DBD for low-dose discharges (70 mW/cm^2) and short treatment times ($\leq$300 s), and the most effective reduction in the number of *S. aureus* cells of 2.77 log was reported after 300 s. *E. coli* biofilm was reduced only by 66.7% [125].

It was shown that ACP is a promising technique but alone cannot achieve complete biofilm inactivation and thus it should be complemented by other surface treatments. Possibility to combine ACP with different biocides such as hydrogen peroxide, sodium hypochlorite, ethylenediaminetetraacetic acid, chlorhexidine, octenidine, and polyhexanide applied before or after the plasma treatment was tested by [126] to reduce biofilms cultivated on titanium discs. Also, Gupta et al. [127] studied the antimicrobial effect of an ACP, plasma jet combined with chlorhexidine, for the sterilization of the biofilms formed by *P. aeruginosa* on titanium surfaces [128].

5.2 Ultrasound-assisted technologies

Ultrasound (US) is a form of energy generated by sound waves at frequencies that are too high to be detected by the human hearing (>16 kHz). The US band is also divided into low frequency (16 kHz–1 MHz) and high frequency (>1 MHz) bands.

US was used as biofilm removal method; however, many studies demonstrated that it should be complemented by other inactivation methods [129, 130]. For example, [130] demonstrated that US removed a significant amount of *E. coli* and *S. aureus* biofilm, up to 4 times higher compared to the swabbing method. Later on, the same researchers [131] showed that two ultrasonic devices developed failed to completely remove *E. coli* and *S. aureus* biofilms for closed surfaces, but they succeeded in biofilm inactivation on opened surfaces (10 s at 40 kHz). The use of chelating agents such as EDTA completely dislodged *E. coli* biofilm but not significantly improved *S. aureus* biofilm removal. A synergistic effect was achieved when US was combined with enzymes (proteolytic or glycolytic) that demonstrated a 2–3 times higher efficacy in biofilm removal compared to sonication.

Combination of US with mild heat and slightly acidic electrolyzed water was used to test the inactivation of *B. cereus* biofilms on green leaf surfaces. Slightly acidic electrolyzed water with 80 mg/L treatment for 15 min combined with US of fixed frequency (40 kHz) and acoustic energy density of 400 W/l at 60°C resulted in a reduction of ~3.0 and ~3.4 log CFU/cm^2 of *B. cereus* reference strains ATCC 10987 and ATCC 14579 [132].

Synergistic effects were registered also for ultrasound (US; 37 kHz, 380 W for 10–60 min) assisted by peroxyacetic acid (PAA; 50–200 ppm) on reducing *Cronobacter sakazakii* biofilms on cucumbers [133].

The efficacy of US (37 kHz, 200 W, for 30 min)-assisted chemical cleaning methods (10% alcohols, 2.5% benzalkonium chloride, and 2.5% didecyl dimethyl ammonium chloride) for the removal of *B. cereus* biofilm from polyurethane conveyor belts in bakeries using US was better compared to each individual method as demonstrated by [134].

5.3 Combined light-based technologies

Ultraviolet (UV) light technology is based on the emission of radiation within the ultraviolet region (100–400 nm). The antimicrobial behavior of UV light is based on the formation of DNA photoproducts that inhibit transcription and replication and can lead to cell death [135]. Since the absorption of the DNA is in the 200–280 nm range with the maximum at 254 nm, this wavelength of the UV-C range is called germicidal UV light [136].

Pulsed light (PL) is the next-generation approach to UV delivery. PL is a technology that can be used to decontaminate surfaces by generating short-time high-energy light pulses (millions or thousands of a second) of an intense broad spectrum (200–1100 nm). PL can be used to decontaminate a great variety of foods as well as to decontaminate contact surfaces, thus improving safety in foods and extending their shelf life [137]. The antimicrobial effect is based on strand breaks that lead to the destruction/chemical modification of the DNA and thus prevent the replication of the bacterial cell [138].

Recently, Rajkovic and coworkers [139] evaluated the efficacy of pulsed UV light treatments to reduce *S.* typhimurium, *E. coli* 0157:H7, *L. monocytogenes*, and *S. aureus* on the surface of dry fermented salami inoculated with 6.3 log CFU/g at 3 J/cm^2 (1 pulse) or 15 J/cm^2 (5 pulses) for 1 or 30 min. The authors found a significant effect of PL treatment time, with the best results after 1 min of applying PL (2.18–2.42 log CFU/g reduction), while after 30 min, the reduction varied from 1.14 to 1.46 log CFU/g.

A comprehensive review in the literature underlined the various researches directed mainly at inactivation of pathogens in food or on surfaces and for preventing biofilm formation [137]. While there are often considerable differences in the rate of microbial inactivation by PL, a maximum reduction of 3-log was typically achieved, which is below the reduction performance standard of 5-log required by HACCP regulation [138].

Regarding the combined methods, synergistic interaction between gallic acid and UV-A light was able to inactivate *E. coli* O157:H7 in spinach biofilm [140]. The UV-A treatment complemented by the gallic acid presence was found to be effective producing a 3-log (CFU/mL) reduction in *E. coli* O157:H7 on the surface of spinach leaves.

However, PL technology limitation related to the inability to effectively treat uneven food surfaces with crevices, the presence of organic material, and large microbial populations generating shading effects should also be taken into account. Future innovation in PL technology will seek to improve fluence efficiency, for example by considering alternative light sources such as LEDs [141], reflective surfaces included in the treatment chamber, using materials such as titanium dioxide to augment irradiation efficacy [138], and other combination of treatments assisted by PL, based on hurdle approach.

5.4 Pulsed electric field

Pulsed electric field (PEF) is a food processing technology that applies short, high-voltage pulses, across a food material placed between two or more electrodes. The pulses enhance cell permeability by damaging the cell membrane, and if the transmembrane potential is sufficiently high, it produces electroporation. Further, if pores are not resealed, it results in cell death. Most of the food applications are designed for liquid flow through pipes where in a certain region the liquid passes in-between the electrodes area that applies the PEFs [142].

Thermosonication (TS) was investigated in combination with PEF to determine its effects on inactivation and sublethal injury of *P. fluorescens* and *E. coli*. While TS was applied at a low (18.6 mm) and high (27.9 mm) wave amplitude, PEF was applied at a low (29 kV cm^{-1}) and high electrical field strength (32 kV cm^{-1}). TS/PEF caused a maximum of 66% inactivation, while sublethally injuring approximately 26% of the *E. coli* population [143].

PEF demonstrated synergistic potential in combination with additives (EDTA or triethyl citrate) to inactivate *Salmonella* serovars in whole liquid eggs [144].

There is a lot of potential demonstrated by PEF and the combination with different other hurdles could contribute to the elimination of persistent clones able to form biofilms.

5.5 High pressure processing

High pressure processing (HPP) is a cutting-edge technology that represents an alternative to conventional processing. HPP has the ability to inactivate microorganisms and enzymes and has a minimal impact on sensorial and nutritional properties of food [145, 146].

Combined with other different hurdles, the pressure-assisted processing could be oriented toward a more targeted inactivation of pathogens and prevention of biofilm formation.

Recent studies were focused on *L. monocytogenes*, a pathogen able to form surface-attached communities that have high tolerance to stress. In order to understand how *agr* gene regulates virulence and biofilm formation, a recent molecular study [147] was conducted. *L. monocytogenes* EGD-e Δ*agrD* showed reduced levels of surface-attached biomass in 0.1 BHI (brain heart infusion) broth.

However, *L. monocytogenes* mutant deficient in *agr* peptide sensing showed no impaired resistance to HPP treatment at 200, 300, and 400 MPa for 1 min compared to wild-type and *L. monocytogenes* EGD-e and thus demonstrating that weakened resistance to cell wall stress is not responsible for the reduced biofilm-forming ability.

Understanding better the molecular mechanisms of stress-related genes will allow to better target pathogen inactivation and to select the right hurdle combination and parameters of unconventional technologies to able to reduce the susceptibility of certain pathogens to form biofilms. These types of studies are just at the beginning and many more researches are expected to focus on these topics in the near future.

6. Conclusions

Pathogenic and toxigenic bacteria are able to form biofilms, structures that protect the cells and allow them to remain postsanitation in the food processing environment.

Specific genes are expressed in all the steps of biofilm formation or are upregulated under influence of different biotic or abiotic factors. Genes codify for cell surface structures and appendages (flagella, curli, fimbriae, and pili) that are facilitating biofilm formation by helping bacteria to move toward surfaces and to adhere to them, for extracellular polymeric substances that stabilize the biofilms and protect the cells and for quorum sensing communication.

Scientists developed novel agents and strategies to control biofilm formation or removal. Their application to the food industry would contribute to eradication of undesirable bacteria from food-processing environments and, subsequently, from food products.

Acknowledgements

This study was supported by a grant of the Executive Agency for Higher Education, Research, Development and Innovation Funding in Romania awarded to the Dunarea de Jos University of Galati (International and European Cooperation—Subprogramme 3.2—Horizon 2020—Contract no: 15/2017), an institution that is a member of the ERA-IB2 consortium "SafeFood" (ID: ERA-IB-16-014).

Florentina Ionela Bucur's work has been funded by the European Social Fund through the Sectoral Operational Programme Human Capital 2014–2020, through the Financial Agreement with the title "Scholarships for entrepreneurial education among doctoral students and postdoctoral researchers (Be Entrepreneur!)," Contract no. 51680/09.07.2019—SMIS code: 124539.

Author details

Leontina Grigore-Gurgu, Florentina Ionela Bucur, Daniela Borda, Elena-Alexandra Alexa, Corina Neagu and Anca Ioana Nicolau*
Dunarea de Jos University of Galati, Faculty of Food Science and Engineering, Galati, Romania

*Address all correspondence to: anca.nicolau@ugal.ro

References

[1] Téllez S. Biofilms and their impact on food industry. VISAVET Outreach Journal. 2010. Available from: https://www.visavet.es/en/articles/biofilms-impact-food-industry.php [Accessed 1.10.2019]

[2] Galié S, García-Gutiérrez C, Miguélez EM, Villar CJ, Lombó F. Biofilms in the food industry: Health aspects and control methods. Frontiers in Microbiology. 2018;**9**:898. DOI: 10.3389/fmicb.2018.00898

[3] Coughlan LM, Cotter PD, Hill C, Alvarez-Ordóñez A. New weapons to fight old enemies: Novel strategies for the (bio)control of bacterial biofilms in the food industry. Frontiers in Microbiology. 2016;**7**:1641. DOI: 10.3389/fmicb.2016.01641

[4] Trachoo N. Biofilms and the food industry. Songklanakarin. Journal of Science and Technology. 2003;**25**(6):807-815

[5] Beloin C, Ghigo JM. Finding gene-expression patterns in bacterial biofilms. Trends in Microbiology. 2005;**13**:16-19. DOI: 10.1016/j.tim.2004.11.008

[6] Chmielewski RAN, Frank JF. Biofilm formation and control in food processing facilities. Comprehensive Reviews in Food Science and Food Safety. 2006;**2**:22-32. DOI: 10.1111/j.1541-4337.2003.tb00012.x

[7] Lemon KP, Higgins DE, Kolter R. Flagellar motility is critical for *Listeria monocytogenes* biofilm formation. Journal of Bacteriology. 2007;**189**:4418-4424. DOI: 10.1128/JB.01967-06

[8] Renier S, Hébraud M, Desvaux M. Molecular biology of surface colonization by *Listeria monocytogenes*: An additional facet of a Gram-positive foodborne pathogen. Environmental Microbiology. 2011;**13**:835-850. DOI: 10.1111/j.1462-2920.2010.02378.x

[9] Barnhart MM, Chapman MR. Curli biogenesis and function. Annual Review of Microbiology. 2006;**60**:131-147. DOI: 10.1146/annurev.micro.60.080805.142106

[10] Giaouris E, Heir E, Desvaux M, Hébraud M, Møretrø T, Langsrud S, et al. Intra- and inter-species interactions within biofilms of important foodborne bacterial pathogens. Frontiers in Microbiology. 2015;**6**:841. DOI: 10.3389/fmicb.2015.00841

[11] Vestby LK, Møretrø T, Langsrud S, Heir E, Nesse LL. Biofilm forming abilities of *Salmonella* are correlated with persistence in fish meal and feed factories. BMC Veterinary Research. 2009;**5**:20. DOI: 10.1186/1746-6148-5-20

[12] Van Houdt R, Michiels CW. Role of bacterial cell surface structures in *Escherichia coli* biofilm formation. Research in Microbiology. 2005;**156**(5 6):626 633. DOI: 10.1016/j.resmic.2005.02.005

[13] Mandin P, Guillier M. Expanding control in bacteria: Interplay between small RNAs and transcriptional regulators to control gene expression. Current Opinion in Microbiology. 2013;**16**(2):125-132. DOI: 10.1016/j.mib.2012.12.005

[14] Vogeleer P, Tremblay YDN, Jubelin G, Jacques M, Harel J. Bioflm-forming abilities of Shiga toxin-producing *Escherichia coli* isolates associated with human infections. Applied and Environmental Microbiology. 2016;**82**(5):1448-1458. DOI: 10.1128/AEM.02983-15

[15] Alvarez-Ordóñez A, Coughlan LM, Briandet R, Cotter PD. Bioflms in food processing environments: Challenges and opportunities. Annual Review of Food Science and Technology.

2019;**10**:10.1-10.23. DOI: 10.1146/annurev-food-032818-121805

[16] Christison CA, Lindsay D, von Holy A. Cleaning and handling implements as potential reservoirs for bacterial contamination of some ready-to-eat foods in retail delicatessen environments. Journal of Food Protection. 2007;**70**:2878-2883. DOI: 10.4315/0362-028X-70.12.2878

[17] De Been M, Francke C, Moezelaar R, Abee T, Siezen RJ. Comparative analysis of two-component signal transduction systems of *Bacillus cereus, Bacillus thuringiensis* and *Bacillus anthracis*. Microbiology. 2006;**152**:3035-3048. DOI: 10.1099/mic.0.29137-0

[18] Houry A, Briandet R, Aymerich S, Gohar M. Involvement of motility and flagella in *Bacillus cereus* biofilm formation. Microbiology. 2010;**156**:1009-1018. DOI: 10.1099/mic.0.034827-0

[19] Houry A, Gohar M, Deschamps J, Tischenko E, Aymerich S, Gruss A, et al. Bacterial swimmers that infiltrate and take over the biofilm matrix. Proceedings of the National Academy of Sciences of the United States of America. 2012;**109**:13088-13093. DOI: 10.1073/pnas.1200791109

[20] Fratamico PM, Annous BA, Guenther NW. Biofilms in the Food and Beverage Industries. Oxford; Cambridge; Philadelphia, New Delhi: Woodhead Publishing Limited; 2012. ISBN: 978-1-84569-716-7

[21] Oniciuc EA, Cerca N, Nicolau AI. Compositional analysis of biofilms formed by *Staphylococcus aureus* isolated from food sources. Frontiers in Microbiology. 2016;7:390. DOI: 10.3389/fmicb.2016.00390

[22] Xue T, Chen X, Shang F. Short communication: Effects of lactose and milk on the expression of biofilm associated genes in *Staphylococcus aureus* strains isolated from a dairy cow with mastitis. Journal of Dairy Science. 2014;**97**(10):6129-6134. DOI: 10.3168/jds.2014-8344

[23] Tirumalai PS, Prakash S. Expression of virulence genes by *Listeria monocytogenes* J0161 in natural environment. Brazilian Journal of Microbiology. 2012;**43**(2):834-843. DOI: 10.1590/S1517-83822012000200050

[24] Alonso AN, Perry KJ, Regeimbal JM, Regan PM, Higgins DE. Identification of *Listeria monocytogenes* determinants required for biofilm formation. PLoS One. 2014;**9**(12):e113696. DOI: 10.1371/journal.pone.0113696

[25] Luo Q, Shang J, Feng X, Guo X, Zhang L, Zhou Q. PrfA led to reduced biofilm formation and contributed to altered gene expression patterns in biofilm-forming *Listeria monocytogenes*. Current Microbiology. 2013;**67**:372-378. DOI: 10.1007/s00284-013-0377-7

[26] Price R, Jayeola V, Niedermeyer J, Parsons C, Kathariou S. The *Listeria monocytogenes* key virulence determinants hly and prfA are involved in biofilm formation and aggregation but not colonization of fresh produce. Pathogens. 2018;7(1):E18. DOI: 10.3390/pathogens7010018

[27] Colagiorgi A, Bruini I, Di Ciccio PA, Zanardi E, Ghidini S, Ianieri A. *Listeria monocytogenes* biofilms in the wonderland of food industry. Pathogens. 2017;**6**(3):41. DOI: 10.3390/pathogens6030041

[28] Mack D, Fischer W, Krokotsch A, Leopold K, Hartmann R, Egge H, et al. The intercellular adhesin involved in biofilm accumulation of *Staphylococcus epidermidis* is a linear beta-1,6-linked glucosaminoglycan: Purification and structural analysis. Journal of Bacteriology. 1996;**178**:175-183. DOI: 10.1128/jb.178.1.175-183.1996

[29] Resch A, Rosenstein R, Nerz C, Götz F. Differential gene expression profiling of *Staphylococcus aureus* cultivated under biofilm and planktonic conditions. Applied and Environmental Microbiology. 2005;**71**(5):2663-2676. DOI: 10.1128/AEM.71.5.2663-2676.2005

[30] van der Woude MW. Re-examining the role and random nature of phase variation. FEMS Microbiology Letters. 2006;**254**:190-197. DOI: 10.1111/j.1574-6968.2005.00038.x

[31] Ranjith K, Arunasri K, Sathyanarayana Reddy G, Adicherla HK, Sharma S, Shivaji S. Global gene expression in *Escherichia coli*, isolated from the diseased ocular surface of the human eye with a potential to form biofilm. Gut Pathogens. 2017;**9**:15. DOI: 10.1186/s13099-017-0164-2

[32] Salazar JK, Deng D, Tortorello ML, Brandl MT, Wang H, Zhang W. Genes ycfR, sirA and yigG contribute to the surface attachment of *Salmonella enterica* Typhimurium and Saint Paul to fresh produce. PLoS One. 2013;**8**(2):e57272. DOI: 10.1371/journal.pone.0057272

[33] Chin KCJ, Duane T, Hebrard TM, Anbalagan K, Dashti MG, Phua KK. Transcriptomic study of *Salmonella enterica* subspecies enterica serovar Typhi biofilm. BMC Genomics. 2017;**18**:836. DOI: 10.1186/s12864-017-4212-6

[34] MacKenzie KD, Palmer MB, Köster WL, White AP. Examining the link between biofilm formation and the ability of pathogenic *Salmonella* strains to colonize multiple host species. Frontiers in Veterinary Science. 2017;**4**:138. DOI: 10.3389/fvets.2017.00138

[35] Gao T, Foulston L, Chai Y, Wang Q, Losick R. Alternative modes of biofilm formation by plant-associated *Bacillus cereus*. Microbiology Open. 2015;**4**:452-464. DOI: 10.1002/mbo3.251

[36] Majed R, Faille C, Kallassy M, Gohar M. *Bacillus cereus* biofilms—Same, only different. Frontiers in Microbiology. 2016;**7**:1054. DOI: 10.3389/fmicb.2016.01054

[37] Caro-Astorga J, Pérez-García A, De Vicente A, Romero D. A genomic region involved in the formation of adhesin fibers in *Bacillus cereus* biofilms. Frontiers in Microbiology. 2015;**5**:745. DOI: 10.3389/fmicb.2014.00745

[38] Álvarez-Ordóñez A, Alvseike O, Omer MK, Heir E, Axelsson L, Holck A, et al. Heterogeneity in resistance to food-related stresses and biofilm formation ability among verocytotoxigenic *Escherichia coli* strains. International Journal of Food Microbiology. 2013;**161**(3):220-230. DOI: 10.1016/j.ijfoodmicro.2012.12.008

[39] Schauder S, Bassler BL. The languages of bacteria. Genes and Development. 2001;**15**(12):1468-1480. DOI: 10.1101/gad.899601

[40] Zhao L, Xue T, Shang F, Sun H, Sun B. *Staphylococcus aureus* AI-2 quorum sensing associates with the KdpDE two-component system to regulate capsular polysaccharide synthesis and virulence. Infection and Immunity. 2010;**78**(8):3506-3515. DOI: 10.1128/IAI.00131-10

[41] Fluckiger U, Ulrich M, Steinhuber A, Döring G, Mack D, Landmann R, et al. Biofilm formation, icaADBC transcription, and polysaccharide intercellular adhesin synthesis by staphylococci in a device-related infection model. Infection and Immunity. 2005;**73**(3):1811-1819. DOI: 10.1128/IAI.73.3.1811-1819.2005

[42] Speziale P, Pietrocola G, Foster TJ, Geoghegan JA. Protein-based biofilm matrices in staphylococci. Frontiers in Cellular and Infection Microbiology. 2014;**4**:171. DOI: 10.3389/fcimb.2014.00171

[43] Holland LM, O'Donnell ST, Ryjenkov DA, Gomelsky L, Slater SR, Fey PD, et al. A staphylococcal GGDEF domain protein regulates biofilm formation independently of cyclic dimeric GMP. Journal of Bacteriology. 2008;**190**:5178-5189. DOI: 10.1128/JB.00375-08

[44] Romilly C, Caldelari I, Parmentier D, Lioliou E, Romby P, Fechter P. Current knowledge on regulatory RNAs and their machineries in *Staphylococcus aureus*. RNA Biology. 2012;**9**:402-413. DOI: 10.4161/rna.20103

[45] Fujii T, Ingham C, Nakayama J, Beerthuyzen M, Kunuki R, Molenaar D, et al. Two homologous Agr-like quorum-sensing systems cooperatively control adherence, cell morphology, and cell viability properties in *Lactobacillus plantarum* WCFS1. Journal of Bacteriology. 2008;**190**:7655-7665. DOI: 10.1128/JB.01489-07

[46] Riedel CU, Monk IR, Casey PG, Waidmann MS, Gahan CGM, Hill C.AgrD-dependent quorum sensing affects biofilm formation, invasion, virulence and global gene expression profiles in *Listeria monocytogenes*. Molecular Microbiology. 2009;**71**:1177-1189. DOI: 10.1111/j.1365-2958.2008.06589.x

[47] Wolska KI, Grudniak AM, Rudnicka Z, Markowska K. Genetic control of bacterial biofilms. Journal of Applied Genetics. 2016;**57**:225-238. DOI: 10.1007/s13353-015-0309-2

[48] Surette MG, Miller MB, Bassler BL. Quorum sensing in *Escherichia coli*, *Salmonella* Typhimurium, and *Vibrio harveyi*: A new family of genes responsible for autoinducer production. Proceeding of the National Academy of Sciences of the USA. 1999;**96**:1639-1644. DOI: 10.1073/pnas.96.4.1639

[49] Challan Belval S, Gal L, Margiewes S, Garmyn D, Piveteau P, Guzzo J. Assessment of the roles of LuxS, S-ribosyl homocysteine, and autoinducer 2 in cell attachment during biofilm formation by *Listeria monocytogenes* EGD-e. Applied and Environmental Microbiology. 2006;**72**:2644-2650. DOI: 10.1128/AEM.72.4.2644-2650.2006

[50] Renier S, Chagnot C, Deschamps J, Caccia N, Szlavik J, Joyce SA. Inactivation of the SecA2 protein export pathway in *Listeria monocytogenes* promotes cell aggregation, impacts biofilm architecture and induces biofilm formation in environmental conditions. Environmental Microbiology. 2014;**16**:1176-1192. DOI: 10.1111/1462-2920.12257

[51] Flemming HC, Wingender J, Szewzyk U, Steinberg P, Rice SA, Kjelleberg S. Biofilms: An emergent form of bacterial life. Nature Reviews Microbiology. 2016;**14**:563-575. DOI: 10.1038/nrmicro.2016.94

[52] Zacharof MP, Lovitt RW. Bacteriocins produced by lactic acid bacteria a review article. APCBEE Procedia. 2012;**2**:50-56. DOI: 10.1016/j.apcbee.2012.06.010

[53] Riley MA, Wertz JE. Bacteriocins: Evolution, ecology, and application. Annual Review of Microbiology. 2002;**56**:117-137. DOI: 10.1146/annurev.micro.56.012302.16

[54] de Arauz LJ, Jozala AF, Mazzola PG, Vessoni Penna TC. Nisin biotechnological production and application: A review. Trends in Food Science & Technology. 2009;**20**:146-154. DOI: 10.1016/j.tifs.2009.01.056

[55] Godoy-Santos F, Pitts B, Stewart PS, Mantovani HC. Nisin penetration and efficacy against *Staphylococcus aureus* biofilms under continuous-flow conditions. Microbiology. 2019;**165**:761-771. DOI: 10.1099/mic.0.000804

[56] Field D, O'Connor R, Cotter PD, Ross RP, Hill C. *In vitro* activities of nisin and nisin derivatives alone and in combination with antibiotics against *Staphylococcus* biofilms. Frontiers in Microbiology. 2016;**7**:1-11. DOI: 10.3389/fmicb.2016.00508

[57] Arevalos-Sánchez M, Regalado C, Martin SE, Domínguez-Domínguez J, García-Almendárez BE. Effect of neutral electrolyzed water and nisin on *Listeria monocytogenes* biofilms, and on listeriolysin O activity. Food Control. 2012;**24**:116-122. DOI: 10.1016/j.foodcont.2011.09.012

[58] Yüksel FN, Buzrul S, Akçelik M, Akçelik N. Inhibition and eradication of *Salmonella* Typhimurium biofilm using P22 bacteriophage, EDTA and nisin. Biofouling. 2018;**34**:1046-1054. DOI: 10.1080/08927014.2018.1538412

[59] Kumar CG, Anand SK. Significance of microbial biofilms in food industry: A review. International Journal of Food Microbiology. 1998;**42**:9-27. DOI: 10.1016/s0168-1605(98)00060-9

[60] Guerra NP, Araujo AB, Barrera AM, Agrasar AT, Macías CL, Carballo J, et al. Antimicrobial activity of nisin adsorbed to surfaces commonly used in the food industry. Journal of Food Protection. 2005;**68**:1012-1019. DOI: 10.4315/0362-028x-68.5.1012

[61] Daeschel MA, Mcguire J, Al-Makhlafi H. Antimicrobial activity of nisin adsorbed to hydrophilic and hydrophobic silicon surfaces. Journal of Food Protection. 1992;**55**:731-735. DOI: 10.4315/0362-028x-55.9.731

[62] Bower CK, McGuire J, Daeschel MA. Suppression of *Listeria monocytogenes* colonization following adsorption of nisin onto silica surfaces. Applied and Environmental Microbiology. 1995;**61**:992-997

[63] Chopra L, Singh G, Kumar Jena K, Sahoo DK. Sonorensin: A new bacteriocin with potential of an anti-biofilm agent and a food biopreservative. Scientific Reports. 2015;**5**:13412. DOI: 10.1038/srep13412

[64] Stiefel P, Mauerhofer S, Schneider J, Maniura-Weber K, Rosenberg U, Ren Q. Enzymes enhance biofilm removal efficiency of cleaners. Antimicrobial Agents and Chemotherapy. 2016;**60**:3647-3652. DOI: 10.1128/AAC.00400-16

[65] Liu X, Tang B, Gu Q, Yu X. Elimination of the formation of biofilm in industrial pipes using enzyme cleaning technique. MethodsX. 2014;**1**:130-136. DOI: 10.1016/j.mex.2014.08.008

[66] Torres CE, Lenon G, Craperi D, Wilting R, Blanco Á. Enzymatic treatment for preventing biofilm formation in the paper industry. Applied Microbiology and Biotechnology. 2011;**92**:95-103. DOI: 10.1007/s00253-011-3305-4

[67] Contesini FJ, de Melo RR, Sato HH. An overview of *Bacillus* proteases: From production to application. Critical Reviews in Biotechnology. 2017;**38**:321-334. DOI: 10.1080/07388551.2017.1354354

[68] Huang H, Ren H, Ding L, Geng J, Xu K, Zhang Y. Aging biofilm from a full-scale moving bed biofilm reactor: Characterization and enzymatic treatment study. Bioresource Technology. 2014;**154**:122-130. DOI: 10.1016/j.biortech.2013.12.031

[69] Lequette Y, Boels G, Clarisse M, Faille C. Using enzymes to remove biofilms of bacterial isolates sampled in the food-industry. Biofouling. 2010;**26**:421-431. DOI: 10.1080/08927011003699535

[70] Kumari P, Jandaik S, Batta S. Role of extracellular proteases in biofilm degradation produced from *Escherichia*

coli and *Pseudomonas mendocina*. International Journal of Current Microbiology and Applied Sciences. 2018;7:1786-1795. DOI: 10.20546/ijcmas.2018.706.212

[71] Nguyen UT, Burrows LL. DNase I and proteinase K impair *Listeria monocytogenes* biofilm formation and induce dispersal of pre-existing biofilms. International Journal of Food Microbiology. 2014;**187**:26-32. DOI: 10.1016/j.ijfoodmicro.2014.06.025

[72] Johansen C, Falholt P, Gram L. Enzymatic removal and disinfection of bacterial biofilms. Applied and environmental microbiology. 1997;**63**:3724-3728

[73] Wang H, Wang H, Xing T, Wu N, Xu X, Zhou G. Removal of *Salmonella* biofilm formed under meat processing environment by surfactant in combination with bio-enzyme. LWT—Food Science and Technology. 2016;**66**:298-304. DOI: 10.1016/j.lwt.2015.10.049

[74] Tajkarimi MM, Ibrahim SA, Cliver DO. Antimicrobial herb and spice compounds in food. Food Control. 2010;**21**:1199-1218. DOI: 10.1016/j.foodcont.2010.02.003

[75] Nazzaro F, Fratianni F, de Martino L, Coppola R, de Feo V. Effect of essential oils on pathogenic bacteria. Pharmaceuticals (Basel, Switzerland). 2013;**6**:1451-1474. DOI: 10.3390/ph6121451

[76] Swamy MK, Akhtar MS, Sinniah UR. Antimicrobial properties of plant essential oils against human pathogens and their mode of action: An updated review. Evidence-Based Complementary and Alternative Medicine. 2016;**2016**:1-21. DOI: 10.1155/2016/3012462

[77] Leonard CM, Virijevic S, Regnier T, Combrinck S. Bioactivity of selected essential oils and some components on *Listeria monocytogenes* biofilms. South African Journal of Botany. 2010;**76**:676-680. DOI: 10.1016/j.sajb.2010.07.002

[78] Sandasi M, Leonard CM, Viljoen AM. The effect of five common essential oil components on *Listeria monocytogenes* biofilms. Food Control. 2008;**19**:1070-1075. DOI: 10.1016/j.foodcont.2007.11.006

[79] de Oliveira MMM, Brugnera DF, das Graças Cardoso M, Alves E, Piccoli RH. Disinfectant action of *Cymbopogon* sp. essential oils in different phases of biofilm formation by *Listeria monocytogenes* on stainless steel surface. Food Control. 2010;**21**:549-553. DOI: 10.1016/j.foodcont.2009.08.003

[80] Borugă O, Jianu C, Mişcă C, Goleţ I, Gruia AT, Horhat FG. *Thymus vulgaris* essential oil: Chemical composition and antimicrobial activity. Journal of Medicine and Life. 2014;**3**:56-60

[81] Kang J, Liu L, Wu X, Sun Y, Liu Z. Effect of thyme essential oil against *Bacillus cereus* planktonic growth and biofilm formation. Applied Microbiology and Biotechnology. 2018;**102**:10209-10218. DOI: 10.1007/s00253-018-9401-y

[82] Mohsenipour Z, Hassanshahian M. The inhibitory effect of *Thymus vulgaris* extracts on the planktonic form and biofilm structures of six human pathogenic bacteria. Avicenna Journal of Phytomedicine. 2015;**5**:309-318

[83] Szczepanski S, Lipski A. Essential oils show specific inhibiting effects on bacterial biofilm formation. Food Control. 2014;**36**:224-229. DOI: 10.1016/j.foodcont.2013.08.023

[84] Amiri H. Essential oils composition and antioxidant properties of three thümus species. Evidence-based Complementary and Alternative

Medicine: eCAM. 2012;**2012**:1-8. DOI: 10.1155/2012/728065

[85] Pérez-Conesa D, Cao J, Chen L, Mclandsborough L, Weiss J. Inactivation of *Listeria monocytogenes* and *Escherichia coli* O157:H7 biofilms by micelle-encapsulated eugenol and carvacrol. Journal of Food Protection. 2011;**74**:55-62. DOI: 10.4315/0362-028x.jfp-08-403

[86] Knowles JR, Roller S, Murray DB, Naidu AS. Antimicrobial action of carvacrol at different stages of dual-species biofilm development by *Staphylococcus aureus* and *Salmonella enterica* serovar Typhimurium. Applied and Environmental Microbiology. 2005;**71**:797-803. DOI: 10.1128/AEM.71.2.797-803.2005

[87] dos Santos Rodrigues JB, de Carvalho RJ, de Souza NT, de Sousa Oliveira K, Franco OL, Schaffner D, et al. Effects of oregano essential oil and carvacrol on biofilms of *Staphylococcus aureus* from food-contact surfaces. Food Control. 2017;**73**:1237-1246. DOI: 10.1016/j.foodcont.2016.10.043

[88] Miladi H, Zmantar T, Kouidhi B, Chaabouni Y, Mahdouani K, Bakhrouf A, et al. Use of carvacrol, thymol, and eugenol for biofilm eradication and resistance modifying susceptibility of *Salmonella enterica* serovar Typhimurium strains to nalidixic acid. Microbial Pathogenesis. 2017;**104**:56-63. DOI: 10.1016/j.micpath.2017.01.012

[89] Bendre RS, Rajput JD, Bagul SD, Karandikar S. Outlooks on medicinal properties of eugenol and its synthetic derivatives. Natural Products Chemistry & Research. 2016;**4**:1-6. DOI: 10.4172/2329-6836.1000212

[90] Li YH, Tian X. Quorum sensing and bacterial social interactions in biofilms. Sensors (Basel, Switzerland). 2012;**12**:32519-32538. DOI: 10.3390/s120302519

[91] Kim YG, Lee JH, Gwon G, Kim SI, Park JG, Lee J. Essential oils and eugenols inhibit biofilm formation and the virulence of *Escherichia coli* O157:H7. Scientific Reports. 2016;**6**:1-11. DOI: 10.1038/srep36377

[92] Principi N, Silvestri E, Esposito S. Advantages and limitations of bacteriophages for the treatment of bacterial infections. Frontiers in Pharmacology. 2019;**10**:1-9. DOI: 10.3389/fphar.2019.00513

[93] Bai J, Kim YT, Ryu S, Lee JH. Biocontrol and rapid detection of food-borne pathogens using bacteriophages and endolysins. Frontiers in Microbiology. 2016;7:1-15. DOI: 10.3389/fmicb.2016.00474

[94] Hagens S, Loessner MJ. Application of bacteriophages for detection and control of foodborne pathogens. Applied Microbiology and Biotechnology. 2007;**76**:513-519. DOI: 10.1007/s00253-007-1031-8

[95] Parasion S, Kwiatek M, Gryko R, Mizak L, Malm A. Bacteriophages as an alternative strategy for fighting biofilm development. Polish Journal of Microbiology. 2014;**63**:137-145

[96] Harper DR, Parracho H, Walker J, Sharp R, Hughes G, Werthén M, et al. Bacteriophages and biofilms. Antibiotics. 2014;**3**:270-284. DOI: 10.3390/antibiotics3030270

[97] Hibma A. Infection and removal of L-forms of *Listeria monocytogenes* with bred bacteriophage. International Journal of Food Microbiology. 1997;**34**:197-207. DOI: 10.1016/s0168-1605(96)01190-7

[98] Soni KA, Nannapaneni R. Removal of *Listeria monocytogenes* biofilms with bacteriophage P100. Journal of Food Protection. 2010;**73**:1519-1524. DOI: 10.4315/0362-028X-73.8.1519

[99] Arachchi GJG, Cridge AG, Dias-Wanigasekera BM, Cruz CD, McIntyre L, Liu R, et al. Effectiveness of phages in the decontamination of *Listeria monocytogenes* adhered to clean stainless steel, stainless steel coated with fish protein, and as a biofilm. Journal of Industrial Microbiology & Biotechnology. 2013;**40**:1105-1116. DOI: 10.1007/s10295-013-1313-3

[100] Garcia KCOD, Corrêa IMO, Pereira LQ, Silva TM, Mioni MSR, Izidoro ACM, et al. Bacteriophage use to control *Salmonella* biofilm on surfaces present in chicken slaughterhouses. Poultry Science. 2017;**96**:3392-3398. DOI: 10.3382/ps/pex124

[101] Sadekuzzaman M, Yang S, Mizan MFR, Ha SD. Reduction of *Escherichia coli* O157:H7 in biofilms using bacteriophage BPECO 19. Journal of Food Science. 2017;**82**:1433-1442. DOI: 10.1111/1750-3841.13729

[102] Donsì F, Ferrari G. Essential oil nanoemulsions as antimicrobial agents in food. Journal of Biotechnology. 2016;**233**:106-120. DOI: 10.1016/j.jbiotec.2016.07.005

[103] Prakash A, Baskaran R, Paramasivam N, Vadivel V. Essential oil based nanoemulsions to improve the microbial quality of minimally processed fruits and vegetables: A review. Food Research International. 2018;**111**:509-523. DOI: 10.1016/j.foodres.2018.05.066

[104] Moghimi R, Ghaderi L, Rafati H, Aliahmadi A, McClements DJ. Superior antibacterial activity of nanoemulsion of *Thymus daenensis* essential oil against *E. coli*. Food Chemistry. 2016;**194**:410-415. DOI: 10.1016/j.foodchem.2015.07.139

[105] Donsì F, Cuomo A, Marchese E, Ferrari G. Infusion of essential oils for food stabilization: Unraveling the role of nanoemulsion-based delivery systems on mass transfer and antimicrobial activity. Innovative Food Science & Emerging Technologies. 2014;**22**:212-220. DOI: 10.1016/j.ifset.2014.01.008

[106] Shah AA, Khan A, Dwivedi S, Musarrat J, Mint AA. Antibacterial and antibiofilm activity of barium titanate nanoparticles. Materials Letters. 2018;**229**:130-133. DOI: 10.1016/j.matlet.2018.06.107

[107] Prakash A, Vadivel V, Rubini D, Nithyanand P. Antibacterial and antibiofilm activities of linalool nanoemulsions against *Salmonella* Typhimurium. Food Bioscience. 2019;**28**:57-65. DOI: 10.1016/j.fbio.2019.01.018

[108] Gonçalves RC, da Silva DP, Signini R, Naves PLF. Inhibition of bacterial biofilms by carboxymethyl chitosan combined with silver, zinc and copper salts. International Journal of Biological Macromolecules. 2017;**105**:385-392. DOI: 10.1016/j.ijbiomac.2017.07.048

[109] Zanetti M, Carniel TK, Dalcanton F, dos Anjos RS, Gracher Riella H, de Araújo PHH, et al. Use of encapsulated natural compounds as antimicrobial additives in food packaging: A brief review. Trends in Food Science and Technology. 2018;**81**:51-60. DOI: 10.1016/j.tifs.2018.09.003

[110] Subhaswaraj P, Barik S, Macha C, Chiranjeevi PV, Siddhardha B. Anti-quorum sensing and anti-biofilm efficacy of cinnamaldehyde encapsulated chitosan nanoparticles against *Pseudomonas aeruginosa* PAO1. LWT. 2018;**97**:752-759. DOI: 10.1016/j.lwt.2018.08.011

[111] Vijayakumar S, Malaikozhundan B, Saravanakumar K, Durán-Lara EF, Wang MH, Vaseeharan B. Garlic clove extract assisted silver nanoparticle—Antibacterial, antibiofilm,

antihelminthic, anti-inflammatory, anticancer and ecotoxicity assessment. Journal of Photochemistry and Photobiology B: Biology. 2019;**198**:111558. DOI: 10.1016/j.jphotobiol.2019.111558

[112] Comin VM, Lopes LQS, Quatrin PM, de Souza ME, Bonez PC, Pintos FG, et al. Influence of *Melaleuca alternifolia* oil nanoparticles on aspects of *Pseudomonas aeruginosa* biofilm. Microbial Pathogenesis. 2016;**93**:120-125. DOI: 10.1016/j.micpath.2016.01.019

[113] Manju S, Malaikozhundan B, Vijayakumar S, Shanthi S, Jaishabanu A, Ekambaram P, et al. Antibacterial, antibiofilm and cytotoxic effects of *Nigella sativa* essential oil coated gold nanoparticles. Microbial Pathogenesis. 2016;**91**:129-135. DOI: 10.1016/j.micpath.2015.11.021

[114] Wu X, Wang Y, Tao L. Sulfhydryl compounds reduce *Staphylococcus aureus* biofilm formation by inhibiting PIA biosynthesis. FEMS Microbiology Letters. 2011;**316**:44-50. DOI: 10.1111/j.1574-6968.2010.02190.x

[115] Liang Z, Qi Y, Guo S, Hao K, Zhao M, Guo N. Effect of AgWPA nanoparticles on the inhibition of *Staphylococcus aureus* growth in biofilms. Food Control. 2019;**100**:240-246. DOI: 10.1016/j.foodcont.2019.01.030

[116] Naskar A, Khan H, Sarkar R, Kumar S, Halder D, Jana S. Anti-biofilm activity and food packaging application of room temperature solution process based polyethylene glycol capped Ag-ZnO-graphene nanocomposite. Materials Science and Engineering: C. 2018;**C91**:743-753. DOI: 10.1016/j.msec.2018.06.009

[117] Akhil K, Jayakumar J, Gayathri G, Khan SS. Effect of various capping agents on photocatalytic, antibacterial and antibiofilm activities of ZnO nanoparticles. Journal of Photochemistry and Photobiology B: Biology. 2016;**160**:32-42. DOI: 10.1016/j.jphotobiol.2016.03.015

[118] Pozo NI, Olmos D, Orgaz B, Božanić DK, González-Benito J. Titania nanoparticles prevent development of *Pseudomonas fluorescens* biofilms on polystyrene surfaces. Materials Letters. 2014;**127**:1-3. DOI: 10.1016/j.matlet.2014.04.073

[119] Chorianopoulos NG, Tsoukleris DS, Panagou EZ, Falaras P, Nychas GJE. Use of titanium dioxide (TiO_2) photocatalysts as alternative means for *Listeria monocytogenes* biofilm disinfection in food processing. Food Microbiology. 2011;**28**:164-170. DOI: 10.1016/j.fm.2010.07.025

[120] Goswami SR, Sahareen T, Singh M, Kumar S. Role of biogenic silver nanoparticles in disruption of cell-cell adhesion in *Staphylococcus aureus* and *Escherichia coli* biofilm. Journal of Industrial and Engineering Chemistry. 2015;**26**:73-78. DOI: 10.1016/j.jiec.2014.11.017

[121] Ranmadugala D, Ebrahiminezhad A, Manley-Harris M, Ghasemi Y, Berenjian A. The effect of iron oxide nanoparticles on *Bacillus subtilis* biofilm, growth and viability. Process Biochemistry. 2017;**62**:231-234. DOI: 10.1016/j.procbio.2017.07.003

[122] Govaert M, Smet C, Vergauwen L, Ećimović B, Walsh JL, Baka M, et al. Influence of plasma characteristics on the efficacy of cold atmospheric plasma (CAP) for inactivation of *Listeria monocytogenes* and *Salmonella* Typhimurium biofilms. Innovative Food Science and Emerging Technologies. 2019;**52**:376-386. DOI: 10.1016/j.ifset.2019.01.013

[123] Ziuzina D, Han L, Cullen PJ, Bourke P. Cold plasma inactivation of internalised bacteria and biofilms for *Salmonella enterica* serovar

Typhimurium, *Listeria monocytogenes* and *Escherichia coli*. International Journal of Food Microbiology. 2015;**210**:53-61. DOI: 10.1016/j.ijfoodmicro.2015.05.019

[124] Patange A, Boehm D, Ziuzina D, Cullen PJ, Gilmore B, Bourke P. High voltage atmospheric cold air plasma control of bacterial biofilms on fresh produce. International Journal of Food Microbiology. 2019;**293**:137-145. DOI: 10.1016/j.ijfoodmicro.2019.01.005

[125] Czapka T, Maliszewska I, Olesiak-Bańska J. Influence of atmospheric pressure non-thermal plasma on inactivation of biofilm cells. Plasma Chemistry and Plasma Processing. 2018;**38**:1181-1197. DOI: 10.1007/s11090-018-9925-z

[126] Koban I, Geisel MH, Holtfreter B, Jablonowski L, Hübner N-O, Matthes R, et al. Synergistic effects of nonthermal plasma and disinfecting agents against dental biofilms *in vitro*. ISRN Dentistry. 2013;**2013**:1-10. DOI: 10.1155/2013/573262

[127] Gupta TT, Karki SB, Matson JS, Gehling DJ, Ayan H. Sterilization of biofilm on a titanium surface using a combination of nonthermal plasma and chlorhexidine digluconate. BioMed Research International. 2017;**2017**:6085741. DOI: 10.1155/2017/6085741

[128] Múgica-Vidal R, Sainz-García E, Álvarez-Ordóñez A, Prieto M, González-Raurich M, López M, et al. Production of antibacterial coatings through atmospheric pressure plasma: A promising alternative for combatting biofilms in the food industry. Food and Bioprocess Technology. 2019;**12**:1251-1263. DOI: 10.1007/s11947-019-02293-z

[129] Cappitelli F, Polo A, Villa F. Biofilm formation in food processing environments is still poorly understood and controlled. Food Engineering Reviews. 2014;**6**:29-42. DOI: 10.1007/s12393-014-9077-8

[130] Oulahal-Lagsir N, Martial-Gros A, Bonneau M, Blum LJ. *Escherichia coli*-milk biofilm removal from stainless steel surfaces: Synergism between ultrasonic waves and enzymes. Biofouling. 2003;**19**(3):159-168. DOI: 10.1080/0892701031000064676

[131] Oulahal N, Martial-Gros A, Bonneau M, Blum LJ. Combined effect of chelating agents and ultrasound on biofilm removal from stainless steel surfaces. Application to "*Escherichia coli* milk" and "*Staphylococcus aureus* milk" biofilms. Biofilms. 2004;**1**:65-73. DOI: 10.1017/s1479050504001140

[132] Hussain MS, Kwon M, Park E, Seheli K, Huque R, Oh DH. Disinfection of *Bacillus cereus* biofilms on leafy green vegetables with slightly acidic electrolyzed water, ultrasound and mild heat. Lwt. 2019;**116**:108582. DOI: 10.1016/j.lwt.2019.108582

[133] Bang HJ, Park SY, Kim SE, Md Furkanur Rahaman M, Ha SD. Synergistic effects of combined ultrasound and peroxyacetic acid treatments against *Cronobacter sakazakii* biofilms on fresh cucumber. LWT—Food Science and Technology. 2017;**84**:91-98. DOI: 10.1016/j.lwt.2017.05.037

[134] Fink R, Oder M, Stražar E, Filip S. Efficacy of cleaning methods for the removal of *Bacillus cereus* biofilm from polyurethane conveyor belts in bakeries. Food Control. 2017;**80**:267-272. DOI: 10.1016/j.foodcont.2017.05.009

[135] Pedrós-Garrido S, Condón-Abanto S, Clemente I, Beltrán JA, Lyng JG, Bolton D, et al. Efficacy of ultraviolet light (UV-C) and pulsed light (PL) for the microbiological decontamination of raw salmon (*Salmo salar*) and food contact surface materials. Innovative Food Science and

Emerging Technologies. 2018;**50**:124-131. DOI: 10.1016/j.ifset.2018.10.001

[136] Turtoi M, Borda D. Decontamination of egg shells using ultraviolet light treatment. World's Poultry Science Journal. 2014;**70**(2):265-278. DOI: 10.1017/S0043933914000282

[137] Mahendran R, Ramanan KR, Barba FJ, Lorenzo JM, López-Fernández O, Munekata PES, et al. Recent advances in the application of pulsed light processing for improving food safety and increasing shelf life. Trends in Food Science and Technology. 2019;**88**:67-79. DOI: 10.1016/j.tifs.2019.03.010

[138] Garvey M, Rowan NJ. Pulsed UV as a potential surface sanitizer in food production processes to ensure consumer safety. Current Opinion in Food Science. 2019;**26**:65-70. DOI: 10.1016/j.cofs.2019.03.003

[139] Rajkovic A, Tomasevic I, De Meulenaer B, Devlieghere F. The effect of pulsed UV light on *Escherichia coli* O157:H7, *Listeria monocytogenes*, *Salmonella* Typhimurium, *Staphylococcus aureus* and staphylococcal enterotoxin A on sliced fermented salami and its chemical quality. Food Control. 2017;**73**:829-837. DOI: 10.1016/j.foodcont.2016.09.029

[140] Cossu A, Ercan D, Wang Q, Peer WA, Nitin N, Tikekar RV. Antimicrobial effect of synergistic interaction between UV-A light and gallic acid against *Escherichia coli* O157:H7 in fresh produce wash water and biofilm. Innovative Food Science and Emerging Technologies. 2016;**37**:44-52. DOI: 10.1016/j.ifset.2016.07.020

[141] Gora SL, Rauch KD, Ontiveros CC, Stoddart AK, Gagnon GA. Inactivation of biofilm-bound *Pseudomonas aeruginosa* bacteria using UVC light emitting diodes (UVC LEDs). Water Research. 2019;**151**:193-202. DOI: 10.1016/j.watres.2018.12.021

[142] Garner AL. Pulsed electric field inactivation of microorganisms: From fundamental biophysics to synergistic treatments. Applied Microbiology and Biotechnology. 2019;**103**:7917-7929. DOI: 10.1007/s00253-019-10067-y

[143] Halpin RM, Duffy L, Cregenzán-Alberti O, Lyng JG, Noci F. The effect of non-thermal processing technologies on microbial inactivation: An investigation into sub-lethal injury of *Escherichia coli* and *Pseudomonas fluorescens*. Food Control. 2014;**41**:106-115. DOI: 10.1016/j.foodcont.2014.01.011

[144] Monfort S, Gayán E, Saldaña G, Puértolas E, Condón S, Raso J, et al. Inactivation of *Salmonella* Typhimurium and *Staphylococcus aureus* by pulsed electric fields in liquid whole egg. Innovative Food Science and Emerging Technologies. 2010;**11**:306-313. DOI: 10.1016/j.ifset.2009.11.007

[145] Bleoancă I, Saje K, Mihalcea L, Oniciuc EA, Smole-Mozina S, Nicolau AI, et al. Contribution of high pressure and thyme extract to control *Listeria monocytogenes* in fresh cheese—A hurdle approach. Innovative Food Science and Emerging Technologies. 2016;**38**:7-14. DOI: 10.1016/j.ifset.2016.09.002

[146] Buckow R, Weiss U, Knorr D. Inactivation kinetics of apple polyphenol oxidase in different pressure–temperature domains. Innovative Food Science & Emerging Technologies. 2009;**10**:441-448. DOI: 10.1016/j.ifset.2009.05.005

[147] Zetzmann M, Bucur FI, Crauwels P, Borda D, Nicolau AI, Grigore-Gurgu L, et al. Characterization of the biofilm phenotype of a *Listeria monocytogenes* mutant deficient in *agr* peptide sensing. MicrobiologyOpen. 2019;**8**:1-9. DOI: 10.1002/mbo3.826

[148] Gross M, Cramton SE, Gotz F, Peschel A. Key role of teichoic acid

net charge in *Staphylococcus aureus* colonization of artificial surfaces. Infection and Immunity. 2001;**69**:3423-3426. DOI: 10.1128/IAI.69.5.3423-3426.2001

[149] Kot B, Sytykiewicz H, Sprawka I. Expression of the biofilm-associated genes in methicillin-resistant *Staphylococcus aureus* in biofilm and planktonic conditions. International Journal of Molecular Science. 2018;**19**(11):3487. DOI: 10.3390/ijms19113487

[150] Hecker M, Mäder U, Völker U. From the genome sequence via the proteome to cell physiology—Pathoproteomics and pathophysiology of *Staphylococcus aureus*. International Journal of Medical Microbiology. 2018;**308**:545-557. DOI: 10.1016/j.ijmm.2018.01.002

[151] Yang YH, Jiang YL, Zhang J, Wang L, Bai XH, Zhang SJ, et al. Structural insights into SraP-mediated *Staphylococcus aureus* adhesion to host cells. PLoS Pathogens. 2014;**10**:e1004169. DOI: 10:e1004169-e1004169

[152] Chastanet A, Losick R. Just-in-time control of spo0A synthesis in *Bacillus subtilis* by multiple regulatory mechanisms. Journal of Bacteriology. 2011;**193**:6366-6374. DOI: 10.1128/JB.06057-11

[153] Mirouze N, Prepiak P, Dubnau D. Fluctuations in spo0A transcription control rare developmental transitions in *Bacillus subtilis*. PLoS Genetics. 2011;7:e1002048. DOI: 10.1371/journal.pgen.1002048

[154] Jers C, Kobir A, Sondergaard EO, Jensen PR, Mijakovic I. *Bacillus subtilis* two-component system sensory kinase DegS is regulated by serine phosphorylation in its input domain. PLoS One. 2011;**6**:E14653-E14653. DOI: 10.1371/journal.pone.0014653

[155] Murray EJ, Kiley TB, Stanley-Wall NR. A pivotal role for the response regulator DegU in controlling multicellular behaviour. Microbiology. 2009;**155**:1-8. DOI: 10.1099/mic.0.023903-0

[156] Spöring I, Felgner S, Preuße M, Eckweiler D, Rohde M, Häussler S, et al. Regulation of flagellum biosynthesis in response to cell envelope stress in *Salmonella enterica* serovar Typhimurium. MBio. 2018;**9**(3):e00736-e00717. DOI: 10.1128/mBio.00736-17

[157] Alikhan NF, Zhou Z, Sergeant MJ, Achtman M. A genomic overview of the population structure of *Salmonella*. PLOS Genetics. 2018;**14**:e1007261. DOI: 10.1371/journal.pgen.1007261

[158] Barbu EM, Ganesh VK, Gurusiddappa S, Mackenzie RC, Foster TJ, Sudhof TC, et al. β-Neurexin is a ligand for the *Staphylococcus aureus* MSCRAMM SdrC. PLoS Pathogens. 2010;**6**:e1000726. DOI: 10.1371/journal.ppat.1000726

[159] Zhang X-S, García-Contreras R, Wood TK. YcfR (BhsA) influences *Escherichia coli* biofilm formation through stress response and surface hydrophobicity. Journal of Bacteriology. 2007;**189**:3051-3062. DOI: 10.1128/JB.01832-06

[160] Weber MM, French CL, Barnes MB, Siegele DA, McLean RJ. A previously uncharacterized gene, yjfO (bsmA), influences *Escherichia coli* biofilm formation and stress response. Microbiology. 2010;**156**:139-147. DOI: 10.1099/mic.0.031468-0

[161] Grantcharova N, Peters V, Monteiro C, Zakikhany K, Römling U. Bistable expression of CsgD in biofilm development of *Salmonella enterica* Serovar Typhimurium. Journal of Bacteriology. 2009;**192**:456-466. DOI: 10.1128/JB.01826-08

[162] Brown PK, Dozois CM, Nickerson CA, Zuppardo A, Terlonge J, Curtiss R 3rd. MlrA, a novel regulator of curli (AgF) and extracellular matrix synthesis by *Escherichia coli* and *Salmonella enterica* serovar Typhimurium. Molecular Microbiology. 2001;**41**:349-363. DOI: 10.1046/j.1365-2958.2001.02529.x

[163] Kearns DB, Chu F, Branda SS, Kolter R, Losick R. A master regulator for biofilm formation by *Bacillus subtilis*. Molecular Microbiology. 2005;**55**:739-749. DOI: 10.1111/j.1365-2958.2004.04440.x

[164] Kodgire P, Dixit M, Rao KKJ. ScoC and SinR negatively regulate epr by corepression in *Bacillus subtilis*. Journal of Bacteriology. 2006;**188**:6425-6428. DOI: 10.1128/JB.00427-06

[165] López D, Vlamakis H, Losick R, Kolter R. Cannibalism enhances biofilm development in *Bacillus subtilis*. Molecular Microbiology. 2009;**74**:609-618. DOI: 10.1111/j.1365-2958.2009.06882.x

[166] Vlamakis H, Chai Y, Beauregard P, Losick R, Kolter R. Sticking together: Building a biofilm the *Bacillus subtilis* way. Nature Reviews. Microbiology. 2013;**11**(3):157-168. DOI: 10.1038/nrmicro2960

[167] Lazarevic V, Soldo B, Médico N, Pooley H, Bron S, Karamata D. *Bacillus subtilis* α-phosphoglucomutase is required for normal cell morphology and biofilm formation. Applied and Environmental Microbiology. 2005;**71**:39-45. DOI: 10.1128/AEM.71.1.39-45.2005

[168] Hobley L, Li B, Wood JL, Kim SH, Naidoo J, Ferreira AS, et al. Spermidine promotes *Bacillus subtilis* bioflm formation by activating expression of the matrix regulator slrR. Journal of Biological Chemistry. 2017;**292**:12041-12053. DOI: 10.1074/jbc.M117.789644

[169] Branda SS, Chu F, Kearns DB, Losick R, Kolter R. A major protein component of the *Bacillus subtilis* biofilm matrix. Molecular Microbiology. 2006;**59**:1229-1238. DOI: 10.1111/j.1365-2958.2005.05020.x

[170] Kobayashi K, Iwano M. BslA(YuaB) forms a hydrophobic layer on the surface of *Bacillus subtilis* biofilms. Molecular Microbiology. 2012;**85**:51-66. DOI: 10.1111/j.1365-2958.2012.08094.x

[171] Hobley L, Ostrowski A, Rao FV, Bromley KM, Porter M, Prescott AR, et al. BslA is a self-assembling bacterial hydrophobin that coats the *Bacillus subtilis* biofilm. Proceedings of the National Academy of Sciences of the United States of America. 2013;**110**:13600-13605. DOI: 10.1073/pnas.1306390110

[172] Zhang J, Poh CL. Regulating exopolysaccharide gene wcaF allows control of *Escherichia coli* biofilm formation. Scientific Reports. 2018;**8**:13127. DOI: 10.1038/s41598-018-31161-7

[173] Pando JM, Karlinsey JE, Lara JC, Libby SJ, Fang FC. The Rcs-regulated colanic acid capsule maintains membrane potential in *Salmonella enterica* serovar Typhimurium. American Society of Microbiology. 2007;**8**(3):e00808-e00817. DOI: 10.1128/mBio.00808-17

[174] Ma Q, Yang Z, Pu M, Peti W, Wood TK. Engineering a novel c-di-GMP-binding protein for biofilm dispersal. Environ. Microbiology. 2011;**13**:631-642. DOI: 10.1111/j.1462-2920.2010.02368.x

[175] Cucarella C, Solano C, Valle J, Amorena B, Lasa I, Penades JR. Bap, a *Staphylococcus aureus* surface protein involved in biofilm formation. Journal of Bacteriology. 2001;**183**:2888-2896. DOI: 10.1128/JB.183.9.2888-2896.2001

[176] Latasa C, Roux A, Toledo-Arana A, Ghigo J-M, Gamazo C, Penadés JR, et al. BapA, a large secreted protein required for biofilm formation and host colonization of *Salmonella enterica* serovar Enteritidis. Molecular Microbiology. 2005;**58**:1322-1339. DOI: 10.1111/j.1365-2958.2005.04907.x

[177] Silva-Hidalgo G, López-Valenzuela M, Cárcamo-Aréchiga N, Cota-Guajardo S, López-Salazar M, Montiel-Vázquez E. Identification of bapA in strains of *Salmonella enterica* subsp. *enterica* isolated from wild animals kept in captivity in Sinaloa, Mexico. Veterinary Medicine International. 2016:3478746. DOI: 10.1155/2016/3478746

[178] Hartford OM, Wann ER, Höök M, Foster TJ. Identification of residues in the *Staphylococcus aureus* fibrinogen-binding MSCRAMM clumping factor a (ClfA) that are important for ligand binding. Jornal of Biological Chemistry. 2001;**276**:2466-2473. DOI: 10.1074/jbc.M007979200

[179] Dunman PM, Murphy E, Haney S, Palacios D, Tucker-Kellogg G, Wu S, et al. Transcription profiling-based identification of *Staphylococcus aureus* genes regulated by the agr and/or sarA loci. Journal of Bacteriology. 2001;**183**:7341-7353. DOI: 10.1128/JB.183.24.7341-7353.2001

[180] Chen J, Xia Y, Cheng C, Fang C, Shan Y, Jin G, et al. Genome sequence of the nonpathogenic *Listeria monocytogenes* Serovar 4a strain M7. Journal of Bacteriology. 2011;**193**:5019-5020. DOI: 10.1128/JB.05501-11

[181] Miller MB, Bassler BL. Quorum sensing in bacteria. Annual Review of Microbiology. 2001;**55**:165-199. DOI: 10.1146/annurev.micro.55.1.165

[182] Gomez MI, Seaghdha MO, Prince AS. *Staphylococcus aureus* protein a activates TACE through EGFR-dependent signaling. The EMBO Journal. 2007;**26**:701-709. DOI: 10.1038/sj.emboj.7601554

[183] Weinmaier T, Riesing M, Rattei T, Bille J, Arguedas-Villa C, Stephan R, et al. The complete genome sequence of *Listeria monocytogenes* LL195—A serotype 4b strain from the 1983 to 1987 listeriosis epidemic in Switzerland. Genome Announcements. 2013;**1**:e00152-e00112. DOI: 10.1128/genomeA.00152-12

[184] Park C, Shin NY, Byun JH, Shin HH, Kwon EY, Choi SM, et al. Down regulation of RNAIII in vancomycin-intermediate *Staphylococcus aureus* strains regardless of the presence of agr mutation. Journal of Medical Microbiology. 2012;**61**:345-352. DOI: 10.1099/jmm.0.035204-0

[185] Lou Z, Chen J, Yu F, Wang H, Kou X, Ma C, et al. The antioxidant, antibacterial, antibiofilm activity of essential oil from *Citrus medica* L. var. *sarcodactylis* and its nanoemulsion. LWT—Food Science and Technology. 2017;**80**:371-377. DOI: 10.1016/j.lwt.2017.02.037

[186] da Silva Gündel S, de Souza ME, Quatrin PM, Klein B, Wagner R, Gündel A, et al. Nanoemulsions containing *Cymbopogon flexuosus* essential oil: Development, characterization, stability study and evaluation of antimicrobial and antibiofilm activities. Microbial Pathogenesis. 2018;**118**:268-276. DOI: 10.1016/j.micpath.2018.03.043

[187] Letsididi K, Lou Z, Letsididi R, Mohammed K, Maguy B. Antimicrobial and antibiofilm effects of trans-cinnamic acid nanoemulsion and its potential application on lettuce. LWT—Food Science and Technology. 2018;**94**:25-32. DOI: 10.1016/j.lwt.2018.04.018

Chapter 15

Dental Biofilm as Etiological Agent of Canine Periodontal Disease

Jana Kačírová, Marián Maďar, Gabriela Štrkolcová, Aladár Maďari and Radomíra Nemcová

Abstract

Periodontal disease is one of the most common health problem affecting dogs. The disease is more prevalent in small breeds and brachycephalic breeds compared to large breeds, and incidence increases with advancing age. In first stage it affects only the gingival tissue and causes gingivitis. It later develops into periodontitis which involves changes in other periodontium tissues. Main etiological agents of periodontal disease are pathogenic bacteria of dental biofilm, and products of their metabolism. In human, *Porphyromonas gingivalis*, *Treponema denticola* and *Tannerella forsythia* play a key role in the etiology of periodontal disease. Also, there are many other candidates as human periodontal pathogens, including *Aggregatibacter actinomycetemcomitans*, *Prevotella intermedia*, *Prevotella nigrescens*, *Fusobacterium nucleatum*, *Parvimonas micra*, *Eikenella corrodens*, *Capnocytophaga gingivalis*, *Eubacterium nodatum* and *Campylobacter rectus*. Since periodontal diseases in dogs are similar to human diseases in terms of disease progression and clinical manifestation, we can assume their common etiology. This chapter is focused on review about canine dental biofilm and about members of biofilm as potential causative agent of canine periodontal disease.

Keywords: dental plaque, biofilm, dog, periodontal disease, gingivitis, periodontitis

1. Introduction

Periodontal disease is a significant veterinary health problem of companion dogs [1]. Periodontal disease refers to a group of inflammatory diseases. In both humans and dogs the initial stages of periodontal disease are observed clinically as red and inflamed gingivae, defined as gingivitis. Without treatment to remove, and disrupt the dental plaque, gingivitis may progress to periodontitis [2]. Periodontitis, the later, irreversible stage of the disease, is an inflammatory disease of supporting teeth tissues [3]. The primary etiological factor in the initiation and progression of periodontal disease is dental plaque [4]. Dental plaque is the community of microorganisms found on a tooth surface as a biofilm, embedded in a matrix of polymers of host and bacterial origin [5]. It is believed that enzymes secreted by dental biofilm bacteria as well as bacterial antigens activate the host inflammatory response initiating disease [6]. Dental calculus that represents mineralized bacterial dental biofilm is considered as secondary etiological factor in periodontal disease [7]. Dental calculus itself is relatively non-pathogenic and, despite its rough surface, is not the direct cause of inflammatory processes, but mainly has an irritant effect.

In addition, the presence of dental calculus leads to greater biofilm accumulation by creating a rough surface [8]. Dental calculus is always covered with a layer of dental biofilm, so it plays an important role as retention factor in the colonization of microorganisms [9].

2. Canine dental biofilm

The oral cavity is a host for a variety of microorganisms including bacteria, viruses, fungi and protozoa that colonize teeth, tongue, oral mucosa, hard palate, caries lesions, periodontal pocket and similarly. The distribution of microorganisms in the oral cavity is not random; most species prefer certain places to others due to the specific local conditions that these sites provide, for example, the anaerobic environment of the gingival sulcus [10, 11]. However, the oral cavity environment is also hostile to microbial life, so only a certain groups of microorganisms entering it are able to colonize it, and survive in this environment. Microorganisms must attach to the surface and form biofilms to remain in oral cavity [12].

Dog oral cavity hides a rich and diverse bacterial community and exceeds the estimates of culture-based studies. Of the cultivable oral microbiota, genera *Actinomyces*, *Streptococcus* and *Granulicatella* are most commonly isolated from saliva. Genera *Porphyromonas*, *Actinomyces* and *Neisseria* are most commonly isolated from plaque [13]. Genera *Porphyromonas*, *Fusobacterium*, *Capnocytophaga*, *Derxia*, *Moraxella*, *Bergeyella*, non-cultivable *Lachnospiraceae*, *Enhydrobacter*, non-classified *Peptostreptococcaceae*, *Xylanibacter*, *Parabacteroides*, *Tannerella*, *Neisseria*, *Treponema* and *Bacteroides* were identified by the pyrosequencing of the 16S rRNA gene [14]. In another oral microbiota study also by pyrosequencing the 16S rRNA gene, the bacterial genera *Actinomyces*, *Porphyromonas*, *Fusobacterium*, *Neisseria*, *Pasteurella*, *Lampropedia*, *Capnocytophaga*, *Frigovirgula*, *Filifactor*, *Conchiformibius*, *Eubacterium*, *Streptococcus*, *Corynebacterium* and *Derxia* have been identified with an abundance >1% [15]. Based on the sequencing of the 16S rRNA gene the presence of other genera in the oral cavity of dogs such as *Abiotrophia*, *Aerococcus*, *Campylobacter*, *Cardiobacterium*, *Clostridium*, *Curtobacterium*, *Dialister*, *Dietzia*, *Dysgonomonas*, *Eikenella*, *Enterococcus*, *Eubacterium*, *Gemella*, *Globicatella*, *Granulicatella*, *Haemophilus*, *Lactobacillus*, *Leptotrichia*, *Leucobacter*, *Micrococcus*, *Micromonas*, *Peptostreptococcus*, *Prevotella*, *Propionibacterium*, *Propionivibrio*, *Rothia*, *Selenomonas*, *Schwartzia*, *Sporocytophaga*, *Wolinella*, *Xanthomonas* and *Xenophilus* was confirmed [16]. Many of them are part of the biofilm formed on teeth surface. In dogs, also in humans, a subgingival biofilm includes colonies of anaerobic, Gram-negative (*Bacteroides* spp., *Capnocytophaga* spp., *Fusobacterium* spp., *Porphyromonas* spp., *Prevotella* spp., *Tannerella* spp. and *Treponema* spp.) as well as Gram-positive bacteria (*Actinomyces* spp., *Corynebacterium* spp., *Eubacterium* spp., *Peptostreptococcus* spp. and *Streptococcus* spp.) [17].

The formation of dental biofilm in the oral cavity is a multi-stage process [18]. It can be divided into four main stages: pellicle formation, initial bacterial adhesion, plaque maturation and finally bacterial dispersion [11]. Initially, a semipermeable layer called pellicle is formed on the tooth surface, which mediates the interaction between tooth, oral fluids and microorganisms [19]. Primary colonizers form biofilm autoaggregation (aggregation between the same species) and coaggregation (aggregation between different species) [20]. In addition, they facilitate the arrival of additional bacteria by providing multiple diverse adhesive sites. They also begin to build a matrix that holds the biofilm together. Some species are incapable of adhering to the surface, but are often able to anchor to a matrix or directly to earlier colonizers [21]. Representatives of the genera *Neisseria*, *Corynebacterium* and *Stenotrophomonas* are involved as primary colonizers in the formation of canine dental biofilm. The most common species of the genus *Neisseria* are *N. zoodegmatis*, *N. animaloris* and *N. weaveri*. Representatives of the genera *Actinomyces*, *Porphyromonas*, *Moraxella*, *Leucobacter*, and the families *Peptostreptococcaceae* and *Pasteurellaceae* probably play the roles of secondary colonizers. Species *Actinomyces canis* and *Porphyromonas gingivicanis* show high levels of biofilm incorporation. The species which featured most frequently in the role of third community member are *Peptostreptococcaceae* spp., *Porphyromonas gingivicanis* and *Leucobacter* spp. [22]. Bacteria from dental biofilm can either be protective, and provide an essential barrier through interactions with the host immune system, or be pathogenic, and cause diseases, such as periodontitis [15]. Oral microbiota varies greatly in healthy dogs and in dogs with disease of oral cavity, and also contains a high proportion of non-cultivable or unexplored species [23]. In healthy dogs, more common species are *Moraxella* spp., *Bergeyella zoohelcum*, *Neisseria shayeganii*, *Pasteurellaceae* spp., *Capnocytophaga* spp. and *Stenotrophomonas* spp. In dogs with periodontitis, species *Peptostreptococcaceae* spp., *Lachnospiraceae* spp. and *Clostridiales* spp. are significantly more prevalent [24].

3. Periodontal disease

Periodontal disease occurs naturally in a wide range of species from rodents to humans [25]. Periodontal disease is one of the most common diseases of adult dogs, with up to 80% of animals affected [23]. All canine breeds are at risk of developing periodontal disease [26]. In general, the disease is more prevalent in small breeds compared to large breeds, and incidence increases with advancing age. In addition, brachycephalic breeds and dogs with teeth overcrowding have been reported to be especially vulnerable to developing the advanced stages of the disease [27]. There are four stages of periodontal disease, each of which is based on the severity of clinical lesions as follows: Stage 1—gingivitis, Stage 2—early periodontitis, Stage 3—moderate periodontitis, Stage 4—advanced periodontitis [28].

Gingivitis is completely reversible, and is recognized by the classic signs of halitosis, bleeding, inflammation, redness and swelling of the gingivae. Periodontitis is irreversible, and attacks the deeper structures that support the teeth, permanently damaging the surrounding bone and periodontal ligament [23]. The breakdown of the collagen fibers of the periodontal ligament results in a periodontal pocket between the gingiva and the tooth. Periodontal pocket deepen due to further destruction of periodontal ligament fibers and alveolar bone resorption. Advanced periodontitis is characterized by gingival erythema and edema, gingival bleeding, gingival recession, tooth mobility, suppuration of periodontal pocket and loss of teeth [29]. We know two main categories of periodontal disease in which loss of supporting structures around the tooth occurs: chronic periodontitis and aggressive periodontitis [30]. Chronic periodontitis is chronic inflammation results in, mostly irreversible, loss of epithelial tissue, bone and ligament. Aggressive periodontitis is characterized by rapid rate of disease progression. It can be present in localized or generalized form; both are early-onset forms of chronic periodontal inflammatory disease. No disease-specific biomarkers exist that differentiate chronic periodontitis from aggressive periodontitis. Although current knowledge suggests that both have similar etiology and histopathology and might indeed be different ends of the same disease spectrum [31].

Periodontal disease is caused by the accumulation of bacterial dental biofilm on the teeth and gingivae, toxic products of the metabolism of these microorganisms, and the host immune response against the infection that triggers the inflammatory process [32]. In case of chronic periodontitis usually have abundance of plaque and calculus, which match with the amount of periodontal destruction. On the other hand, in case of aggressive periodontitis, there is usually a mismatch between the amount of local factors and the periodontal destruction [33]. Periodontal disease affects more frequently and more severely regions of premolars and molars than regions of maxillary and mandibular incisors. Missing of teeth is observed at a high and increasing incidence with age. The tooth most commonly lost is the first premolar, followed by the other premolars and molars, where severe periodontitis is frequently found [34]. Periodontitis is a serious infection that can have medical consequences such as anorexia and weight loss, chronic pain, swollen gums, dental caries, breakage or loss of teeth and breakage of the maxillary or mandibular bone [35]. Unfortunately, the damage from periodontal disease is not confined to just loss of teeth. Oral infection, especially periodontitis, may affect the course and pathogenesis of a number of systemic diseases, such as chronic bronchitis, pulmonary fibrosis, endocarditis, interstitial nephritis, glomerulonephritis and hepatitis [1].

4. Periodontal pathogens

Although there is sufficient evidence that biofilm accumulation and maturation is essential for initiation and progression of periodontal disease, studies show that bacterial species colonizing periodontal pocket have different roles in the pathogenesis of this disease [36]. Microbial density is considered to be critical for the development of gingivitis, and some types of chronic periodontitis, while the species of the microorganisms may be of greater importance in the initiation of aggressive periodontitis [35]. Subgingival microbiota in periodontitis may contain hundreds of bacterial species, but only a small number is associated with disease progression, and is considered to be of importance etiologically [37]. The presence of *Mycobacterium tuberculosis* is an indication of tuberculosis, and *Treponema pallidum* a positive diagnosis of syphilis, but there is no single microorganism, which is attributable to chronic periodontitis [38].

As with any other infection, identification of the microbial pathogens associated with the etiology of periodontitis is the first step towards the development of effective therapeutic approaches. The establishment of a microorganism as a true pathogen should be based on two main levels of evidence: (1) the organism should be present in higher prevalence and/or levels in disease than in health, and (2) its suppression or elimination should reduce or stop disease progression [39]. In human, the presence of three species of Gram-negative anaerobic bacteria within subgingival biofilm, *Porphyromonas gingivalis*, *Treponema denticola* and *Tannerella forsythia*, described by Socransky et al. as the "red complex," show a strong association to periodontitis, and some studies have indicated their involvement also in dogs [40]. There are many others candidates as human periodontal pathogens, including *Aggregatibacter actinomycetemcomitans*, *Prevotella intermedia*, *Prevotella nigrescens*, *Fusobacterium nucleatum*, *Parvimonas micra*, *Eikenella corrodens*, *Capnocytophaga gingivalis*, *Treponema socranskii*, *Eubacterium nodatum* and *Campylobacter rectus* [41]. Some of them are also associated with canine periodontal disease. *Tannerella forsythia*, *Campylobacter rectus* and *Porphyromonas gulae* were detected in almost all dogs with periodontitis. *Prevotella intermedia* and *Eikenella corrodens* were detected only in few dogs with periodontitis, *Treponema denticola*, *Capnocytophaga ochracea* and *Porphyromonas gingivalis* were detected in only one specimen. In addition, *Prevotella nigrescens* and *Aggregatibacter actinomycetemcomitans* were not detected in any of the specimens [42]. *Actinomyces canis* and *Corynebacterium canis* are significantly more prevalent in dogs with periodontitis than in healthy dogs [23, 24].

Species of the genera *Actinomyces* and *Corynebacterium* can play the same role in periodontitis in dogs that *P. gingivalis* plays in periodontitis in humans. This assumption is based on the finding that the ratio of these genera that have trypsin-like activity is increased in areas with periodontitis compared to their ratio in non-periodontal areas and may partly explain the absence of *P. gingivalis* [43]. Protist organisms, such as *Entamoeba gingivalis* and *Trichomonas tenax*, are also present in canine dental biofilm, and correlate with periodontal disease. These findings provide the evidence for the presence of oral protozoa in canine dental biofilm, and suggest a possible role for protozoa in the periodontal disease process [44].

4.1 Genus *Porphyromonas*

The genus *Porphyromonas* is phylogenetically classified in the family *Porphyromonadaceae*, order *Bacteroidales*, class *Bacteroides* and phylum *Bacteroidetes*. Representatives of this genus are Gram-negative, obligate anaerobic, non-motile and non-spore-forming rods. Several species of *Porphyromonas*, including *P. endodontalis*, *P. circumdentaria* and *P. gingivalis* were isolated from the biofilm of adult dogs, but not from any oral sites of puppies and adolescent dogs [45]. Also, several new *Porphyromonas* species (*P. gulae*, *P. macacae*, *P. cangingivalis*, *P. cansulci*, *P. creviocanis*, *P. gingivacanis*, *P. canoris*, *P. denticanis*) associated with periodontal disease have been described [23]. In humans, the major periodontal pathogen is *P. gingivalis* [46]. There are several differences between isolates *Porphyromonas* from humans and from dogs. For example, *P. gingivalis* isolates of canine origin are catalase positive, but isolates of human origin are catalase negative. These catalase positive organisms like *P. gingivalis* may represent the species *P. gulae* [47].

P. gulae is rarely found in humans and healthy animals, usually is isolated from animals, including dogs, with active periodontitis [48]. From dogs with periodontal disease are most frequently isolated three *Porphyromonas* species: *P. gulae*, *P. salivosa* (current name, *P. macacae*) and *P. denticanis* [49]. Of these only, *P. gulae* exhibits virulence characteristics similar to those of the human periodontal pathogen *P. gingivalis* such as lysyl- and arginyl-specific proteolytic activity of the gingipains. This finding suggested that *P. gulae* may play a key role in the development of periodontitis in dogs [50]. In addition, fimbrial protein with the same size and antigenicity similar the 41-kDa fimbrial subunit protein (fimbrillin, FimA) of *P. gingivalis* was identified in *P. gulae* [51]. The fimbria is an important cell structure involved in the adherence and invasion of host's cells, and stimulates the production of inflammatory cytokines by macrophages and fibroblasts. This adhesive ability is considered to be a major pathogenic characteristic of *Porphyromonas* that causes periodontal tissue destruction [52].

P. cangingivalis is the most prevalent canine oral bacterial species in both plaque from healthy gingiva and plaque from dogs with early periodontitis. The ability of *P. cangingivalis* to predominate in both health and disease environments suggests that it is both metabolically flexible enough to colonize in health and also able to compete against other *Porphyromonas* spp. in a disease environment [24]. *P. cangingivalis* has a complete protoporphyrin IX synthesis pathway potentially allowing it to synthesize its own heme unlike pathogenic Porphyromonads such as *P. gingivalis* that acquire heme predominantly from blood. The ability to synthesize siroheme and vitamin B_{12} point to enhanced metabolic flexibility for *P. cangingivalis*, which may underlie its prevalence in the canine oral cavity [53].

4.2 Genus *Tannerella*

The genus *Tannerella* is phylogenetically classified in the family *Porphyromonadaceae*, order *Bacteroidales*, class *Bacteroides* and phylum *Bacteroidetes*.

Representatives of this genus are Gram-negative, anaerobic, non-motile rods. The primary periodontal pathogen is *Tannerella forsythia* originally described as *Bacteroides forsythus*, and reclassified to *Tannerella forsythia* based on 16S rRNA phylogenetic analysis [54]. *T. forsythia* should be regarded as common member of oral microbiota in dogs [42], but dogs with gingivitis or periodontitis are more likely to be infected with *T. forsythia* than healthy animals [26]. *T. forsythia* has several virulence factors, including surface antigen BspA, cell surface proteolytic enzymes, hemagglutinin, cell envelope lipoproteins, glycosidases and cell surface (S)-layer, which contribute to pathogenic potential. The surface protein BspA can bind extracellular matrix components as well as other oral bacteria, and is partly responsible for alveolar bone resorption [55].

4.3 Genus *Campylobacter*

The genus *Campylobacter* is phylogenetically classified in the family Campylobacteraceae, order *Campylobacterales*, class *Epsilonproteobacteria* and phylum *Proteobacteria*. Species *Campylobacter rectus* (formerly *Wolinella recta*), Gram-negative, microaerophilic and motile bacterium, is regarded as common member of oral microbiota in dogs. *C. rectus*, *Tannerella forsythia* and *Porphyromonas gulae* are three major species present in dogs with periodontitis base on study using molecular biological approaches [42]. Several possible factors of virulence have been described for *C. rectus*, such as flagellum, surface layer proteins (S-layer), RTX-type toxins, GroELlike proteins and lipopolysaccharide [56]. *C. rectus* may be an important indicator of periodontal disease. Together with other oral anaerobic bacteria, *C. rectus* is associated with the initiation and progression of periodontal disease [57].

4.4 Genus *Treponema*

The genus *Treponema* is phylogenetically classified in the family *Spirochaetaceae*, order *Spirochaetales*, class *Spirochaetesa* and phylum *Spirochaetes*. *Treponemes* are Gram-negative, obligate anaerobic, motile spirochetes. *Treponemes* are involved in the development of chronic domestic animal diseases, including periodontal diseases in dogs [58]. Dogs harbor several different *Treponema* spp. in their oral cavity, and they can be common in both healthy and periodontitis affected dogs, indicating they are part of the normal oral microbiota [40]. Canine dental biofilm include species *T. denticola*, *T. socranskii*, *T. vincentii*, *T. maltophilum*, *T. medium* and *T. pectinovorum* [59]. Prevalence *T. denticola* and *T. socranskii* is significantly higher in dogs with periodontitis than in dogs without periodontitis. In addition, *Treponema* spp. are not only in the microbial biofilm but also within the gingival tissue [60].

Treponemes, including *T. denticola*, are found on the surface of dense subgingival bacterial biofilms, particularly at the interface of biofilms and gingival epithelium. *T. denticola* has been shown to adhere to fibroblasts and epithelial cells as well as extracellular components of the matrix present in periodontal tissues, and produces several harmful factors that can contribute to virulence of bacteria [61]. The main virulence factors of *T. denticola* in chronic periodontitis include motility and chemotaxis, the ability to interact synergistically with other periodontal pathogens, the ability to produce cytotoxic metabolites, the ability to form biofilms and a variety of cell surface proteins. Motility and chemotaxis allow the bacterium to rapidly colonize new sites, penetrate deep periodontal pocket and penetrate into epithelial layers. Cell surface proteins cause dysregulation of host defense, thereby helping to protect the subgingival biofilm and causing host tissue destruction [58].

4.5 Genus *Fusobacterium*

The genus *Fusobacterium* is phylogenetically classified in the family *Fusobacteriaceae*, order *Fusobacteriales*, class *Fusobacteriia* and phylum *Fusobacteria*. Species *Fusobacterium nucleatum* and *Fusobacterium canifelinum* were identified in subgingival plaque from dogs with and without periodontitis [62]. Based on phenotypic and genotypic differences, *F. nucleatum* is divided into five subspecies, namely *F. nucleatum* subspp. *nucleatum*, *F. nucleatum* subspp. *polymorphum*, *F. nucleatum* subspp. *fusiforme*, *F. nucleatum* subspp. *vincentii* and *F. nucleatum* subspp. *animalis*, whose prevalence varies with disease [63]. At present, the mechanisms of pathogenicity of *F. nucleatum* are unclear. Butyrate production is considered a virulence factor. The association of *F. nucleatum* with periodontal disease is probably through its role as a transient colonizer between Gram-positive and Gram-negative species, mainly in humans. Consequently, *F. nucleatum* can serve as a bridge between species that can colonize exposed tooth surfaces (early colonizers), and species that require interactions with other species (late colonizers). Since late colonizers tend to be species associated with periodontal destruction, bridging with *F. nucleatum* could play an important role in determining the pathogenicity of a mature oral biofilm community [64, 65]. Of the large number of periodontal pathogens, *F. nucleatum* is most frequently involved in infections outside the oral cavity, including pneumonia, pyogenic liver abscess, sepsis, infectious endocarditis, brain abscesses and caecal inflammation [66].

4.6 Genus *Parvimonas*

The genus *Parvimonas* is phylogenetically classified in the family *Peptoniphilaceae*, order *Tissierelliales*, class *Tissierellia* and phylum *Firmicutes*. The species *Parvimonas micra* originally classified as *Peptostreptococcus micros* was first reclassified in 1999 to *Mircomonas micros*, and the second time reclassified in 2006 to *Parvimonas micra* [67]. *P. micra* is anaerobic, asaccharolytic Gram-positive coccus found in dogs with periodontitis but not in the healthy dogs [68]. The virulence factors produced by *P. micra*, which may play a role in the pathogenesis of periodontitis, are poorly characterized. *P. micra* may modulate the inflammatory response in the host and contribute to the destruction of periodontal tissue. In addition, *P. micra* is capable of adhering to epithelial cells, also to other periodontal pathogens [69], and is able to form biofilms in conjunction with *Frederiksenia canicola* and *P. gulae*. *P. micra* might provide a catalyst for progressive tissue destruction, inflammation and alveolar bone loss in canine periodontal disease, in keeping with the keystone-pathogen hypothesis [68].

4.7 Genus *Prevotella*

The genus *Prevotella* is phylogenetically classified in the family *Prevotellaceae*, order *Bacteroidales*, class *Bacteroidia* and phylum *Bacteroidetes*. Representatives of this genus are Gram-negative, anaerobic, non-motile rods. The primary periodontal pathogen is species *Prevotella intermedia*. Within the *P. intermedia* strains, heterogeneity was found in terms of serology and DNA homology. In 1992, based on complex DNA-DNA hybridization, it was suggested that *P. intermedia* be reclassified into two species, *P. intermedia* and *P. nigrescens* [70]. *P. intermedia* and *P. nigrescens*, members of the "orange complex" described by Socransky et al., are among the most common species in subgingival plaque in humans. *P. intermedia* may under certain conditions increase the activity of degradation enzymes and promote the progression of periodontitis [71]. *P. intermedia* is also present in canine dental plaque. In dogs, the counts of *P. intermedia* correlated with the amount of plaque and the degree of gingivitis [72]. *Prevotella dentalis* is also associated with periodontitis. *P. dentalis* (formerly *Mitsuokella dentalis*) was originally named after Japanese bacteriologist Mitsuok, who described this organism for the first time [38]. Mitsuoka isolates a large number of *P. dentalis* strains from humans, dogs and pigs that seem to be closely related to the *Bacteroides* genus [73].

4.8 Oral protozoa

For several decades, research in periodontology is focused on the characterization of bacterial communities thought to be involved in canine periodontal diseases. However, other microorganisms are known to inhabit the oral cavity and could also influence the process of periodontal disease. There were identified two oral protozoa, *Entamoeba gingivalis* and *Trichomonas tenax*, which can inhabit the canine oral periodontium. Both were statistically associated to animals with periodontal disease [74].

The species *Entamoeba gingivalis* is phylogenetically classified in the genus *Entamoeba*, class *Archamoebae* and phylum *Amoebozoa*. The protozoan *E. gingivalis* resides in the oral cavity and is frequently observed in the periodontal pockets of humans and pets. The parasite *E. gingivalis* is more prevalent and more abundant in periodontal pockets, suggesting that this ecological niche is either propitious for its survival, or that the parasite induces changes leading to this environment [75]. *E. gingivalis* is an opportunistic pathogen, which, together with synergistic symbiotic bacteria, can cause periodontal diseases in hosts with low immunity [76]. Pathogenicity of protozoa *E. gingivalis* in the oral cavity is not completely understood [77].

The species *Trichomonas tenax* is phylogenetically classified in the genus *Trichomonas*, family *Trichomonadidae* and order *Trichomonadida*. *T. tenax* inhabits the oral cavities of various mammals, including humans, dogs, cats and horses [78]. *T. tenax*, an anaerobic motile-flagellated protozoan, is 12–20 μm long and 5–6 μm wide organism. It is either ellipsoidal or ovoid in shape and has four anterior flagella of unequal lengths [79]. *T. tenax* can ingest bacteria and various particles by phagocytosis necessary for their development. *T. tenax*, detected in periodontal cases, is likely to be related to the onset and evolution of periodontal disease [80]. This parasite has been reported to be involved in a number of cases of pulmonary trichomoniasis. Besides bronchopulmonary exudates the trichomonads have also been found in pleural fluid, submaxillary gland and infra-auricular lymph node [81]. Several mechanisms may explain the deleterious effects of the *T. tenax* parasite towards periodontal tissues. Recent studies have emphasized the ability of parasites to induce changes in some features of microbial communities. *T. tenax* can escape the host immune response via a complex strategy caused by an imbalance of the oral cavity microbiocenosis. Pathogenic bacteria involved in periodontal host colonization and immune subversion use complement and toll-like receptor (TLR) signaling pathways. Like bacteria, parasites are recognized by TLR. *T. tenax* also produce fibronectin-like proteins, responsible for tissue adhesion. Given this pathogenic property, host-tissue disruption and lysis may be induced by *T. tenax* secretion of peptidases such as cathepsin B-like proteinases for matricial type 1 collagen and gelatin hydrolyses or haemolysins for erythrolysis [79].

5. The possibility of transferring bacteria from the oral microbiome of dogs to human

Except for to the impact on animal health, bacteria from the oral cavity of animals may also have harmful effects on human health in the case of microbial transmissibility, for example, through dog bites. Dog bite wounds are polymicrobial, with a broad combination of aerobic and anaerobic microorganisms. The microbiology of infected bite wounds from dogs is similar to that of the organisms that colonize the dog's oral cavity. Less frequently, isolates may also come from the environment and patients' skin [82]. On average, a dog bite wound contains two to five different species of bacteria [83]. *Pasteurella* species are the most frequent isolates of dog bites (50%), especially *Pasteurella canis* is the most common isolate of dog bites [84]. Other common aerobic organisms include *P. multocida*, *P. dagmatis*, *Staphylococcus* spp. (including MRSA), *Streptococcus* spp. (including *S. pyogenes*), *Neisseria* spp., *Capnocytophaga canimorsus*, *Corynebacterium* spp., *Moraxella* spp., *Enterococcus* spp. and *Bacillus* spp. [82, 83]. The most common anaerobic organism isolated from infected dog bite wounds is *Fusobacterium nucleatum* [82]. *Fusobacterium canifelinum* was also isolated from wounds in humans after dog bites [85]. Other common anaerobes include *Prevotella* spp., *Bacteroides* spp., *Porphyromonas* spp., *Propionibacterium* spp. and *Peptostreptococcus* spp. [82, 83]. Several *Porphyromonas* species (*P. macacae*, *P. canoris*, *P. circumdentaria*, *P. cangingivalis*, and *P. cansulci*) [84] and other periodontal pathogens (*Tannerella forsythia*, *Prevotella intermedia* and *Prevotella dentalis*) were also isolated from infected dog bite wounds [82]. In addition, some pathogens such as *Leptospira*, *Rabies virus*, *Clostridium tetani* or *Francisella tularensis*, which can cause systemic infection after bites by dogs, were isolated from wounds in humans [83].

As transmission of oral bacteria during normal contacts between dogs and humans is also feasible one might expect correlations between the oral microbiota of dogs and humans [17]. Oral-to-oral transfer of *Neisseria shayeganii*, *Porphyromonas canigingivalis*, *Tannerella forsythia* and *Streptococcus* minor from dogs to humans is suspected. The finding of potentially zoonotic and periodontopathic bacteria in the canine oral microbiome may be a public health concern [15].

6. Conclusion

Review of literature showed that some bacterial species like *Tannerella forsythia*, *Campylobacter rectus*, *Treponema denticola*, *Fusobacterium nucleatum*, *Parvimonas micra* and *Prevotella intermedia* are the important pathogens for periodontitis in both humans and dogs. On the other hand, *Porphyromonas gulae* is specifically associated with canine periodontal disease. In addition, it is assumed, that oral protozoa such as *Entamoeba gingivalis* and *Trichomonas tenax* play role in canine periodontal disease. In summary, periodontal disease is polymicrobial disease and further analyses of the associated species of periodontitis and their virulence factors in dogs are needed.

Acknowledgements

This publication was supported by the project of the Ministry of education science Research and sport of the Slovak Republic VEGA 1/0788/19 and VEGA 1/0081/17.

Author details

Jana Kačírová*, Marián Maďar, Gabriela Štrkolcová, Aladár Maďari
and Radomíra Nemcová
University of Veterinary Medicine and Pharmacy in Košice, Košice, Slovakia

*Address all correspondence to: kacirova.jana@gmail.com

References

[1] Stella JL, Bauer AE, Croney CC. A cross-sectional study to estimate prevalence of periodontal disease in a population of dogs (*Canis familiaris*) in commercial breeding facilities in Indiana and Illinois. PLoS ONE. 2018;**13**(1):e0191395. DOI: 10.1371/journal.pone.0191395

[2] Davis IJ et al. Longitudinal quantification of the gingival crevicular fluid proteome during progression from gingivitis to periodontitis in a canine model. Journal of Clinical Periodontology. 2016;**43**(7):584-594. DOI: 10.1111/jcpe.12548

[3] Saini R et al. Periodontitis, a true infection. Journal of Global Infectious Diseases. 2009;**1**(2):149-150. DOI: 10.4103/0974-777X.56251

[4] Lovegrove JM. Dental plaque revisited: Bacteria associated with periodontal disease. Journal of the New Zealand Society of Periodontology. 2004;**87**:7-21. ISSN: 0111-1485

[5] Marsh PD. Dental plaque as a biofilm and a microbial community: Implications for health and disease. BMC Oral Health. 2006;**6**(Suppl 1):S14. DOI: 1472-6831-6-S1-S14

[6] Wallis C et al. A longitudinal assessment of periodontal health status in 53 Labrador retrievers. The Journal of Small Animal Practice. 2018;**59**(9): 560-569. DOI: 10.1111/jsap.12870

[7] Niklaus PL, Lindhe J, editors. Clinical Periodontology and Implant Dentistry. 2nd ed. Danvers: Wiley-Blackwell; 2015. p. 1480. ISBN: 978-0-470-67248-8

[8] Niemiec BA. Veterinary Periodontology. Danvers: Wiley-Blackwell; 2013. p. 372. DOI: 10.1002/9781118705018

[9] Hellwig E, Klimek J, Attin T. Záchovná stomatologie a parodontologie. Praha: Grada Publishing; 2003. p. 332. ISBN: 80-247-0311-4

[10] Wade WG. New aspects and new concepts of maintaining "microbiological" health. Journal of Dentistry. 2010;**38**(Suppl 1):S21-S25. DOI: S0300-5712(10)70007-5

[11] Huang R, Li M, Gregory RL. Bacterial interactions in dental biofilm. Virulence. 2011;**2**(5):435-444. DOI: 10.4161/viru.2.5.16140

[12] Marsh PD. Dental plaque: Biological significance of a biofilm and community life-style. Journal of Clinical Periodontology. 2005;**32**(Suppl 6):7-15. DOI: 10.1111/j.1600-051X.2005.00790.x

[13] Elliott DR et al. Cultivable oral microbiota of domestic dogs. Journal of Clinical Microbiology. 2005;**43**(11):5470-5476. DOI: 10.1128/JCM.43.11.5470-5476.2005

[14] Sturgeon A et al. Metagenomic analysis of the canine oral cavity as revealed by high-throughput pyrosequencing of the 16S rRNA gene. Veterinary Microbiology. 2012;**162**(2-4):891-898. DOI: S0378-1135(12)00638-4

[15] Oh C et al. Comparison of the oral microbiomes of canines and their owners using next-generation sequencing. PLoS ONE. 2015;**10**(7):e0131468. DOI: 10.1371/journal.pone.0131468

[16] Dewhirst FE et al. The canine oral microbiome. PLoS ONE. 2012;**7**(4):e36067. DOI: 10.1371/journal.pone.0036067

[17] Golynska M et al. Molecular-level evaluation of selected periodontal

pathogens from subgingival regions in canines and humans with periodontal disease. Journal of Veterinary Science. 2016;**18**(1):51-58. DOI: 10.4142/jvs.2017.18.1.51

[18] Dhir S. Biofilm and dental implant: The microbial link. Journal of Indian Society of Periodontology. 2013;**17**(1): 5-11. DOI: 10.4103/0972-124X.107466

[19] Hannig C, Hannig M. The oral cavity: A key system to understand substratum-dependent bioadhesion on solid surfaces in man. Clinical Oral Investigations. 2009;**13**(2):123-139. DOI: 10.1007/s00784-008-0243-3

[20] Chandki R, Banthia P, Banthia R. Biofilms: A microbial home. Journal of Indian Society of Periodontology. 2011;**15**(2):111-114. DOI: 10.4103/0972-124X.84377

[21] Sauer K et al. *Pseudomonas aeruginosa* displays multiple phenotypes during development as a biofilm. Journal of Bacteriology. 2002;**184**(4):1140-1154. ISSN: 0021-9193

[22] Holcombe LJ et al. Early canine plaque biofilms: Characterization of key bacterial interactions involved in initial colonization of enamel. PLoS ONE. 2014;**9**(12):e113744. DOI: 10.1371/journal.pone.0113744

[23] Riggio MP et al. Molecular identification of bacteria associated with canine periodontal disease. Veterinary Microbiology. 2011;**150**(3-4):394-400. DOI: S0378-1135(11)00139-8

[24] Davis IJ et al. A cross-sectional survey of bacterial species in plaque from client owned dogs with healthy gingiva, gingivitis or mild periodontitis. PLoS ONE. 2013;**8**(12):e83158. DOI: 10.1371/journal.pone.0083158

[25] Hennet PR, Harvey CE. Natural development of periodontal disease in the dog: A review of clinical, anatomical and histological features. Journal of Veterinary Dentistry. 1992;**9**(3):13-19. ISSN: 0898-7564

[26] Di Bello A et al. Periodontal disease associated with red complex bacteria in dogs. The Journal of Small Animal Practice. 2014;**55**(3):160-163. DOI: 10.1111/jsap.12179

[27] Marshall MD et al. A longitudinal assessment of periodontal disease in 52 Miniature Schnauzers. BMC Veterinary Research. 2014;**10**:166. DOI: 1746-6148-10-166

[28] Carvalho CM et al. Mandibulectomy for treatment of fractures associated with severe periodontal disease. The Canadian Veterinary Journal. 2015;**56**(3):292-294. ISSN: 0008-5286

[29] Preshaw PM et al. Periodontitis and diabetes: A two-way relationship. Diabetologia. 2011;**55**(1):21-31. DOI: 10.1007/s00125-011-2342-y

[30] Armitage GC. Development of a classification system for periodontal diseases and conditions. Annals of Periodontology. 1999;**4**(1):1-6. DOI: 10.1902/annals.1999.4.1.1

[31] Kinane DF, Stathopoulou PG, Papapanou PN. Periodontal diseases. Nature Reviews. Disease Primers. 2017;**3**:17038. DOI: 10.1038/nrdp.2017.38

[32] Albuquerque C et al. Canine periodontitis: The dog as an important model for periodontal studies. Veterinary Journal. 2011;**191**(3):299-305. DOI: S1090-0233(11)00299-1

[33] Ramachandra SS et al. Differential diagnosis between chronic versus aggressive periodontitis and staging of aggressive periodontitis: A cross-sectional study. Contemporary Clinical Dentistry. 2018;**8**(4):594-603. DOI: 10.4103/ccd.ccd_623_17

[34] Isogai H et al. Epidemiological study on periodontal diseases and some other dental disorders in dogs. Nihon Juigaku Zasshi. 1989;**51**(6):1151-1162. ISSN: 0021-5295

[35] Stephan B et al. Activity of pradofloxacin against *Porphyromonas* and *Prevotella* spp. implicated in periodontal disease in dogs: Susceptibility test data from a European multicenter study. Antimicrobial Agents and Chemotherapy. 2008;**52**(6):2149-2155. DOI: 10.1128/AAC.00019-08

[36] Wolff L, Dahlen G, Aeppli D. Bacteria as risk markers for periodontitis. Journal of Periodontology. 1994;**65**(5 Suppl):498-510. DOI: 10.1902/jop.1994.65.5s.498

[37] AlJehani YA. Risk factors of periodontal disease: Review of the literature. International Journal of Dentistry. 2014;**2014**:182513. DOI: 10.1155/2014/182513

[38] Arora N, Mishra A, Chugh S. Microbial role in periodontitis: Have we reached the top? Some unsung bacteria other than red complex. Journal of Indian Society of Periodontology. 2014;**18**(1):9-13. DOI: 10.4103/0972-124X.128192

[39] Perez-Chaparro PJ et al. Newly identified pathogens associated with periodontitis: A systematic review. Journal of Dental Research. 2014;**93**(9):846-858. DOI: 0022034514542468

[40] Nises J et al. The occurrence of *Treponema* spp. in gingival plaque from dogs with varying degree of periodontal disease. PLoS ONE. 2018;**13**(8):e0201888. DOI: 10.1371/journal.pone.0201888

[41] Teles R et al. Lessons learned and unlearned in periodontal microbiology. Periodontology 2000. 2013;**62**(1):95-162. DOI: 10.1111/prd.12010

[42] Kato Y et al. Molecular detection of human periodontal pathogens in oral swab specimens from dogs in Japan. Journal of Veterinary Dentistry. 2011;**28**(2):84-89. DOI: 10.1177/089875641102800204

[43] Takada K, Hirasawa M. Expression of trypsin-like activity by the genera *Corynebacterium* and *Actinomyces* in canine periodontitis. Journal of Medical Microbiology. 2000;**49**(7):621-625. DOI: 10.1099/0022-1317-49-7-621

[44] Patel N et al. The prevalence of canine oral *Protozoa* and their association with periodontal disease. The Journal of Eukaryotic Microbiology. 2016;**64**(3):286-292. DOI: 10.1111/jeu.12359

[45] Isogai H et al. Ecology of genus *Porphyromonas* in canine periodontal disease. Zentralblatt für Veterinärmedizin. Reihe B. 1999;**46**(7):467-473. ISSN: 0514-7166

[46] Mysak J et al. *Porphyromonas gingivalis*: Major periodontopathic pathogen overview. Journal of Immunology Research. 2014;**2014**:476068. DOI: 10.1155/2014/476068

[47] Harvey CE, Thornsberry C, Miller BR. Subgingival bacteria: Comparison of culture results in dogs and cats with gingivitis. Journal of Veterinary Dentistry. 1995;**12**(4):147-150. ISSN: 0898-7564

[48] Fournier D et al. *Porphyromonas gulae* sp. nov., an anaerobic, gram-negative *coccobacillus* from the gingival sulcus of various animal hosts. International Journal of Systematic and Evolutionary Microbiology 2001;**51**(Pt 3):1179-1189. DOI: 10.1099/00207713-51-3-1179

[49] Holden JA et al. *Porphyromonas gulae* activates unprimed and gamma interferon-primed macrophages via the

pattern recognition receptors toll-like receptor 2 (TLR2), TLR4, and NOD2. Infection and Immunity. 2017;**85**(9):1-15. DOI: 10.1128/IAI.00282-17

[50] Lenzo JC et al. *Porphyromonas gulae* has virulence and immunological characteristics similar to those of the human periodontal pathogen *Porphyromonas gingivalis*. Infection and Immunity. 2016;**84**(9):2575-2585. DOI: 10.1128/IAI.01500-15

[51] Hamada N et al. Molecular and antigenic similarities of the fimbrial major components between *Porphyromonas gulae* and *P. gingivalis*. Veterinary Microbiology. 2008;**128**(1-2):108-117. DOI: S0378-1135(07)00472-5

[52] do Nascimento Silva A et al. Pathogenicity and genetic profile of oral *Porphyromonas* species from canine periodontitis. Archives of Oral Biology. 2017;**83**:20-24. DOI: S0003-9969(17)30213-3

[53] O'Flynn C et al. Comparative genomics of the genus *Porphyromonas* identifies adaptations for heme synthesis within the prevalent canine oral species *Porphyromonas cangingivalis*. Genome Biology and Evolution. 2015;**7**(12):3397-3413. DOI: 10.1093/gbe/evv220

[54] Sharma A. Virulence mechanisms of *Tannerella forsythia*. Periodontology 2000. 2010;**54**(1):106-116. DOI: 10.1111/j.1600-0757.2009.00332.x

[55] Chukkapalli SS et al. Chronic oral infection with major periodontal bacteria *Tannerella forsythia* modulates systemic atherosclerosis risk factors and inflammatory markers. Pathogens and Disease. 2015;**73**(3):1-12. DOI: 10.1093/femspd/ftv009

[56] Arce RM et al. Characterization of the invasive and inflammatory traits of oral *Campylobacter rectus* in a murine model of fetoplacental growth restriction and in trophoblast cultures. Journal of Reproductive Immunology. 2010;**84**(2):145-153. DOI: 10.1016/j.jri.2009.11.003

[57] Ihara H et al. Detection of *Campylobacter rectus* in periodontitis sites by monoclonal antibodies. Journal of Periodontal Research. 2003;**38**(1): 64-72. DOI: 1o627

[58] Dashper SG et al. Virulence factors of the oral spirochete *Treponema denticola*. Journal of Dental Research. 2010;**90**(6):691-703. DOI: 10.1177/0022034510385242

[59] Valdez M et al. Isolation of oral spirochetes from dogs and cats and provisional identification using polymerase chain reaction (PCR) analysis specific for human plaque *Treponema* spp. Journal of Veterinary Dentistry. 2000;**17**(1):23-26. ISSN: 0898-7564

[60] Nordhoff M et al. Association of *Treponema* spp. with canine periodontitis. Veterinary Microbiology. 2008;**127**(3-4):334-342. DOI: 10.1016/j.vetmic.2007.09.011

[61] Sela MN. Role of *Treponema denticola* in periodontal diseases. Critical Reviews in Oral Biology and Medicine. 2001;**12**(5):399-413. ISSN: 1045-4411

[62] Senhorinho GN et al. Occurrence and antimicrobial susceptibility of *Porphyromonas* spp. and *Fusobacterium* spp. in dogs with and without periodontitis. Anaerobe. 2012;**18**(4):381-385. DOI: 10.1016/j.anaerobe.2012.04.008

[63] Han YW. *Fusobacterium nucleatum*: A commensal-turned pathogen. Current Opinion in Microbiology. 2015;**23**: 141-147. DOI: S1369-5274(14)00180-5

[64] Merritt J et al. Autoaggregation response of *Fusobacterium nucleatum*.

Applied and Environmental Microbiology. 2009;**75**(24):7725-7733. DOI: 10.1128/AEM.00916-09

[65] Signat B et al. *Fusobacterium nucleatum* in periodontal health and disease. Current Issues in Molecular Biology. 2011;**13**(2):25-36. DOI: v13/25

[66] Henne K et al. Sex-specific differences in the occurrence of *Fusobacterium nucleatum* subspecies and *Fusobacterium periodonticum* in the oral cavity. Oncotarget. 2018;**9**(29):20631-20639. DOI: 10.18632/oncotarget.25042

[67] Uemura H et al. *Parvimonas micra* as a causative organism of spondylodiscitis: A report of two cases and a literature review. International Journal of Infectious Diseases. 2014;**23**:53-55. DOI: 10.1016/j.ijid.2014.02.007

[68] Sanguansermsri P et al. Interspecies dynamics among bacteria associated with canine periodontal disease. Molecular Oral Microbiology. 2017;**33**(1):59-67. DOI: 10.1111/omi.12199

[69] Tanabe S, Bodet C, Grenier D. *Peptostreptococcus* micros cell wall elicits a pro-inflammatory response in human macrophages. Journal of Endotoxin Research. 2007;**13**(4):219-226. DOI: 10.1177/0968051907081869

[70] Fukui K et al. Incidence of *Prevotella* intermedia and *Prevotella nigrescens* carriage among family members with subclinical periodontal disease. Journal of Clinical Microbiology. 1999;**37**(10):3141-3145. ISSN: 0095-1137

[71] Zhang Y et al. Population-genomic insights into variation in *Prevotella* intermedia and *Prevotella nigrescens* isolates and its association with periodontal disease. Frontiers in Cellular and Infection Microbiology. 2017;7:409. DOI: 10.3389/fcimb.2017.00409

[72] Allaker RP et al. Prevalence of *Porphyromonas* and *Prevotella* species in the dental plaque of dogs. The Veterinary Record. 1997;**140**(6):147-148. ISSN: 0042-4900

[73] Hiranmayi KV et al. Novel pathogens in periodontal microbiology. Journal of Pharmacy & Bioallied Sciences. 2017;**9**(3):155-163. DOI: 10.4103/jpbs.JPBS_288_16

[74] Patel N, Holcombe L, Andrew P. Oral protists: Importance to canine periodontal disease. Protistology. 2016;**10**:57-58. ISSN: 1680-0826

[75] Bonner M et al. Reassessing the role of *Entamoeba gingivalis* in periodontitis. Frontiers in Cellular and Infection Microbiology. 2018;**8**:379. DOI: 10.3389/fcimb.2018.00379

[76] Liu GY et al. Experimental study on the pathogenesis of *Entamoeba gingivalis*. Zhongguo Ji Sheng Chong Xue Yu Ji Sheng Chong Bing Za Zhi. 2001;**19**(4):229-232. ISSN: 1000-7423

[77] Mielnik-Blaszczak M et al. *Entamoeba gingivalis*: Prevalence and correlation with dental caries in children from rural and urban regions of Lublin Province, Eastern Poland. Annals of Agricultural and Environmental Medicine. 2018;**25**(4):656-658. DOI: 10.26444/aaem/80403

[78] Dybicz M et al. Molecular identification of *Trichomonas tenax* in the oral environment of domesticated animals in Poland: Potential effects of host diversity for human health. Annals of Agricultural and Environmental Medicine. 2018;**25**(3):464-468. DOI: 10.26444/aaem/92309

[79] Marty M et al. *Trichomonas tenax* and periodontal diseases: A concise review. Parasitology. 2017;**144**(11):1417-1425. DOI: 10.1017/S0031182017000701

[80] Benabdelkader S et al. Specific clones of *Trichomonas tenax* are associated with periodontitis. PLoS ONE. 2019;**14**(3):e0213338. DOI: 10.1371/journal.pone.0213338

[81] Kutisova K et al. Tetratrichomonads from the oral cavity and respiratory tract of humans. Parasitology. 2005;**131**(Pt 3):309-319. ISSN: 0031-1820

[82] Abrahamian FM, Goldstein EJ. Microbiology of animal bite wound infections. Clinical Microbiology Reviews. 2011;**24**(2):231-246. DOI: 10.1128/CMR.00041-10

[83] Rothe K, Tsokos M, Handrick W. Animal and human bite wounds. Deutsches Ärzteblatt International. 2015;**112**(25):433-442; quiz 443. DOI: 10.3238/arztebl.2015.0433

[84] Talan DA et al. Bacteriologic analysis of infected dog and cat bites. Emergency medicine animal bite infection study group. The New England Journal of Medicine. 1999;**340**(2):85-92. DOI: 10.1056/NEJM199901143400202

[85] Conrads G et al. *Fusobacterium canifelinum* sp. nov., from the oral cavity of cats and dogs. Systematic and Applied Microbiology. 2004;**27**(4):407-413. DOI: 10.1078/0723202041438509

Chapter 16

Biofilm, a Cozy Structure for *Legionella pneumophila* Growth and Persistence in the Environment

Arwa Abu Khweek and Amal O. Amer

Abstract

Legionella pneumophila (*L. pneumophila*) is the causative agent of Legionnaires' disease. Transmission to humans is mediated via inhalation of contaminated water droplets. *L. pneumophila* is widely distributed in man-made water systems, multiple species of protozoa, and nematodes. *L. pneumophila* persist within multi-species biofilms that cover surfaces within water systems. Virulence, spread, and resistance to biocides are associated with survival of *L. pneumophila* within multi-organismal biofilm. Outbreaks of Legionellosis are correlated with the existence of *L. pneumophila* in biofilms, even after the intensive chemical and physical treatments. Several factors negatively or positively modulate the persistence of *L. pneumophila* within the microbial consortium-containing *L. pneumophila*. Biofilm-forming *L. pneumophila* continue to be a public health and economic burden and directly influence the medical and industrial sectors. Diagnosis and hospitalization of patients and prevention protocols cost governments billions of dollars. Dissecting the biological and environmental factors that promote the persistence and physiological adaptation in biofilms can be fundamental to eliminating and preventing the transmission of *L. pneumophila*. Herein, we review different factors that promote persistence of *L. pneumophila* within the biofilm consortium, survival strategies used by the bacteria within biofilm community, gene regulation, and finally challenges associated with biofilm resistance to biocides and *anti-Legionella* treatments.

Keywords: *legionella pneumophila*, biofilm, Legionellosis, protozoa, *Caenorhabditis elegans*

1. Introduction

L. pneumophila, the causative agent of Legionellosis, as being pathogenic to humans was following an outbreak of pneumonia at a convention of the American Legion in Philadelphia, USA in July 1976 [1]. This pathogen causes a severe form of pneumonia termed Legionnaires' disease (LD), and less frequently, Pontiac fever, a self-limited flu-like illness. Approximately 90% of LD cases are caused by *L. pneumophila*. Transmission of *L. pneumophila* occur primarily through the spread of contaminated aerosols present in cooling towers, condensers, faucets, showers, and hot tubs [2, 3]. Although stringent water quality examinations, the formation of contaminated aerosols remains to be a major problem associated with disease spread [4].

Multiple mechanisms of persistence are harbored by *L. pneumophila* in various environmental conditions and in humans. Following invasion of amoeba or human macrophages, *L. pneumophila* form the *Legionella*-containing vacuole (LCV), which acquires vesicles from early and late endosomes, mitochondria and the endoplasmic reticulum (ER), thus escaping the microbicidal endocytic pathway. Hijacking the endocytic pathway by LCV is fundamental in initiating and maintaining a niche that secure *L. pneumophila* replication [5, 6]. Importantly, a battery of effector proteins produced by the Dot/Icm type IV secretion system of *L. pneumophila.* The Dot/Icm secreted effectors are required for successful intracellular replication of *L. pneumophila* [7–13]. Like other intracellular bacteria, *L. pneumophila* switch between a transmissive (virulent) and replicative (non-virulent) biphasic cycles. This switch is essential to ensure bacterial replication in nutrient starved or rich environments and transmit between different niches [14]. Nutrient rich environment is conducive of the replicative phase, where *L. pneumophila* express few virulence factors. While nutrient deprived environment is promotive of the transmissive phase, especially when the phagosome is unable to support the replication phase of *L. pneumophila*. Hallmark features of the transmissive phase include, increased motility, expression of plethora of virulence factors, resistance to stressors and egress from the infected host [14]. In the environment, *L. pneumophila* survive as free living (planktonic) or form bacterial biofilms with other organisms that adhere to surfaces [15–20]. Moreover, *L. pneumophila* is able to differentiate into inert, cyst-like but extremely infectious mature intracellular form (MIF) [21, 22]. Resilience of *L. pneumophila* extracellularly and under harsh environmental settings is attributed to its ability to exist in viable non-culturable (VBNC) state [23, 24]. Harboring a VBNC mode hinders the detection of many *Legionella* species. In nature, colonization and persistence is promoted via biofilm formation [25], and survival within freshwater amoeba and *C. elegans* [5, 26].

Herein, we review factors that mediate biofilm persistence, strategies utilized by the bacteria to become a member of the biofilm consortium and modes of eradicating *L. pneumophila* biofilm.

1.1 Constituents of *L. pneumophila* biofilm

L. pneumophila is found as sessile cells associated with biofilms in freshwater environments, [19, 27, 28]. Biofilms mediate bacterial attachments to surfaces and to other pre-attached bacterial communities. Attachment is attained via forming an extracellular matrix (ECM) that is composed mainly of water, proteins, exopolysaccharides, lipids, DNA and RNA, and inorganic compounds [29–32]. Three developmental phases are required for biofilm formation. (I) initial attachments to a surface, (II) maturation and extracellular matrix formation, and (III) detachments and dispersion of the bacteria. Biofilms eventually develop into three-dimensional structures containing water channels, which allow bacteria to obtain nutrients, oxygen and get rid of waste products. The behavior of *L. pneumophila* has mainly been studied in the context of mono- or mixed species biofilms, due to the complexity of biofilm formed in natural environment [17–19, 33, 34]. Interestingly, *L. pneumophila* exhibit minor representation among other species in freshwater and environmental biofilms, [27, 28], and the existence of *L. pneumophila* may be influenced by other microorganisms in complex biofilms [35]. Some bacterial species positively or negatively affect the persistence of *L. pneumophila* biofilm [19]. Intriguingly, *Klebsiella pneumoniae (K. pneumoniae)*, *Flavobacterium* sp., *Empedobacter breve, Pseudomonas putida* and *Pseudomonas fluorescens* positively associated with the long-term persistence of *L. pneumophila* in biofilms [18, 19, 36]. Other species within biofilms seem to be the provider of capsular polysaccharides, extracellular matrix that support the adherence [37–39], or the contributor of growth factors that stimulate growth of *L. pneumophila* [19]. *Pseudomonas aeruginosa* (*P. aeruginosa*), *Aeromonas hydrophila*, *Burkholderia cepacia*, *Acidovorax* sp., and *Sphingomonas* sp. [40] are among species that antagonize the persistence of *L. pneumophila* within the biofilm [19]. Inhibition of *L. pneumophila* biofilm by *P. aeruginosa* could be a consequence of the effect of homoserine lactone quorums sensing (QS) molecule [41], or production of bacteriocin [40]. Interestingly, *L. pneumophila* can coexist in biofilm formed by *P. aeruginosa* and *K. pneumoniae* indicating that the inhibition of *L. pneumophila* biofilm formation by *P. aeruginosa* can be alleviated by the permissive *K. pneumoniae* [19]. The authors suggest that the growth provided by *K. pneumoniae* to promote survival of *L. pneumophila* can at the same time lessen the inhibitory effect by *P. aeruginosa* [19]. Therefore, the identity, number and nature of interactions between bacterial species (commensalism or interference) can directly affect growth of *L. pneumophila* within biofilms.

Biofilm formation of *L. pneumophila* in the laboratory is achieved by growing the bacteria under stringent conditions in nutrient-rich Buffered Yeast Extract medium (BYE) [18, 34]. Different temperatures correlated with different amount, degree of attachment and rate of biofilm formation. Mushroom like structure containing water channels is the hallmark features of biofilms formed at 25°C. In contrast, at 37°C *L. pneumophila* biofilm is thicker and deficient of water channels observed at 25°C. However, filamentous appearance with mat-like structure has been observed with *L. pneumophila* grown at 42°C. Studies in our laboratory showed that in contrast to the *dotA* mutant that lacks the type IV secretion, WT *L. pneumophila* form biofilm when grown statically at 37°C for 7 days.

Our knowledge is lacking regarding the factors encoded by *L. pneumophila* that promote the attachment and persistence within multispecies biofilms created by other bacteria.

1.2 Formation of biofilms as a survival niche in oligotrophic environment

Biofilm is extremely nutritious environment that harbors a mixture of living, dead organisms as well as protozoa and bacteria. To be a productive member of the microbial consortium, *L. pneumophila* has to compete with other bacteria for nutrients in a multispecies biofilm. Therefore, it is essential for *L. pneumophila* to strive in an environment adjacent to bacterial neighbors that best sustains their growth and survival [42]. Given the fastidious and auxotrophic nature of *L. pneumophila*, supplementation of the laboratory media with amino acids and iron is essential for growth [43, 44]. However, the ability of *L. pneumophila* to survive in oligotrophic environments is puzzling and suggests that the bacteria can live on a diet provided by other members in the biofilm community. To overcome the starvation mode in oligotrophic environment, *L. pneumophila* incorporate in two- and multispecies biofilms. Instead of attaching as a primary colonizer, *L. pneumophila* use a strategic mode where they dock to a pre-established biofilm, thus mediating bacterial survival and association in the biofilm community [19, 42].

Obtaining the required carbon, nitrogen, and amino acids for replication of *L. pneumophila* seems to be primarily reliant on necrotrophic feeding on the products of dead bacteria and tissues within the biofilm [35, 36]. Moreover, heterotrophic bacteria support growth of *L. pneumophila* on media that does not usually support growth because it is deficient in L-cysteine and ferric pyrophosphate [45]. Consistent with this, *L. pneumophila* showed satellite colonies around some aquatic bacteria including *Flavobacterium breve*, *Pseudomonas spp*., *Alcaligenes spp*., and *Acinetobacter spp*. Further, *L. pneumophila* are able to obtain nutrients directly from algae and to grow on the extracellular products produced by cyanobacteria under laboratory conditions [46]. Further, several algae such as *Scenedesmus* spp., *Chlorella* spp., and *Gloeocystis* spp., supported the growth of *L. pneumophila* in basal salt media [28].

The second mechanism by which *L. pneumophila* obtain nutrient in biofilms is through amoeba. Amoeba serve as a secure niche that provides the environmental host for survival and replication of *Legionella* species in the environment [47, 48], and protect the bacteria from antibacterial agents [49]. Importantly, pathogenesis of *L. pneumophila* is correlated with persistence and adaptation of *L. pneumophila* in various amoebal hosts, and the nature of protozoal species can directly affect biofilm colonization with *L. pneumophila* [50, 51]. Indeed, *L. pneumophila* can parasitize more than 20 species of amoebae, three species of ciliated protozoa and one species of slime mold [52, 53]. Further, multiplication within amoeba mediated increase production of polysaccharides by *L. pneumophila,* thus enhancing its capacity to establish biofilm [54]. Further, debris from dead amoeba has been shown to support *L. pneumophila* growth [55], and the biomass of protozoa is directly correlated with outbreaks of *L. pneumophila*. Moreover, absence of amoeba did not result to an increase in the number of biofilm-associated *L. pneumophila*. Instead, *L. pneumophila* can enter the VBNC state to mediate their survival [28]. It has been suggested that metazoan such as the *C. elegans* could provide a natural host for *L. pneumophila* [56, 57]. Moreover, *L. pneumophila* survive within biofilm containing protozoan and *C. elegans* [58]. Therefore, harnessing nutrient from mixed species biofilms as well as survival in the amoeba and *C. elegans* enhances the persistence of *L. pneumophila*. Therefore, diversity of biofilm-associated organisms would provide a various means of nutrient acquisition in oligotrophic environment for such a fastidious organism.

1.3 Factors influencing biofilm formation by *L. pneumophila*

1.3.1 Cyclic-di-GMP

Regulation of bacterial pathogenesis and biofilm formation has been associated with the bacterial second messenger Cyclic-dimeric diguanylate (c-di-GMP) [59–62]. Biofilm regulation for several bacteria has been shown to be reliant on c-di-GMP [63–65]. Two main enzymes have been implicated in regulating the synthesis of the c-di-GMP. (I) A diguanylate cyclases (DGCs) containing GGDEF domain mediates the production of c-di-GMP from two GTPs molecules [66]. (II) A phosphodiesterases (PDEs) proteins containing EAL domain that mediate the degradation of c-di-GMP [66].

The *L. pneumophila* genome encodes for 22–24 GGDEF/EAL-containing proteins that vary between strains, suggesting that c-di-GMP signaling plays a role in the *L. pneumophila* life style [67–69]. Furthermore, *L. pneumophila* replication within amoeba and macrophages as well as virulence is influenced by the expression of GGDEF/EAL-containing proteins [68, 69]. Three GGDEF/EAL-containing proteins positively regulate biofilm formation in *L. pneumophila* Lens, [67]. *L. pneumophila* lacking these proteins showed reduced biofilm formation, however the level of c-di-GMP was not different when compared to the wild type (WT) bacteria [67]. However, two GGDEF/EAL-containing proteins have been shown to negatively regulate biofilm formation and deletion of these proteins resulted in overproduction of biofilm but surprisingly a decrease in the level of the c-di-GMP [67]. Therefore, GGDEF/EAL-containing proteins utilize different mechanisms to regulate biofilm by *L. pneumophila* when compared to other bacteria.

The Haem Nitric oxide/Oxygen (H-NOX) binding domains family of hemoprotein sensors have been demonstrated to play a role in regulating biofilm formation and the c-di-GMP activity [70]. Intriguingly, *L. pneumophila* is the only prokaryote found to encode two H-NOX proteins and show widespread of the H-NOX proteins in their genomes. Hyper-biofilm formation phenotype is attributed to deletion of *hnox1* without influencing growth of *L. pneumophila* in nutrient proficient media (BYE), mouse macrophages or *Acanthamoeba castellanii*. Importantly, a diguanylate cyclase is adjacent to *hnox1* and when overexpressed, *L. pneumophila* exhibits a hyper-biofilm phenotype. Presence of the H-NOX in the NO-bound state inhibited the diguanylate cyclase activity; suggesting that the diguanylate cyclase activity is regulated by NO [70]. Exposure to NO did not result in dispersing the adherent bacteria, but instead the biofilm intensity was increased. The reduced level of c-di-GMP has been associated with the excessive biofilm formation and the c-di-GMP degrading ability could enhance biofilm formation [67]. In the aquatic environment, exposure to NO occurs when *L. pneumophila* is in close contact to denitrifying bacteria, or when exposed to NO produced by macrophages or protozoa. Therefore, biofilm formation can be regulated by NO sensing.

1.3.2 Iron

Even though it is essential for *L. pneumophila* growth and replication [71–73], the concentration of iron must be stringently regulated, to overcome the toxic effect associated with production of reactive oxygen species (ROS), when used in excessive amount [74, 75]. Biofilm formation is inhibited when a fivefold increase in the concentration of iron pyrophosphate was used [17]. In addition, iron salt has been shown to disturb biofilm formation by other bacteria such including *P. aeruginosa* [76]. Recently, the effect of iron pyrophosphate and several iron chelators on the persistence of *L. pneumophila* in mixed biofilm were tested [77]. Chelating ferrous iron dipyridyl, DIP, enhanced the growth of (WT or mutant in iron uptake), suggesting that DIP positively contributes to the persistence of *L. pneumophila* [77]. Interestingly, DIP has no effect on the bacterial population in biofilm or survival of free-living amoeba in the biofilm and is independent of iron acquisition systems as mutants in iron uptake were not affected by DIP. These data suggest that contribution of DIP to the persistence of *L. pneumophila* in biofilm is via protecting *L. pneumophila* from the adverse effects of iron due to a decrease in ROS production [77].

1.3.3 Genetic control

Even though biofilm formation plays a role in the colonization, survival, dissemination and likely the pathogenesis of *L. pneumophila* [78], the genetic factors and molecular mechanisms involved in this process need to be elucidated. Genes that belong to the putative twin-arginine translocation pathway, which is required for transport of folded proteins across the cytoplasmic membrane, have been shown to be required for biofilm formation. Biofilm formation is reduced in mutants with insertional inactivation of the *tatB* and *tatC* genes [79]. Further, biofilm formation in static microtiter plates is impaired in a strain lacking the flagellar sigma factor FliA (σ^{28}) [18]. Expression of genes associated with the transmissive phase of *L. pneumophila* is controlled by FliA [80, 81]. Biofilm-derived *L. pneumophila* down-regulate FliA expression compared to planktonic bacteria in mouse macrophages infection, [82]. Production of flagella is controlled by *L. pneumophila* quorum sensing (Lqs) signaling compound LAI-1(3-hydroxypentadecane-4-one) as well as the stationary phase regulatory network, sensing availability of nutrient

[83]. However, the flagella are not required for attachment and persistence of *L. pneumophila* biofilm formed by *K. pneumonia* [19]. This is consistent with our observation showing the down-regulation of the flagella during biofilm formation in mouse macrophage [82].

Binding to sulfated glycosaminoglycans (CAGs) of the host extracellular matrix is mediated via the *Legionella* collagen-like (LcI) adhesin. Even though LcI is widely distributed in different *L. pneumophila* environmental and clinical isolates, it is lacking in poor biofilm producers; indicating the acquisition of this gene by horizontal gene transfer to *L. pneumophila* [84]. The GC content of *lpg2644* is different from the rest of *L. pneumophila* genome [84], indicating the acquisition of this gene by horizontal gene transfer to *L. pneumophila* [84]. Further, biofilm formation, cell–cell adhesion and cell-matrix interactions is reduced in strains with mutation in *lpg2644* [84]. The *L. pneumophila lpg2644* gene is differentially regulated during growth phases and biofilm formation [41]. Regulation of late stages of biofilm formation is mediated by *P. aeruginosa* quorum sensing (3OC12-HSL). Therefore, regulation of biofilm formation promotes dispersion of bacteria and mediates initiation of another biofilm cycle to another surface [41]. These events are crucial for the proliferation and transmission of *L. pneumophila* [78].

1.3.4 Quorum sensing

In Gram-negative bacteria, gene expression of several bacterial processes, including virulence, sporulation, bioluminescence, competence and biofilm formation is regulated by quorum sensing (QS) [85, 86]. Quorum sensing bacteria are usually identified in man-made water systems and it is well appreciated that QS signaling regulate environmental biofilm production [87]. The LAI-1 (3-hydroxypentadecane-4-one) QS autoinducer is the only (*Legionella* quorum sensing) Lqs system identified up to date [88–91]. The *L. pneumophila* LAI-1 is detected by the Lqs system which is composed of the autoinducer synthase LqsA, the homologous sensor kinases LqsS and the response regulator LqsR [88–90]. The Lqs system of *L. pneumophila* is homologous to the *cqsAS* QS of *Vibrio cholera*, which regulates cell-density, virulence and biofilm formation [85, 92]. Importantly, the *L. pneumophila* biofilm formation is inhibited by the *P. aeruginosa* quorum sensing autoinducer (3-oxo-C12-HSL), which down-regulate the expression of *lqsR* [41, 93]. Therefore, QS could potentially disperse *L. pneumophila* biofilm during later stages.

1.4 Modulation of gene expression in biofilms

Differential gene expression between planktonic and biofilm forming *L. pneumophila* was shown through transcriptomic analysis [17]. The gene expression pattern was compared with the replicative and transmissive phases during growth of *L. pneumophila* in *A. castellanii* [94]. Importantly, gene expression profile of sessile bacteria is similar to the replicative phase of *L. pneumophila*. Furthermore, genes that are involved in repressing the transmissive phase were well expressed in the sessile bacteria [17], suggesting that biofilm is a secure niche for *L. pneumophila* [17]. The *pvcAB* gene cluster (which is regulated by iron) is among the genes that were highly expressed in the sessile form [17]. The *L. pneumophila pvcA* and *pvcB* genes are homologous exhibit homology to the *P. aeruginosa* proteins PvcA and PvcB and are required for the production of the iron binding protein (siderophore). The *pvcA* and *pvcB* in *L. pneumophila* encode for a siderophore-like molecule, which promote iron sequestration at a sub toxic level. The second gene cluster, including *ahpC2* and *ahpD*, encodes for alkyl hydroperoxide reductases and play a role in protection against oxidative stress [95, 96] displayed the highest induction in biofilm cells [95]. Iron plays a role in the production of reactive oxygen species and the metabolism of iron and oxidative stress is related. Induction of both *pvcAB* and *ahpC2D* genes in sessile cells could be utilized to overcome the toxic environment associated with high iron level concentrations.

Further, examining the expression of the macrophage infectivity potentiator (*mip*) to transcriptionally active *L. pneumophila* infected in cell culture was used to evaluate the virulence of biofilm-associated *L. pneumophila* [16]. Expression of *mip* is required for growth in protozoa and human macrophages [97]. Further, *mip* expression is up-regulated during the transmissive stages of *L. pneumophila* life cycle, but downregulated at early stages of infection [98]. At early stages of biofilm formation, which is similar to the replicative phase, expression of *mip* was constant. However, at later stages of biofilm formation, which is similar to the replicative phase, *mip* expression was predominately up-regulated [16]. Upregulation of *mip* expression could be correlated with the switch to the transmissive phase observed in the planktonic form and suggests that biofilm could protect the replicative form of *L. pneumophila*.

1.5 Biocides treatments of *L. pneumophila* biofilm and bacterial resistance

L. pneumophila survive in biofilms covering environmental and artificial water systems such as ventilation and conditioning systems [78]. In addition, biofilm-containing *L. pneumophila* can become a transient or permanent habitat for other relevant microorganisms. Therefore, biofilm-associated organisms can survive for days, weeks or even months depending on the substratum and the environmental factors that stimulate biofilm formation [99, 100]. To restrict *L. pneumophila* growth, numerous chemical, physical and thermal disinfection methods have been used against *L. pneumophila* [101]. However, these treatments generally do not result in total elimination of the bacterium, and after a lag period, recolonization occurs as quickly as the treatments are discontinued [35]. Biofilm-associated *L. pneumophila* is extremely resistant to disinfectants and biocides [101, 102]. Further, exposure of biofilm-encased bacteria to biocides could lead to entry into a viable non-culturable status [103]. Chlorine and its derivatives are the most common biocides used in disinfection protocols and have been shown to be appropriate in eliminating planktonic *L. pneumophila* but not biofilms [104]. Resistance of *L. pneumophila* to disinfection is due not only to its capacity to survive within biofilm, but also the bacteria exhibit the intra-amoebal life-style [105, 106]. Therefore, amoeba- associated *L. pneumophila* are more resistant to disinfection possibly due to differences in membrane chemistry or life cycle stages of this primitive organism [35, 107]. It has been shown that vesicles containing intracellular *L. pneumophila* released by amoeba are resistant to biocide treatments [108]. Importantly, these vesicles remained viable for few months [109]. Understanding the molecular mechanisms that governs the intra-amoeba related resistance should pave the way for development of new strategies to eradicate *L. pneumophila*.

Other methods have been used to limit *L. pneumophila* such as applying heat which has been shown to be effective in reducing the number of bacteria and protozoan trophozoites, but infective against killing cysts [110, 111]. UV radiation is also effective when the bacteria are in direct contact with the radiation [112]. However, higher UV intensities are required to inactivate the protozoa [113]. Other methods have been proposed to control *L. pneumophila* growth such as controlling the carbon source within anthropogenic water system [114], or addition of phages to control bacterial or specifically *L. pneumophila* growth. The phage is capable of degrading polysaccharides and therefore destabilizing the biofilm [115, 116]. Furthermore, nanoparticles have been shown to be effective in reduction of *L. pneumophila* biofilm volume and showed some efficacy against *Staphylococcus aureus* and *Staphylococcus epidermidis* biofilms [117–119]. Moreover, several natural compounds (biosurfactants, antimicrobial peptides, protein and essential oil) have been shown to exhibit anti-*Legionella* properties [120]. Collectively, it is necessary to control *L. pneumophila* growth and their natural hosts to optimize eradication of the bacteria.

2. Conclusions

Several chemical and physical parameters can influence the behavior of *L. pneumophila* in biofilms, including the surface, the temperature, carbon and metal concentrations, and the presence of biocides [17, 18, 34, 114, 121–128]. Biological factors such as being a member of mixed species biofilm or parasitizing free-living amoeba or nematodes influence biofilm formation by *L. pneumophila*. Biofilm-associated *L. pneumophila* is resistant to biocides and Legionellosis outbreaks have been attributed to biofilms. Therefore, it is essential to design new remedies for eradication of *L. pneumophila* biofilm in different environmental settings. Treatment studies should be performed when the bacterium is in its natural host to determine how the bacterium is protected inside the amoeba and if the passages through the natural hosts modify the resistance. Thus, preventing biofilm formation appears as one strategy to reduce water system contamination.

Acknowledgements

Studies in Dr. Amer's laboratory are supported by The Ohio State University Center for Clinical and Translational Science Longitudinal Pilot Award (CCTS), R21 AI113477, R01 AI24121 and R01 HL127651. Studies in Dr. Abu Khweek's laboratory are supported by Birzeit University.

Conflict of interest

The authors of the manuscript declare that the submitted work was carried out in the absence of any personal, professional or financial relationships that could potentially be construed as a conflict of interest.

Author contributions

Arwa Abu Khweek wrote the book chapter, and Amal O. Amer edited the manuscript.

Author details

Arwa Abu Khweek[1] and Amal O. Amer[2]*

1 Department of Biology and Biochemistry, Birzeit University, West Bank, Palestine

2 Department of Microbial Infection and Immunity, Infectious Disease Institute, The Ohio State University, Columbus, OH, United States

*Address all correspondence to: amal.amer@osumc.edu

References

[1] Fraser DW et al. Legionnaires' disease: Description of an epidemic of pneumonia. The New England Journal of Medicine. 1977;**297**(22):1189-1197

[2] Wagner C et al. Collagen binding protein Mip enables legionella pneumophila to transmigrate through a barrier of NCI-H292 lung epithelial cells and extracellular matrix. Cellular Microbiology. 2007;**9**(2):450-462

[3] Steinert M, Hentschel U, Hacker J. Legionella pneumophila: an aquatic microbe goes astray. FEMS Microbiology Reviews. 2002;**26**(2):149-162

[4] Fields BS, Benson RF, Besser RE. Legionella and Legionnaires' disease: 25 years of investigation. Clinical Microbiology Reviews. 2002;**15**(3):506-526

[5] Isberg RR, O'Connor TJ, Heidtman M. The legionella pneumophila replication vacuole: Making a cosy niche inside host cells. Nature Reviews. Microbiology. 2009;**7**(1):13-24

[6] de Felipe KS et al. Evidence for acquisition of legionella type IV secretion substrates via interdomain horizontal gene transfer. Journal of Bacteriology. 2005;**187**(22):7716-7726

[7] Abu Khweek A et al. The Sphingosine-1-phosphate Lyase (LegS2) contributes to the restriction of legionella pneumophila in murine macrophages. PLoS One. 2016;**11**(1):e0146410

[8] Khweek AA et al. A bacterial protein promotes the recognition of the legionella pneumophila vacuole by autophagy. European Journal of Immunology. 2013;**43**(5):1333-1344

[9] Losick VP, Isberg RR. NF-kappaB translocation prevents host cell death after low-dose challenge by legionella pneumophila. The Journal of Experimental Medicine. 2006;**203**(9):2177-2189

[10] de Felipe KS et al. Legionella eukaryotic-like type IV substrates interfere with organelle trafficking. PLoS Pathogens. 2008;**4**(8):e1000117

[11] Price CT et al. Host proteasomal degradation generates amino acids essential for intracellular bacterial growth. Science. 2011;**334**(6062):1553-1557

[12] Belyi Y et al. Legionella pneumophila glucosyltransferase inhibits host elongation factor 1A. Proceedings of the National Academy of Sciences of the United States of America. 2006;**103**(45):16953-16958

[13] Laguna RK et al. A legionella pneumophila-translocated substrate that is required for growth within macrophages and protection from host cell death. Proceedings of the National Academy of Sciences of the United States of America. 2006;**103**(49):18745-18750

[14] Newton HJ et al. Molecular pathogenesis of infections caused by legionella pneumophila. Clinical Microbiology Reviews. 2010;**23**(2):274-298

[15] O'Toole G, Kaplan HB, Kolter R. Biofilm formation as microbial development. Annual Review of Microbiology. 2000;**54**:49-79

[16] Andreozzi E et al. Role of biofilm in protection of the replicative form of legionella pneumophila. Current Microbiology. 2014;**69**(6):769-774

[17] Hindre T et al. Transcriptional profiling of legionella pneumophila

biofilm cells and the influence of iron on biofilm formation. Microbiology. 2008;**154**(Pt 1):30-41

[18] Mampel J et al. Planktonic replication is essential for biofilm formation by legionella pneumophila in a complex medium under static and dynamic flow conditions. Applied and Environmental Microbiology. 2006;**72**(4):2885-2895

[19] Stewart CR, Muthye V, Cianciotto NP. Legionella pneumophila persists within biofilms formed by Klebsiella pneumoniae, Flavobacterium sp., and Pseudomonas fluorescens under dynamic flow conditions. PLoS One. 2012;**7**(11):e50560

[20] Atlas RM. Legionella: From environmental habitats to disease pathology, detection and control. Environmental Microbiology. 1999;**1**(4):283-293

[21] Berk SG et al. Packaging of live legionella pneumophila into pellets expelled by Tetrahymena spp. does not require bacterial replication and depends on a dot/Icm-mediated survival mechanism. Applied and Environmental Microbiology. 2008;**74**(7):2187-2199

[22] Faulkner G, Garduno RA. Ultrastructural analysis of differentiation in legionella pneumophila. Journal of Bacteriology. 2002;**184**(24):7025-7041

[23] Steinert M et al. Resuscitation of viable but nonculturable legionella pneumophila Philadelphia JR32 by Acanthamoeba castellanii. Applied and Environmental Microbiology. 1997;**63**(5):2047-2053

[24] Garcia MT et al. Acanthamoeba polyphaga resuscitates viable non-culturable legionella pneumophila after disinfection. Environmental Microbiology. 2007;**9**(5):1267-1277

[25] Valster RM, Wullings BA, van der Kooij D. Detection of protozoan hosts for legionella pneumophila in engineered water systems by using a biofilm batch test. Applied and Environmental Microbiology. 2010;**76**(21):7144-7153

[26] Horwitz MA. Formation of a novel phagosome by the Legionnaires' disease bacterium (legionella pneumophila) in human monocytes. The Journal of Experimental Medicine. 1983;**158**(4):1319-1331

[27] Declerck P et al. Replication of legionella pneumophila in biofilms of water distribution pipes. Microbiological Research. 2009;**164**(6):593-603

[28] Declerck P. Biofilms: The environmental playground of legionella pneumophila. Environmental Microbiology. 2010;**12**(3):557-566

[29] Shirtliff ME, Mader JT, Camper AK. Molecular interactions in biofilms. Chemistry & Biology. 2002;**9**(8):859-871

[30] Sutherland IW. The biofilm matrix--an immobilized but dynamic microbial environment. Trends in Microbiology. 2001;**9**(5):222-227

[31] Costerton JW. Overview of microbial biofilms. Journal of Industrial Microbiology. 1995;**15**(3):137-140

[32] Costerton JW et al. Bacterial biofilms in nature and disease. Annual Review of Microbiology. 1987;**41**:435-464

[33] Pecastaings S et al. Sessile legionella pneumophila is able to grow on surfaces and generate structured monospecies biofilms. Biofouling. 2010;**26**(7):809-819

[34] Piao Z et al. Temperature-regulated formation of mycelial mat-like biofilms

by legionella pneumophila. Applied and Environmental Microbiology. 2006;**72**(2):1613-1622

[35] Taylor M, Ross K, Bentham R. Legionella, protozoa, and biofilms: Interactions within complex microbial systems. Microbial Ecology. 2009;**58**(3):538-547

[36] Vervaeren H et al. Introduction of a boost of legionella pneumophila into a stagnant-water model by heat treatment. FEMS Microbiology Ecology. 2006;**58**(3):583-592

[37] Wu MC et al. Isolation of genes involved in biofilm formation of a Klebsiella pneumoniae strain causing pyogenic liver abscess. PLoS One. 2011;**6**(8):e23500

[38] Basson A, Flemming LA, Chenia HY. Evaluation of adherence, hydrophobicity, aggregation, and biofilm development of Flavobacterium johnsoniae-like isolates. Microbial Ecology. 2008;**55**(1):1-14

[39] Kives J, Orgaz B, Sanjose C. Polysaccharide differences between planktonic and biofilm-associated EPS from Pseudomonas fluorescens B52. Colloids and Surfaces. B, Biointerfaces. 2006;**52**(2):123-127

[40] Guerrieri E et al. Effect of bacterial interference on biofilm development by legionella pneumophila. Current Microbiology. 2008;**57**(6):532-536

[41] Mallegol J et al. Essential roles and regulation of the legionella pneumophila collagen-like adhesin during biofilm formation. PLoS One. 2012;**7**(9):e46462

[42] Watnick P, Kolter R. Biofilm, city of microbes. Journal of Bacteriology. 2000;**182**(10):2675-2679

[43] George JR et al. Amino acid requirements of legionella pneumophila. Journal of Clinical Microbiology. 1980;**11**(3):286-291

[44] Edelstein PH. Comparative study of selective media for isolation of legionella pneumophila from potable water. Journal of Clinical Microbiology. 1982;**16**(4):697-699

[45] Wadowsky RM, Yee RB. Satellite growth of legionella pneumophila with an environmental isolate of Flavobacterium breve. Applied and Environmental Microbiology. 1983;**46**(6):1447-1449

[46] Tison DL et al. Growth of legionella pneumophila in association with blue-green algae (cyanobacteria). Applied and Environmental Microbiology. 1980;**39**(2):456-459

[47] Rowbotham TJ. Preliminary report on the pathogenicity of legionella pneumophila for freshwater and soil amoebae. Journal of Clinical Pathology. 1980;**33**(12):1179-1183

[48] Newsome AL et al. Isolation of an amoeba naturally harboring a distinctive legionella species. Applied and Environmental Microbiology. 1998;**64**(5):1688-1693

[49] Loret JF, Greub G. Free-living amoebae: Biological by-passes in water treatment. International Journal of Hygiene and Environmental Health. 2010;**213**(3):167-175

[50] Murga R et al. Role of biofilms in the survival of legionella pneumophila in a model potable-water system. Microbiology. 2001;**147**(Pt 11):3121-3126

[51] Rowbotham TJ. Pontiac fever, amoebae, and legionellae. Lancet. 1981;**1**(8210):40-41

[52] Hagele S et al. Dictyostelium discoideum: A new host model system for intracellular pathogens of the genus

legionella. Cellular Microbiology. 2000;**2**(2):165-171

[53] Kikuhara H et al. Intracellular multiplication of legionella pneumophila in Tetrahymena thermophila. Journal of UOEH. 1994;**16**(4):263-275

[54] Bigot R et al. Intra-amoeba multiplication induces chemotaxis and biofilm colonization and formation for legionella. PLoS One. 2013;**8**(10):e77875

[55] Temmerman R et al. Necrotrophic growth of legionella pneumophila. Applied and Environmental Microbiology. 2006;**72**(6):4323-4328

[56] Brassinga AK et al. Caenorhabditis is a metazoan host for legionella. Cellular Microbiology. 2010;**12**(3):343-361

[57] Hellinga JR et al. Identification of vacuoles containing extraintestinal differentiated forms of legionella pneumophila in colonized Caenorhabditis elegans soil nematodes. Microbiology. 2015;**4**(4):660-681

[58] Rasch J et al. Legionella-protozoa-nematode interactions in aquatic biofilms and influence of Mip on Caenorhabditis elegans colonization. International Journal of Medical Microbiology. 2016;**306**(6):443-451

[59] Tamayo R, Pratt JT, Camilli A. Roles of cyclic diguanylate in the regulation of bacterial pathogenesis. Annual Review of Microbiology. 2007;**61**:131-148

[60] Romling U, Galperin MY, Gomelsky M. Cyclic di-GMP: The first 25 years of a universal bacterial second messenger. Microbiology and Molecular Biology Reviews. 2013;**77**(1):1-52

[61] Martinez-Gil M, Ramos C. Role of cyclic di-GMP in the bacterial virulence and evasion of the plant immunity. Current Issues in Molecular Biology. 2017;**25**:199-222

[62] Abu Khweek A, Fetherston JD, Perry RD. Analysis of HmsH and its role in plague biofilm formation. Microbiology. 2010;**156**(Pt 5):1424-1438

[63] Conner JG et al. The ins and outs of cyclic di-GMP signaling in vibrio cholerae. Current Opinion in Microbiology. 2017;**36**:20-29

[64] Valentini M, Filloux A. Biofilms and cyclic di-GMP (c-di-GMP) Signaling: Lessons from Pseudomonas aeruginosa and other bacteria. The Journal of Biological Chemistry. 2016;**291**(24):12547-12555

[65] Bobrov AG et al. Systematic analysis of cyclic di-GMP signalling enzymes and their role in biofilm formation and virulence in Yersinia pestis. Molecular Microbiology. 2011;**79**(2):533-551

[66] Simm R et al. GGDEF and EAL domains inversely regulate cyclic di-GMP levels and transition from sessility to motility. Molecular Microbiology. 2004;**53**(4):1123-1134

[67] Pecastaings S et al. New insights into legionella pneumophila biofilm regulation by c-di-GMP signaling. Biofouling. 2016;**32**(8):935-948

[68] Levi A et al. Cyclic diguanylate signaling proteins control intracellular growth of legionella pneumophila. MBio. 2011;**2**(1):e00316-e00310

[69] Allombert J et al. Three antagonistic cyclic di-GMP-catabolizing enzymes promote differential dot/Icm effector delivery and intracellular survival at the early steps of legionella pneumophila infection. Infection and Immunity. 2014;**82**(3):1222-1233

[70] Carlson HK, Vance RE, Marletta MA. H-NOX regulation of c-di-GMP metabolism and biofilm formation

in legionella pneumophila. Molecular Microbiology. 2010;77(4):930-942

[71] Radtke AL, O'Riordan MX. Intracellular innate resistance to bacterial pathogens. Cellular Microbiology. 2006;**8**(11):1720-1729

[72] Reeves MW et al. Metal requirements of legionella pneumophila. Journal of Clinical Microbiology. 1981;**13**(4):688-695

[73] Schaible UE, Kaufmann SH. Iron and microbial infection. Nature Reviews. Microbiology. 2004;**2**(12):946-953

[74] Andrews SC, Robinson AK, Rodriguez-Quinones F. Bacterial iron homeostasis. FEMS Microbiology Reviews. 2003;**27**(2-3):215-237

[75] Lemire JA, Harrison JJ, Turner RJ. Antimicrobial activity of metals: Mechanisms, molecular targets and applications. Nature Reviews. Microbiology. 2013;**11**(6):371-384

[76] Musk DJ, Banko DA, Hergenrother PJ. Iron salts perturb biofilm formation and disrupt existing biofilms of pseudomonas aeruginosa. Chemistry & Biology. 2005;**12**(7):789-796

[77] Portier E et al. Iron availability modulates the persistence of legionella pneumophila in complex biofilms. Microbes and Environments. 2016;**31**(4):387-394

[78] Lau HY, Ashbolt NJ. The role of biofilms and protozoa in legionella pathogenesis: Implications for drinking water. Journal of Applied Microbiology. 2009;**107**(2):368-378

[79] De Buck E et al. Legionella pneumophila Philadelphia-1 tatB and tatC affect intracellular replication and biofilm formation. Biochemical and Biophysical Research Communications. 2005;**331**(4):1413-1420

[80] Heuner K et al. Influence of the alternative sigma(28) factor on virulence and flagellum expression of legionella pneumophila. Infection and Immunity. 2002;**70**(3):1604-1608

[81] Molofsky AB, Shetron-Rama LM, Swanson MS. Components of the legionella pneumophila flagellar regulon contribute to multiple virulence traits, including lysosome avoidance and macrophage death. Infection and Immunity. 2005;**73**(9):5720-5734

[82] Abu Khweek A et al. Biofilm-derived legionella pneumophila evades the innate immune response in macrophages. Frontiers in Cellular and Infection Microbiology. 2013;**3**:18

[83] Schell U, Simon S, Hilbi H. Inflammasome recognition and regulation of the legionella flagellum. Current Topics in Microbiology and Immunology. 2016;**397**:161-181

[84] Duncan C et al. Lcl of legionella pneumophila is an immunogenic GAG binding adhesin that promotes interactions with lung epithelial cells and plays a crucial role in biofilm formation. Infection and Immunity. 2011;**79**(6):2168-2181

[85] Zhu J et al. Quorum-sensing regulators control virulence gene expression in vibrio cholerae. Proceedings of the National Academy of Sciences of the United States of America. 2002;**99**(5):3129-3134

[86] Ng WL, Bassler BL. Bacterial quorum-sensing network architectures. Annual Review of Genetics. 2009;**43**:197-222

[87] Shrout JD, Nerenberg R. Monitoring bacterial twitter: Does quorum sensing determine the behavior of water and wastewater treatment biofilms?

Environmental Science & Technology. 2012;**46**(4):1995-2005

[88] Tiaden A et al. The legionella pneumophila response regulator LqsR promotes host cell interactions as an element of the virulence regulatory network controlled by RpoS and LetA. Cellular Microbiology. 2007;**9**(12):2903-2920

[89] Tiaden A et al. Synergistic contribution of the legionella pneumophila lqs genes to pathogen-host interactions. Journal of Bacteriology. 2008;**190**(22):7532-7547

[90] Spirig T et al. The legionella autoinducer synthase LqsA produces an alpha-hydroxyketone signaling molecule. The Journal of Biological Chemistry. 2008;**283**(26):18113-18123

[91] Tiaden A et al. The autoinducer synthase LqsA and putative sensor kinase LqsS regulate phagocyte interactions, extracellular filaments and a genomic island of legionella pneumophila. Environmental Microbiology. 2010;**12**(5):1243-1259

[92] Miller MB et al. Parallel quorum sensing systems converge to regulate virulence in vibrio cholerae. Cell. 2002;**110**(3):303-314

[93] Kimura S et al. Pseudomonas aeruginosa las quorum sensing autoinducer suppresses growth and biofilm production in legionella species. Microbiology. 2009;**155**(Pt 6):1934-1939

[94] Bruggemann H et al. Virulence strategies for infecting phagocytes deduced from the in vivo transcriptional program of legionella pneumophila. Cellular Microbiology. 2006;**8**(8):1228-1240

[95] Rocha ER, Smith CJ. Role of the alkyl hydroperoxide reductase (ahpCF) gene in oxidative stress defense of the obligate anaerobe bacteroides fragilis. Journal of Bacteriology. 1999;**181**(18):5701-5710

[96] LeBlanc JJ, Davidson RJ, Hoffman PS. Compensatory functions of two alkyl hydroperoxide reductases in the oxidative defense system of legionella pneumophila. Journal of Bacteriology. 2006;**188**(17):6235-6244

[97] Cianciotto NP, Fields BS. Legionella pneumophila mip gene potentiates intracellular infection of protozoa and human macrophages. Proceedings of the National Academy of Sciences of the United States of America. 1992;**89**(11):5188-5191

[98] Wieland H et al. Intracellular multiplication of legionella pneumophila depends on host cell amino acid transporter SLC1A5. Molecular Microbiology. 2005;**55**(5):1528-1537

[99] Blasco MD, Esteve C, Alcaide E. Multiresistant waterborne pathogens isolated from water reservoirs and cooling systems. Journal of Applied Microbiology. 2008;**105**(2):469-475

[100] Buse HY et al. Microbial diversities (16S and 18S rRNA gene pyrosequencing) and environmental pathogens within drinking water biofilms grown on the common premise plumbing materials unplasticized polyvinylchloride and copper. FEMS Microbiology Ecology. 2014;**88**(2):280-295

[101] Kim BR et al. Literature review--efficacy of various disinfectants against legionella in water systems. Water Research. 2002;**36**(18):4433-4444

[102] Borella P et al. Water ecology of legionella and protozoan: Environmental and public health perspectives. Biotechnology Annual Review. 2005;**11**:355-380

[103] Giao MS et al. Incorporation of natural uncultivable legionella

pneumophila into potable water biofilms provides a protective niche against chlorination stress. Biofouling. 2009;**25**(4):335-341

[104] Cooper IR, Hanlon GW. Resistance of legionella pneumophila serotype 1 biofilms to chlorine-based disinfection. The Journal of Hospital Infection. 2010;**74**(2):152-159

[105] Hilbi H, Hoffmann C, Harrison CF. Legionella spp. outdoors: Colonization, communication and persistence. Environmental Microbiology Reports. 2011;**3**(3): 286-296

[106] Steinert M et al. Regrowth of legionella pneumophila in a heat-disinfected plumbing system. Zentralbl Bakteriol. 1998;**288**(3): 331-342

[107] Dupuy M et al. Efficiency of water disinfectants against legionella pneumophila and Acanthamoeba. Water Research. 2011;**45**(3):1087-1094

[108] Berk SG et al. Production of respirable vesicles containing live legionella pneumophila cells by two Acanthamoeba spp. Applied and Environmental Microbiology. 1998;**64**(1):279-286

[109] Bouyer S et al. Long-term survival of legionella pneumophila associated with Acanthamoeba castellanii vesicles. Environmental Microbiology. 2007;**9**(5):1341-1344

[110] Storey MV et al. The efficacy of heat and chlorine treatment against thermotolerant Acanthamoebae and legionellae. Scandinavian Journal of Infectious Diseases. 2004;**36**(9):656-662

[111] Farhat M et al. Effects of disinfection on legionella spp., eukarya, and biofilms in a hot water system. Applied and Environmental Microbiology. 2012;**78**(19):6850-6858

[112] Schwartz T, Hoffmann S, Obst U. Formation of natural biofilms during chlorine dioxide and u.v. disinfection in a public drinking water distribution system. Journal of Applied Microbiology. 2003;**95**(3):591-601

[113] Hijnen WA, Beerendonk EF, Medema GJ. Inactivation credit of UV radiation for viruses, bacteria and protozoan (oo)cysts in water: A review. Water Research. 2006;**40**(1):3-22

[114] Pang CM, Liu WT. Biological filtration limits carbon availability and affects downstream biofilm formation and community structure. Applied and Environmental Microbiology. 2006;**72**(9):5702-5712

[115] Lammertyn E et al. Evidence for the presence of legionella bacteriophages in environmental water samples. Microbial Ecology. 2008;**56**(1):191-197

[116] Hughes KA, Sutherland IW, Jones MV. Biofilm susceptibility to bacteriophage attack: The role of phage-borne polysaccharide depolymerase. Microbiology. 1998;**144**(Pt 11):3039-3047

[117] Raftery TD et al. Discrete nanoparticles induce loss of legionella pneumophila biofilms from surfaces. Nanotoxicology. 2014;**8**(5):477-484

[118] Subbiahdoss G et al. Magnetic targeting of surface-modified superparamagnetic iron oxide nanoparticles yields antibacterial efficacy against biofilms of gentamicin-resistant staphylococci. Acta Biomaterialia. 2012;**8**(6):2047-2055

[119] Taylor EN et al. Superparamagnetic iron oxide nanoparticles (SPION) for the treatment of antibiotic-resistant biofilms. Small. 2012;**8**(19):3016-3027

[120] Berjeaud JM et al. Legionella pneumophila: The paradox of a highly sensitive opportunistic waterborne pathogen able to persist in the

environment. Frontiers in Microbiology. 2016;7:486

[121] Wright JB, Ruseska I, Costerton JW. Decreased biocide susceptibility of adherent legionella pneumophila. The Journal of Applied Bacteriology. 1991;**71**(6):531-538

[122] Bezanson G et al. In situ colonization of polyvinyl chloride, brass, and copper by legionella pneumophila. Canadian Journal of Microbiology. 1992;**38**(4):328-330

[123] Turetgen I, Cotuk A. Monitoring of biofilm-associated legionella pneumophila on different substrata in model cooling tower system. Environmental Monitoring and Assessment. 2007;**125**(1-3):271-279

[124] Rogers J et al. Influence of plumbing materials on biofilm formation and growth of legionella pneumophila in potable water systems. Applied and Environmental Microbiology. 1994;**60**(6):1842-1851

[125] Donlan RM et al. Legionella pneumophila associated with the protozoan Hartmannella vermiformis in a model multi-species biofilm has reduced susceptibility to disinfectants. Biofouling. 2005;**21**(1):1-7

[126] Liu Z et al. Effect of flow regimes on the presence of legionella within the biofilm of a model plumbing system. Journal of Applied Microbiology. 2006;**101**(2):437-442

[127] Lehtola MJ et al. Survival of Mycobacterium avium, legionella pneumophila, Escherichia coli, and caliciviruses in drinking water-associated biofilms grown under high-shear turbulent flow. Applied and Environmental Microbiology. 2007;**73**(9):2854-2859

[128] van der Kooij D, Veenendaal HR, Scheffer WJ. Biofilm formation and multiplication of legionella in a model warm water system with pipes of copper, stainless steel and cross-linked polyethylene. Water Research. 2005;**39**(13):2789-2798

Chapter 17

Oral Microbiota from the Stomatology Perspective

Andrea Stašková, Radomíra Nemcová, Stanislav Lauko and Andrej Jenča

Abstract

Besides the properties typical of body cavities, the oral cavity exhibits many differentiating features that allow it to occupy position of an autonomous functional and biological unit, a characteristic ecosystem. An appropriate homeostasis of oral biocenosis and balanced conditions for microorganisms concerning proportions of physiological and pathogenic or potentially pathogenic microbiota play an important role with regard to the oral cavity health and eventually the overall health of an individual. The oral cavity is a constantly changing habitat. The current market offers a number of relevant preparations supporting oral health, and alternative approaches serving these purposes are also available. Results of the studies that focused on microbiocenosis of the dental plaque and interactions between individual bacterial species indicate a probiotic potential of some oral bacteria and their prospective use in prevention of oral cavity diseases. This chapter deals with the state of physiological microbiota found in oral biofilms, with the most important infections of the oral cavity and the potential use of probiotics as a prospective alternative approach to prevention and therapy of oral cavity diseases.

Keywords: oral cavity, microbiome, biofilm, focal infection, probiotics

1. Introduction

Microbiological analysis of oral microbiota is still a challenge the science has to face. Up to this day, we have knowledge of only a portion of microorganisms living in the oral cavity. Their research is very important from the point of view of prevention, diagnostic and treatment of oral and general diseases [1]. Dental caries is the most common chronic disease in the world affecting people regardless of sex, age and ethnic origin, although it affects more the individuals with low social-economic status. *Streptococcus mutans* was identified as the causative agent of this disease. Presented were also results indicating participation of acidogenic bacteria in the process of its development [1]. These bacteria are generally called cariogenic bacteria. However, no pathogen is the direct and only cause of the development of dental caries or periodontitis. More profound knowledge of microbial composition of the oral biofilm of humans on the surface of teeth or in the subgingival space can help to understand better the complexity of pathogenesis of the development of dental diseases, and find new ways how to affect positively the oral health through balanced, physiologically beneficial microbiota [2]. The oral cavity is a constantly changing habitat. Traditional methods intended for the studies of diversity of mirobiocenoses are based on conventional isolation of bacteria by cultivation, their morphology and identification by means of their biochemical properties. These methods do not suffice to ensure concise characterisation and quantification of microbiota, are time demanding, provide results not earlier than after 48 hours and involve only cultivable bacteria. High percentage of bacteria is cultivated only with difficulties due to unknown requirements on their growth [2]. Currently, a number of genetic techniques intended for quantification, identification and characterisation of bacterial communities are available. The study of the external influence on oral cavity microbiocenosis is inevitable due to high incidence and prevalence of dental caries or periodontopathies, despite the current widespread use of oral hygiene preparations [3]. Today's market offers a multitude of such preparations, and also, alternative approaches for the improvement of oral health are available. Scientific studies presented interesting knowledge about beneficial bacteria capable of inhibiting the growth of pathogenic bacteria by their bioactive products. This concerns, for example, the proof of the suppression of oral pathogens by *Streptococcus salivarius* K12 probiotic bacteria, or their bioactive compounds can serve as a basis for the development of new strategies contributing to prevention and treatment of oral diseases [4].

2. Oral cavity microbiome

Immediately after birth, the sterile mouth cavity of the newborn individual mediates the contact between the internal and external environments and, at this time, also its colonisation by microorganisms commences. After several days, the microbiota characteristic of the oral cavity becomes stabilised [5]. In the process of colonisation of the mouth cavity of newborns, streptococci are acquired the first. Over time, the diversity of populating microorganisms grows until the individual acquires certain microbiota the stability of which depends on compensation mechanisms ensuring suitable conditions in the mouth cavity. Mucosa and teeth in the oral cavity come into constant contact with the exogenous microbiota, and the health state of dentition is also affected by proportions of individual groups of microorganisms. Some factors, for example unsuitable diet, can irreversibly affect the homeostasis of the oral ecosystem and subsequently lead to propagation of pathological changes in the oral cavity [6].

Although the oral microbiota contains bacteria, fungi, viruses and archaea, research has focused mostly on oral bacterial populations present in the highest numbers [7]. Fungi as one of the components of the oral microbiota were identified by pyrosequencing with focus on RNA, which exhibits high species variability. Peterson et al. [8] reported that the number of fungal species in the oral microbiota ranges from 9 to 23.

Molecular microbiology techniques based on 16S rRNA allowed scientists to describe more than 700 bacterial species present in the oral cavity of humans. More than 50% of bacterial species were not cultivated and thus their role in the oral microbial ecology has not been explained. It was assessed that approximately 1000 bacterial species are capable of stable existence in the mouth of humans, while each man can harbour 50–200 species of this diverse spectrum [9]. Many species are found temporarily in the saliva or as a part of biofilms formed on teeth or mucosa. Analysis of biodiversity in the mouth cavity showed that the number of oral phylotypes is considerably undervalued. Quantification of oral microbiota of humans was performed by metagenomics of unique phylotypes using pyrosequencing 454 and sequencing by Ilumina technology. This quantification method confirmed 668 bacterial phylotypes in microbiota of one plaque, which is considerably more than the numbers published in the previous studies. Similar sequencing technique detected

3621 phylotypes in the saliva and 6888 phylotypes in a subgingival plaque [7]. Oral cavity bacteria identified by modern sequencing methods are classified in various strains the majority of which belongs to strains (phyla) *Firmicutes*, *Fusobacteria*, *Bacteroidetes*, *Actinobacteria*, *Proteobacteria*, *Spirochaetes* and *Synergistetes*.

In addition to its principal function—intake of food and water—the mouth cavity fulfils a number of other important functions that include the primary protective function against microorganisms entering the gastrointestinal tract and various functions involving discrimination of taste, temperature and pressure [10]. In addition to properties typical of body cavities, the oral cavity possesses many differentiation features owing to which it acquires a position of separate functional and biological unit, a characteristic ecosystem [11].

Microbiota of the oral cavity is not uniform and changes according to anatomical and physiological conditions; it is different at the orifices of salivary glands, on the surface of teeth, in sulcus gingivalis, on the tongue, at tonsils or at the buccal mucosa [12]. The growth of oral microorganisms depends on temperature, pH, oxidation-reduction potential, availability of nutrients and water, morphology of oral structures, flow of saliva and the presence of antimicrobial compounds. Each of these factors puts a selection pressure on the oral ecosystem and helps to maintain balance between populations of microorganisms (**Figure 1**).

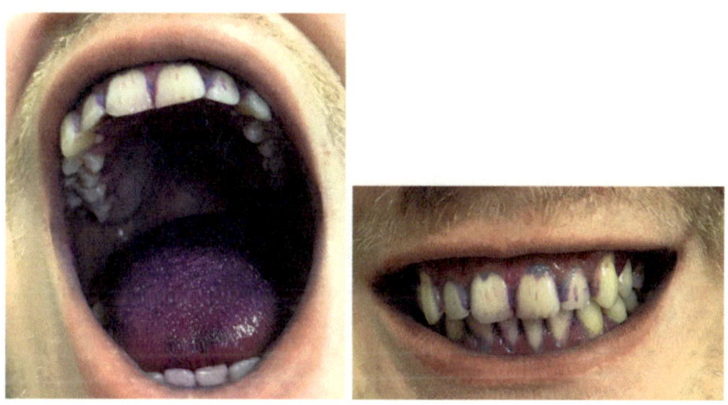

Figure 1.
Detection of oral biofilm by means of a plaque-finder, the new dental plague is coloured red, the older one is coloured blue.

2.1 Oral microbiota in sulcus gingivalis

Sulcus gingivalis is one of the sites where the microorganisms from the external environment begin to act as first. The total count of cultivable bacteria in sulcus gingivalis of healthy people is relatively low and amounts to about 103–106 CFU (colony forming units) per gingival slit. Sulcus gingivalis supplies nutrients to bacteria, exhibits low redox potential and thus is colonised mostly by obligate anaerobic rods. The subgingival plaque is also dominated by *Actinomyces* and streptococci that belong among Gram-positive microorganisms. It has been assumed that microbiota of sulcus gingivalis is related to the composition of the supragingival plaque with frequent occurrence of black-pigmented rods of *Porphyromonas gingivalis*, *Porphyromonas endodontalis*, *Prevotella melaninogenica*, *Prevotella intermedia*, *Prevotella loescheii* and *Prevotella denticola*.

The most frequent bacterial populations in the sulcus gingivalis are the following: *Streptococcus sanguis*, *Streptococcus mitis*, *Staphylococcus epidermidis*, *Micrococcus* spp., *Mycoplasma* spp., *Trichomonas tenax*, *Entamoeba gingivalis*, *Streptococcus intermedius*, *Veillonella parvula*, *Streptococcus mobillorum*, *Streptococcus constellatus*, *Peptostreptococcus micros*, *Lactobacillus casei*, *Lactobacillus acidophilus*, *Eubacterium lentum*, *Propionibacterium acnes*, *Catonella* spp., *Johnsonella* spp., *Rothia dentocariosa*, *Actinomyces viscosus*, *Actinomyces odontolyticus*, *Actinomyces naeslundii*, *Capnocytophaga gingivalis*, *Capnocytophaga ochracea*, *Prevotella oralis*, *Prevotella denticola*, *Bacteroides melaninogenicus*, *Fusobacterium nucleatum*, *Eikenella corrodens*, *Wolinella* spp., *Campylobacter sputorum*, *Selenomonas sputigena*, *Treponema* spp., and *Leptotrichia* spp., *Granulicatella* spp. [13]. Of the more noteworthy representatives, one should mention parasitic protozoa *Entamoeba gingivalis* and *Trichomonas tenax* [14].

2.2 Oral microbiota on the surface of teeth

Dental plaque consists of microorganisms producing a complex matrix composed of extracellular products of microorganisms and salivary components. Bacteria isolated from supragingival plaques include mostly Gram-positive, facultatively anaerobic species, particularly streptococci, and members of the genus *Actinomyces.* Bacteria of the genera *Veillonella*, *Haemophilus* and *Bacterioides* are usually isolated from deeper layers.

Formation of the dental plaque can be divided to several stages: formation of pellicle, initial bacterial adhesion, bacterial colonisation and plaque maturation and finally its mineralisation and calcification (**Figures 2** and **3**), i.e., formation of dental calculus (*calculus dentis*) [9].

Some bacteria are able to adhere to the tooth surface and by their factors of virulence and metabolic products are capable of causing dental caries or other bacterial diseases of additional parts of the oral cavity [15]. After disturbance of the balance between the original microbiota and the propagated potentially pathogenic microorganisms, various diseases frequently occur in the oral cavity. Therefore, these microorganisms may exhibit some pathogenicity, but only under certain conditions, and therefore, we refer to them as facultative or opportunistic pathogens [16]. From the surface of teeth, we may isolate *Streptococcus sanguis*, *Streptococcus mutans* and bacteria of the genera *Neisseria*, *Haemophilus*, *Lactobacillus*, *Propionibacterium*, *Actinomyces*, *Leptotrichia*, *Fusobacterium*, *Veillonella*, *Bacteroides* and *Bacterionema*, described in **Table 1**.

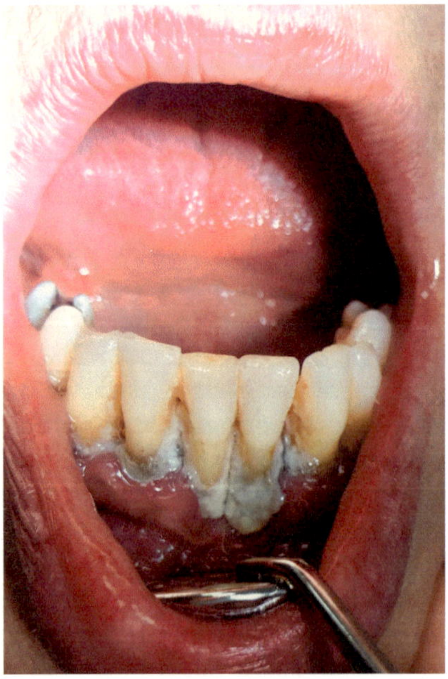

Figure 2.
Deposit of supragingival dental calculus on the vestibular area of teeth in the mandible.

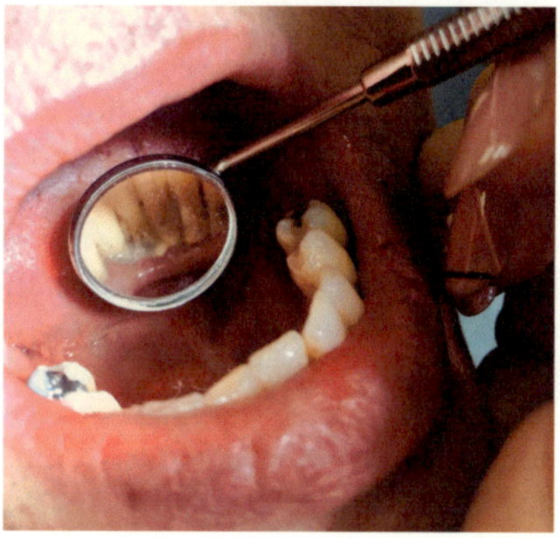

Figure 3.
Deposit of dental calculus on the lingual area of teeth in the mandible.

	Anaerobic microorganisms	Aerobic microorganisms
Teeth surface	*Prevotella buccalis*	*Neisseria* spp.
	Actinomyces viscosus	*Protozoa*
	Propionibacterium spp.	*Streptococcus mutans*
	Lactobacillus spp.	*Aggregatibacter Actinomycetemcomitans*
	Actinomyces israelii	*Mycoplasma* spp.
	Actinomyces naeslundii	*Streptococcus sanguis*
	Nocardia spp.	
	Rothia dentocariosa	
	Peptostreptococcus	
	Actinomyces israelii	
	Veillonella spp.	
	Fusobacterium spp.	
	Leptotrichia spp.	
	Prevotella oralis	
	Actinomyces odontolyticus	

Table 1.
Oral microbiota on the surface of teeth.

2.3 Oral microbiota of the tongue

From the tongue, there was isolated particularly *Streptococcus salivarius,* while *Streptococcus mutans* and *Streptococcus sanguis* appeared in the oral cavity only after eruption of teeth [17]. The tongue may become a reservoir of microorganisms participating in periodontal diseases. Bacteria that occur in the saliva may originate from various parts of the oral cavity and the microbial composition of saliva resembles that of the tongue (**Tables 2** and **3**).

2.4 Oral microbiota of the saliva

Free fluoride ions, found in the saliva in concentrations ranging from 0.01 to 0.05 ppm, are an important factor of remineralisation of enamel [18]. Individual proportions of calcium, fluorine and phosphates indicate potential remineralisation effect of the saliva on the dental tissue. Saliva has a positive suppression effect on the development of dental caries. This effect results from the content of unsaturated ions of phosphates, fluorine and calcium while there is a continuous exchange of these ions between the tooth crown and the saliva. At neutral pH, a balance is established between enamel minerals and the saliva. When the action of organic acids produced by bacteria disturbs this balance, pH in the oral cavity decreases and demineralisation of tooth surface occurs. Some components of the saliva neutralise the acidic environment and reduce the demineralisation rate and thus prevent the dental caries. This buffering capacity of the saliva is ensured by phosphate, bicarbonate and proteinaceous buffers [19].

Glycoprotein mucin acts as a lubricant of the oral cavity surface, produces a protective barrier against the external environment and, at the same time, facilitates chewing, swallowing and speech. It is one of the agglutination factors of the saliva that causes aggregation of bacteria. It can interact with *Streptococcus sanguis*, *Streptococcus mitis*, *Streptococcus gordonii*, *Aggregatibacter actinomycetemcomitans*, *Pseudomonas aeruginosa* and *Escherichia coli*. Saliva also contains other biologically active compounds, such as hormones, glucose, cholesterol, fatty acids and urea [20].

Microorganisms do not tolerate large variations in the level of pH. The pH in the oral cavity is close to neutral and ranges between 6.75 and 7.25. Saliva exhibits remineralisation abilities but the remineralisation process requires some time [21]. Increased frequency of easily metabolizable saccharides at the presence of plaques increases the risk of development of caries [22]. In this respect, saccharose plays a significant role as it easily diffuses into the plaque and is highly soluble [21]. Saccharides use microorganisms as a source of energy and a building material. Organic acids synthesised by microorganisms during metabolic processing of saccharides cause a decrease in the level of pH and a subsequent loss of minerals from the teeth surface [23].

	Anaerobic microorganisms	Aerobic microorganisms
Tongue	*Campylobacter (Campylobacter sputorum)*	*Streptococcus mitis*
	Propionibacterium	*Streptococcus salivarius*
	Actinomyces	*Staphylococcus* spp.
	Veillonella	*Enterobacteriaceae*
	Bacteroides (Bacteroides melaninogenicus)	*Streptococcus sanguis*
	Peptococcus	*Corynebacterium* spp.
	Prevotella (Prevotella oralis)	*Candida* and other microscopic fungi
	Peptostreptococcus	*Micrococcus* spp.
		Staphylococcus spp.
		Neisseriaceae

Table 2.
Oral microbiota of the tongue.

	Microorganisms
Saliva	*Streptococcus milleri*
	Streptococcus salivarius
	Actinomyces spp.
	Veillonella spp.
	Streptococcus sanguis
	Streptococcus mitior
	Lactobacillus spp.
	Streptococcus mutans

Table 3.
Oral microbiota of the saliva.

3. Oral cavity diseases

A variety of diseases involve the oral cavity including dentition problems, maxillary and mandibular disorders and diseases, gingivitis, diseases of the tongue, palate, internal mucosa and lips [24]. In the oral cavity, there are also salivary glands that fulfil very important functions within the digestive system and these paired glands may be afflicted with various inflammatory and noninflammatory diseases that can cause additional complications in the oral cavity [25]. Due to the diversity of anatomical structures and varied microbiota in the oral cavity, this part of the body can be affected by a great number of diseases such as tumour and benign diseases, inflammatory and noninflammatory and inherent or acquired diseases [22]. They are caused by infectious and noninfectious agents. The infectious agents include viruses, bacteria and fungi, and other may be caused by hormonal changes, systemic diseases, hypersensitive responses, immunodeficiency states or tumours [11].

3.1 Focal infection

Focal infection of dentogenic origin is defined as a secondary or total infection caused by spreading of microorganisms to distant organs, while the primary infection is located in the tissues of apical and marginal periodontium. Oral focus is a focus of the chronic inflammatory process of primary infection localised in the tissues of the oral cavity, which is the source of infection. From the point of view of focal infection, the most serious etiological agents are *Streptococcus viridans*, *Streptococcus mitis*, *Streptococcus milleri* and *Streptococcus sanguis* [26]. As a matter of fact, this involves a numerous group of diseases or states that include also periodontitis or periodontitis marginalis. Focal infection is a focus from which the infection spreads to the entire organism and causes damage to tissues and organs [27]. In the course of several years, the opinion about the source of focal infection in the oral cavity gradually changed. In the past, mostly foci in the area of teeth roots, the so-called dead teeth, were considered the sources of focal infection [28]. Due to insufficient possibilities of treatment of root canals, many teeth were extracted [29]. Currently, this very practice is the subject of increasingly serious discussions within professional circles as a potential massive source of infection of an organism. Endodontics is a branch of dentistry dealing with diagnosis and treatment of pathological conditions of dental pulp and periapical tissues [30]. Endodontic treatment means the treatment of the dental pulp, in the majority of cases its complete removal and perfect filling of root canal using correct techniques and treatment procedures.

The role of root filling is to close hermetically the entry to foramen apicale dentis and fill up completely the infection-free tooth canal [31]. Such treatment will prolong functionality and life of inflammation-affected teeth pillars. Imperfect removal of the infected tooth pulp or transfer of infection to the periapical space and incomplete filling of the root canal turns such tooth into a source of focal infection.

There are many foci in the oral cavity that can become potential sources of focal odontogenic infection. Origin of these foci may be attributed to neglected care of the oral cavity, pathological action of some microorganisms or unfavourable anatomic conditions in this cavity.

3.2 Sources of focal infection

3.2.1 Dental pulp necrosis and gangrene

Dental pulp necrosis may develop after injury or as a result of degenerative processes in the dental pulp, and can be affected as a whole or only its part. The principal cause is a pronounced damage to vascular supply. Colliquative necrosis results in decomposition of the dental pulp tissue. At coagulation necrosis, the infected dental pulp produces fluid rich in proteins. Such condition may occur during preparation close to the dental pulp [22]. Dental pulp gangrene is a secondarily altered necrosis that develops after infection of the necrotic pulp and can be of two types, dry or wet. Dry gangrene develops after partial infection of the necrotic pulp and the remnant pulp dries up. Wet gangrene is more frequent—it develops by the action of multiple microbiota from the carious dentin. Necrotic dental pulp tissue has a strong offensive smell due to accumulated gases such as skatole and indole [31].

3.2.2 Teeth with chronic dental pulp inflammation

Chronic-closed dental pulp inflammations, *pulpitis chronica clausa* in Latin, occur frequently in teeth with caries that penetrated into the dental pulp. The consequence is a chronic abscess with clinically mutedental pulp. During preparation, small amount of pus or dark blood is sometimes discharged from the pulp cavity [22]. Chronic-closed dental pulp inflammations are also frequently clinically mute, and in such cases, the dental pulp shows fibrocystic or at atrophic changes.

The affected pulp tissue is prone to calcification or denticles. The residual pulp shows chronic inflammatory infiltration. Such condition may result in partial or complete obliteration of the root canal [28]. Internal granuloma (*pulpitis chronica granulomatosa interna*) is a chronic productive inflammation with typical finding of considerably hyperaemic granular tissue. A characteristic feature of this process is fibroblasts that form capillaries and cells of chronic inflammatory cellularization [32]. Injury is the most frequent cause of this type of chronic inflammation, also chronic traumatization of the tooth may contribute to damage to the dental pulp [33].

3.2.3 Teeth with periapical findings

Inflammations in the periodontium region affect several types of tissues such as parts of the suspension apparatus of teeth, compacta, spongiosis of alveolar bone and root surface cementum. Such changes are collectively referred to as periodontitis [34]. The causes of periapical inflammation may include infections, chemical irritation and acute or chronic trauma. The most frequent cause of the development of periapical focus is necrotic, passively infected tooth pulp in the root canal. This way altered dental pulp contains compound microbiota with predominance of Gram-positive streptococci, but also enterococci, lactobacilli, *Candida* and *Neisseria* species and anaerobic bacteria such as *Fusobacteria* and *Bacteroides* [35]. Infection causes softening of the dentin wall of the root canal and the metabolic products of microorganisms induce inflammatory conditions in the periodontium region (**Figure 4**). The most frequent site of the development is the apex of the tooth root, but the inflammation process is observed also in the areas of lateral ramifications or sub-pulpal tooth canal. The inflammation is acute or primarily chronic, or chronic with acute exacerbation.

3.2.4 Periodontal abscesses

Abscess is a collection of pus in a newly formed cavity. Periodontitis may be associated with development of periodontal abscesses [34]. They are divided into soft tissue and hard tissue abscesses. They manifest themselves by oedemas and pain, the more advanced forms also by the presence of yellowish pus. Retraction of gingiva may result in evacuation of pus. Bone abscess affects bone spongiosa and is manifested by intense strong pain upon tapping a tooth close to the abscess. Sometimes even shivers may occur and pus is not evacuated after retraction of gingiva [36]. Untreated bone abscess may result in sequestration of the affected bone, but this form is very rare [31].

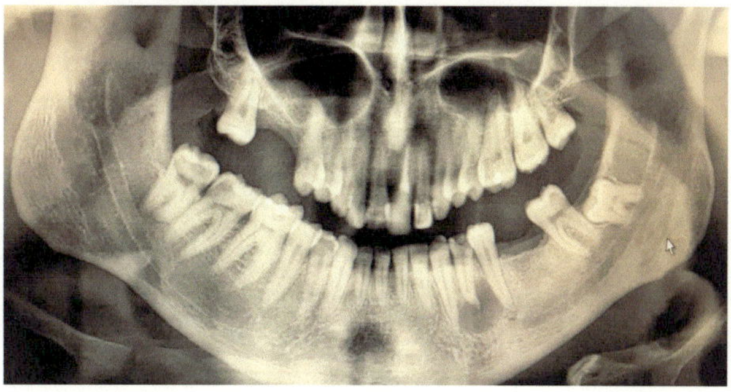

Figure 4.
Periapical finding in tooth No. 34, X-ray - opg 2D image.

3.2.5 Periodontal pockets

Periodontal pockets develop by extension of periodontal fissure, most frequently with approximal localisation. It can be located by one tooth but can affect all teeth in the maxilla and in the mandible. The periodontal pocket mostly contains subgingival dental plaque, subgingival dental calculus, dead microorganisms, leukocytes, proliferating nonspecific granular tissue and inflammatory exudate [37]. Periodontal pockets are classified as true, false, active and nonactive. The false periodontal pockets develop by enlargement of the marginal gingiva without shift of the dento-gingival connection, and the alveolar bone remains intact [11]. The true periodontal pockets are associated with resorption of the alveolar bone. The true periodontal pocket has been described as a space between the gingiva and tooth, coronary delimited by the edge of the marginal gingiva and apically delimited by the base of the periodontal pocket [38]. The difference between the true and false periodontal pockets is diagnosed by X-ray examination [22]. In the active pocket, one may find signs of inflammation, purulent exudations and postprobe haemorrhage. These active periodontal pockets require treatment. The nonactive pockets are free of marked findings. It suffices to carry out regular monitoring of these quiescent forms of periodontal pockets [34].

3.2.6 Gingivitis

Gingivitis is the most frequent microbial inflammation in the human body induced by microbiota of the dental plaque. It can occur as a constant symptom of periodontitis. According to its course, gingivitis may be classified as acute or chronic. Acute gingivitis is painful, the gingiva is red to red-violet and haemorrhage occurs upon stimulus but also spontaneously. Chronic gingivitis manifests itself by a red-pink colour, haemorrhage upon probing and stimulus-induced pain. The shape of the gingiva is altered and large false pockets are frequently observed. The causes are varied and can be divided to local and general [34].

We recognise several types of acute gingivitis. Gingivae affected by gingivitis acuta simplex are slightly reddened while those affected by gingivitis catarrhalis acuta are hyperaemic and swollen. If this process is limited to one or two papillae, we refer to it as papilitis [37]. Gingivitis vesiculosa is manifested by production of vesicles with a clear content and reddened surrounding of vesicles. Gingivae affected by gingivitis pseudomembranosa are red, swollen, associated with production of pseudomembranes—this is fibrinous purulent inflammation. The most frequent form of gingivitis is ulcerous gingivitis that affects younger people [35]. The causes of this disease are many—weakened organism due to infectious disease, vitamin deficit, stress and drugs. The symptoms include swollen gums and the apexes of papillae truncated by necrosis. It is localised mostly in the zone of frontal teeth and molars [22]. Chronic gingivitis is classified as gingivitis cattarrhalis chronica, gingivitis gravidarum, gingivitis pubertalis, gingivitis scurbutica, gingivitis at epilepsies and leukaemia and elephantiasis fibromatosis gingivae [34].

3.2.7 Retained radices (radices relictae)

This condition occurs in patients with neglected hygiene. If the crown portion of the tooth disintegrates due to untreated caries, the roots of teeth are retained in the gums. Failure to ensure timely treatment of root canal may result in infection of the root pulp and thus in potential dental focal infection [36].

3.3 Diseases of the lips-cheilitis

Inflammation of lips extending to or beyond the border of lips can occur as acute or chronic. The factors most frequently involved in cheilitis are external factors. The currently known forms of cheilitis are actinic, angular, allergic, exfoliative, glandular and granulomatous [39].

Actinic cheilitis is referred to as solar cheilosis or solar keratosis of the lips that develops due to excessive exposure to UV radiation. It is localised in the lower lip in men and in the upper one in women. The risk group are fair-skinned (Caucasian) types of people. The clinical symptoms include dryness and scaliness of lips, their greyish colouration, swelling, ulceration, deepened folds and coarse lesions. Histological examination will confirm hyperkeratosis as a consequence of thickening of the epithelial cells and epithelial dysplasia. The potential ways of treatment include cryosurgery, electro-surgery, laser, and 5-fluorouracyl [40, 41].

Angular cheilitis is also referred to as angular cheilosis, commissural cheilitis or angular stomatitis. It is an inflammation of one or eventually of both angles of the mouth. The causes include bacterial (*Staphylococcus aureus*, haemolytic *Streptococcus*) or yeast infections (*Candida albicans*) mechanical damage to lips by denture prosthesis of fixation apparatus. Also, malnutrition involving deficiency of group B vitamins should be considered. Granulomatos cheilitis presents as swelling of the upper and lower lips and, at the same time, as one of the manifestations of orofacial granulomatosis, which is a separate disease, or as a monosymptomatic form of the Melkersson-Rosethal syndrome. Three symptoms are characteristic of this disease—recurrent orofacial swelling, recurrent facial paralysis and fissured tongue. One can also observe chapped, red-brown lips or buccal nerve paralysis. It is induced by allergic response to cinnamon or various benzoates and can represent also early manifestation of Crohn disease, mycobacterial infection or sarcoidosis. Aetiology of the disease is unknown. It has been assumed that sudden inflammation or random aggregation of inflammatory cells may be involved. Diagnosis is very difficult, important are histological results, which may indicate presence of granulomas and the positive findings may imply the Melkersson-Rosethal syndrome. This finding was obtained also in patients with Crohn disease and affected mouth.

3.4 Diseases of the tongue

The most important diseases of the tongue include atrophy of the tongue fur, rhomboid glossitis, geographic tongue, fissured tongue, herpetic geometric glossitis, black hairy tongue, oral leucoplakia and macroglossia [37].

Rhomboid glossitis also known as central papillary atrophy presents as typical loss of tongue papillae along the midline posterior dorsal tongue, caused by oral candidiasis (**Figure 5**).

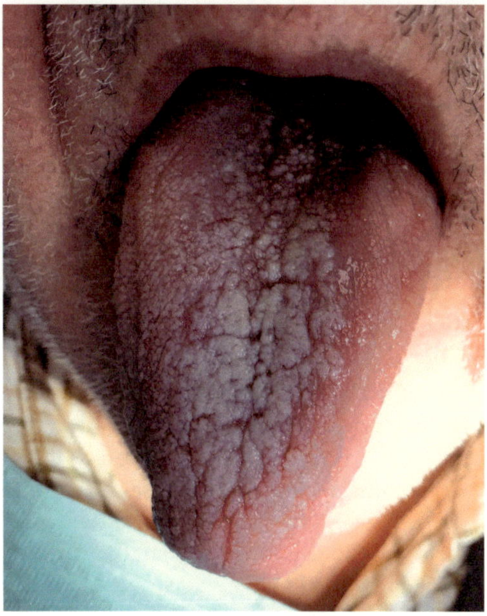

Figure 5.
A white coating on the tongue caused by an overgrowth of Candida albicans.

The tongue lesion is shiny, frequently symmetrical, well delineated, depapillated. The risk factors include smoking, inadequate oral hygiene, use of unsuitable prosthesis and HIV infection. The treatment is based on the use of corticosteroid inhalators and sprays. It occurs worldwide and affects men, women and children. Diagnosis is based on clinical examination and laboratory confirmation of *Candida* spp. The most effective prevention/treatment, especially in smokers, is giving up smoking and the use of antimycotics [39].

The term geographic tongue, lingua geographica, is used to describe inflammation affecting the dorsal surface of the tongue. Its characteristic feature is depapillation of some parts of the tongue resulting in the alternation of depapillated and normal-structure areas producing a map-like (geographic) pattern.

The depapillated areas are smooth and more intensively red coloured, and except this colour differences, the condition mostly causes no other symptoms. However, it may cause burning mouth syndrome after consumption of some foods [34]. The exact aetiology is unknown but association with smoking, stress and genetic association with human leukocyte antigens (HLAs), diabetes or psoriasis has been assumed. Diagnosis is made on the basis of clinical and histological examination. Differential diagnosis must distinguish this condition from oral lichenic planus, erythematous candidiasis and leucoplakia. Effective drug therapy is based on antihistaminics and corticosteroids.

Fissured tongue affects 5–10% of population with higher susceptibility occurring in older individuals. The exact aetiology is unknown but imbalance of the level of salivary electrolytes and haematological abnormalities were observed. This condition affects the dorsal side of the tongue. In the central part, a central fissure (groove) is observed with multiple smaller fissures branching off the central one. Patients with Down, Melkersson-Rosethal and Sjögren syndromes are at risk. Improvement in oral hygiene, particularly the tongue, may result in the recovery from this disease [22].

Black hairy tongue is the term used to refer to the hypertrophy of filiform papillae of the tongue that acquire black colour. This disease affects the dorsal part of the tongue. There are several causes that induce this disease: keratinization of cells, restoration of the epithelial layer without complete exfoliation of the old layer, change in pH in the oral cavity, use of oxidation agents, smoking, antibiotics, bacterial and yeast infections and radiotherapy. Complication of this disease involves papillae that are markedly elongated and thus can cause tickling sensation, which may result in vomiting. Therapy consists in intensive cleaning of the tongue and administration of antimycotics [42].

3.5 Diseases of the salivary glands

The diseases afflicting salivary glands include xerostomia, siallorhoea, inflammation of salivary glands—sialadenitis, Sjögren syndrome, calculi in salivary glands—sialolithiasis, cysts, sialadenosis and tumours of salivary glands [43].

Xerostomia or dry mouth syndrome is associated with reduced production of saliva, and this condition is also termed hyposalivation. It is caused by carcinomas or unsuitable therapy. An extensive group of diseases are inflammations of the salivary glands—sialadenitis. They are classified as acute bacterial sialadenitis, chronic sialoadenitis, viral sialadenitis, specific sialadenitis and autoimmmune sialadenitis—the Sjögren syndrome [34].

Acute bacterial sialadenitis is most frequently caused by pathogenic bacteria *Streptococcus aureus*, *Streptococcus viridans* and *Streptococcus pneumoniae*. The principal pathways of spreading of this infection are haematogenic and lymphogenic. The risk factors that support the development of infection include decreased production of saliva, cachexia sialolihtiasis and malignancies. Clinical manifestations include purulent and abscess forms.

3.6 Dental caries and periodontitis

Dental caries is the most frequent dental and oral disease. It occurs worldwide [31]. Root caries is caused by *Streptococcus mutans*, *Lactobacillus acidophilus*, *Actinomyces* sp. and *Nocardia* sp. Bacteria first pass through the enamel, then through dentin, and finally, they reach the cementum layer. During clinical examination of dentition, black or dark yellow lesions are observed on teeth [44]. The most important mineral in teeth is hydroxyapatite. Remineralisation of teeth is ensured by prolin and minerals contained in the saliva [45]. Residues of food in the oral cavity, sweet beverages, beverages with high concentration of acids and citrus fruit are sources of bacterial nutrition.

The ability of bacteria *Streptococcus mutans* to form biofilms is important from the clinical point of view, particularly in relation to the development of dental caries. Dental caries has a multispecies aetiology. Mutant streptococci are referred to as a cluster of acidogenic streptococci species inhabiting dental plaques. *Streptococcus mutans* and *Streptococcus sobrinus* are the bacteria most frequently isolated from dental carious lesions. There were published individual case reports involving infectious endocarditis with participation of these bacteria [46]. The development of dental caries starts with dissolution of the mineral portion of tooth manifested by lesions and white sports on teeth, followed by local destruction of the enamel and dentin. If this process is left alone without treatment, inflammation of the dental pulp and periapical tissues follows. Many strategies focused on reduction of the occurrence of dental caries and their specific effect consisting in reduction of counts or acidogenic activity of *Streptococcus mutans* in the dental plaque [47].

In 2011, information about new bacterial species *Scardovia wiggsiae* appeared in professional microbiological and stomatological literature. The authors reported that in addition to *Streptococcus mutans,* this bacterium participates in the development of dental plaques and acute early age dental caries affecting dentition of children [44]. The relevant investigations were carried out by a team of scientists from Forsyth Institute, Cambridge, headed by A. C. R. Tanner, and involved bacterial population in samples of dental plaques and from the depth of cavities in primary dentition of 2–6-year-old children. The results were compared with the findings in the plaques of children without dental caries or white spots indicating demineralisation of enamel [48]. Because dental caries develops with participation of acidotolerant bacteria, the laboratory cultivation was carried out in anaerobic environment on blood agar of pH 5.0. In this way, the authors selected species that may play an important role in cariogenesis. Partial 16S rRNA sequences obtained from 5608 isolates were characterised on the basis of species. Subsequently, the findings of individual bacterial species from children with and without caries were compared. The species most frequently isolated from children with acute dental caries were *Streptococcus mutans*, *Scardovia wiggsiae*, *Veilonella parvula*, *Streptococcus cristatus* and *Actinomyces gerensceriae*. According to Human Oral Microbiome Database, the authors identified 198 taxons and 45 of them were until then characterised as noncultivable. The results showed that both *Streptococcus mutans* and the new bacteria *Scardovia wiggsiae* were isolated from 80% of the children with dental caries, but these bacteria were absent in 80% of children free from dental caries. The microorganism most frequently present in progressing dental caries was *Streptococcus mutans* and the newly discovered species *Scardovia wiggsiae* co-participated in the development of dental caries but was cultivated also independently from the cases of progressing dental caries. Many saccharolytic bacteria participate in reduction of pH, but their growth is selectively restricted at low pH at which the cariogenic acidotolerant species that include also the newly discovered *Scardovia wiggsiae* are able to multiply.

Periodontitis is a serious infection of gingiva that damages soft tissues and degrades the osseous tissue, can cause looseness of teeth or result in their loss (**Figure 6**). It affects approximately 10% of the world population. It is a subject to internal and external factors and the influence of bacteria, particularly the Gram-positive ones, referred to sometimes as the “red complex”, namely *Treponema denticola*, *Porphiromonas gingivalis* and *Tanerella forsythia* [39].

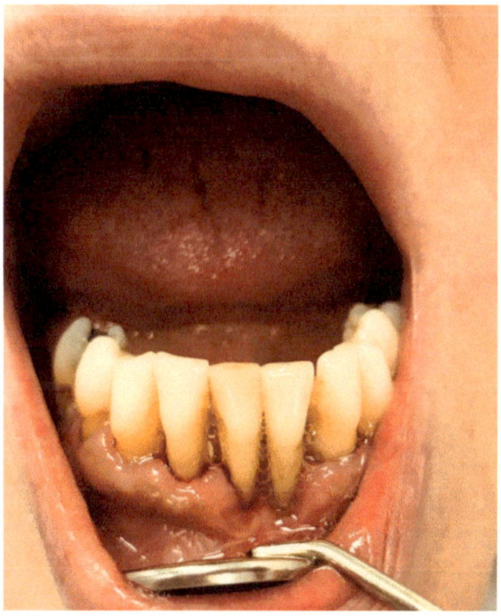

Figure 6.
Periodontitis afflicted lower front teeth in the mandible.

4. Oral probiotics and their influence on oral cavity diseases

4.1 Importance of probiotics to the oral cavity health

The authorship of the concept of probiotics has been attributed to the Russian scientist and Nobel prize winner Elie Metchnikoff, who at the turn of the nineteenth and twentieth centuries theorised that Bulgarian pheasants own their long life to the consumption of fermented milk products. Since then, the scientists confirmed that the use of probiotic strains, particularly those of the genera *Lactobacillus* and *Bifidobacterium*, can support gastrointestinal, genitourinary and oral health by maintaining the microbial balance of these ecosystems [49]. According to the up-to-date definition, probiotics are live microorganisms, which when administered in adequate amounts confer a health benefit on the host [50].

In the past decade, the awareness of probiotics and their contribution to human and animal populations increased and a wide range of probiotic products appeared on the market. Research activities focused on the search for new probiotics help to understand the process of development of probiotic products and their potential role in prevention or treatment of diseases [51]. Composition of microbiota of today's man differs from that in the past. Modern people are exposed to a number of negative influences that affect their microbiological balance. As long as the harmony and balance is maintained, we speak about symbiosis. The imbalance of microbiota is referred to as dysbiosis, which involves changes in proportions and heterogeneity of commensal species resulting in disturbed functioning of protective barriers and subsequent development of diseases [52]. The adverse influences that cause dysbiosis include particularly the use of antibiotics and chemotherapeutics, stress situations resulting from the lifestyle of the modern man, unsuitable eating habits or drinking regimen and changes in composition of food or the environment. Searching for amendment of the developed dysbiosis became the prime stimulus of the study of probiotics.

The increasing bacterial resistance to antibiotics and the demands of the wide public on natural therapy resulted in decreased use of conventional antimicrobials and raised the need for development of new ways of treatment [53]. A separate issue is the probiotics intended for oral cavity. People associate the term probiotic with the health of the intestinal tract and necessity to use them during antibiotic treatment, which became a common practice but oral antibiotics also play an important role in the overall health of an individual. It was demonstrated that probiotics have the potential for modification of the oral microbiota and are effective in the prevention and treatment of oral cavity diseases, such as dental caries and periodontal diseases associated with dysbiosis [54]. Today, the global market already offers some probiotic preparations that prevent formation of dental plaques, support health of gingivae and teeth and help to fight the bad breath [55]. The most frequently investigated bacteria include representatives of the genera *Lactobacillus*, *Streptococcus* and *Bifidobacterium*. Species of these taxons are members of normal microbiota found in the gastrointestinal tract, while some of them prefer to colonise the oral cavity [56]. Potentially, pathogenic microorganisms enter the body through the mouth or nose and thus the oral probiotics constitute and excellent first-line protective barrier of the mouth and throat. Clinical studies in humans that investigated treatment of periodontal diseases by probiotics reported overall contributions such as the decreased bleeding of gums. The studies that involved the use of probiotics as a supplement to clinical periodontal treatment showed a more pronounced improvement of the clinical status of patients in comparison with the clinical treatment alone [57]. One of the preparations used in Slovakia is ProDentis [58], a preparation containing mostly *Lactobacillus reuteri*. One clinical study was based on the use of pastilles containing *Lactobacillus reuteri* strains as a supplement of therapy of chronic periodontitis. Results of this study revealed a marked decrease of occurrence of *Porphyromonas gingivalis* in the saliva and in subgingival and supragingival plaques [59].

4.2 Properties of oral probiotics and mechanism of their effect

The effectiveness of probiotic microorganisms in the oral cavity depends on their ability to resist to the environmental conditions and protective mechanisms, to adhere to the surfaces coated by saliva, easily colonise the mouth and grow in it and inhibit oral pathogens without harming the host [55]. Ideal properties of oral probiotics are presented as follows [60]:

1. binding to dental surfaces,

2. production of antimicrobial substances against oral pathogens,

3. aberration of environmental conditions in the mouth, and

4. reduction of the inflammatory response.

The mechanism of effect of probiotics in the oral cavity (**Figure 7**) is almost identical with that in the gastrointestinal tract, i.e., modulation of the immune response, metabolic effects and harmonisation of the intestinal or oral microbiota.

Probiotic bacteria excrete various antimicrobial compounds such as organic acids, hydrogen peroxide and bacteriocins [61]. In addition, they compete with pathogens for the adhesive sites on mucous membranes. They can also modify their environment by modulation of its pH or the oxidation–reduction potential, which can interfere with the ability of pathogens to establish themselves on the mucosa. The beneficial effects of probiotics may include stimulation of the nonspecific immunity and modulation of humoral and cellular immune responses [14, 62].

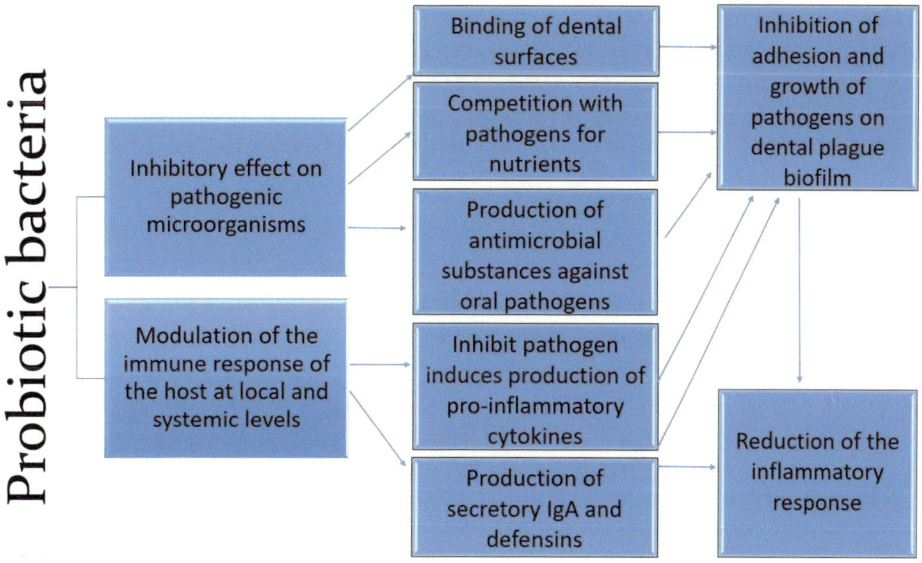

Figure 7.
Mechanism of the effect of probiotics in the oral cavity.

4.3 Potential risks of the use of probiotics

Before introduction of any probiotic product to market, it has to comply with very strict conditions. The product should be a result of thorough research of the relevant strain and each dose should contain sufficient bacteria even after storage of the preparation [63]. Probiotics are considered safe as they contain nonpathogenic microorganisms, and this is one of the reasons for high willingness of patients to use them. Risk to health may occur during a long-term use of probiotic preparation at diseases or states that enable their potential passage to the body at the development of secondary infection. Such states include bloody diarrhoea, immunosuppressive treatment or irradiation. Additional potential risks include transfer of vancomycin resistance by strains of *Enterococcus faecium*, administration of high doses of probiotics to autoimmune patients, infants and newborns with immunity and intestinal permeability disorders and administration to patients with immature or markedly disturbed immune system or patients with AIDS [64].

Probiotics can be routinely used as a food supplement, and their positive health claims were well described [65]. Despite that, some undesirable effects of the use of probiotics can also occur [66]. Usually, this involves only mild reactions that affect small percentage of users. When using probiotic products, it is necessary to consult a doctor about potential indications and undesirable effects [67]. Clinical indications of the use of probiotics are very extensive. One of their unwanted effects are digestion problems that may involve tympany and increased thirst [68]. Biogenic amines are low molecular weight organic compounds produced by degradation of amino acids, which may affect negatively the human organism [69]. The biogenic amines produced from the accepted food by fermentation activity of probiotic bacteria have excitation effect on the nervous system and decrease blood flow through organs, which can result in headaches [70]. Biogenic amines are histamine, tyramine, tryptamine, putrescin, spermidine and phenyl ethylamine [71]. In some groups of people, the use of probiotic products results in increased risk of infections, such as in immunosuppressed individuals or patients after surgeries hospitalised for long time. One should not forget to mention allergic reactions associated with the use of probiotic components. Probiotic products contain various additives such as lactose, eggs, soya or other generally known allergens. The consumers should avoid components that may induce in them hypersensitivity or allergic reactions [72]. Basically, such cases are rare and the probiotic treatment can be referred to as the treatment on a natural basis. At the same time, it is recommended to increase gradually the doses of probiotics until reaching the full dose in order to prevent potential side effects that occur particularly in weakened individuals.

5. Conclusion

There is an increasing concern about the fact that oral diseases put a systemic load on the organism. This stresses the importance of oral health for the overall health of an individual and the population. Predictions have been made in the past that the scientific and technological advances in the field of molecular biology, immunology and genetics, together with ageing of the population, will require future complex health service measures within which the care of the oral health will become important from the point of view of management of overall health and economy, and thus will necessitate novel oral health approaches. Bioactive compounds, as substances capable of affecting the microbiocenosis environment, are considered an alternative when searching for replacement for antibiotics. Results of the studies focused on microbiocenosis of the dental biofilm and interactions between individual bacterial species indicate a probiotic potential of some oral bacteria and their potential to prevent oral cavity diseases. Qualitative influence on pathogenic bacterial microbiota of the oral cavity, exerted by probiotic bacteria such as *Streptococcus salivarius*, brings not only health but also economic benefits. One should only hope that additional evidence of beneficial effects of probiotics and increased knowledge about biochemical and immunological mechanisms of their action will improve the potential of treatment and prevention of oral diseases and result in more rational and targeted use of bacterial supplements under specific clinical conditions.

Acknowledgements

This publication was supported by the project of the Ministry of Education Science Research and Sport of the Slovak Republic VEGA 1/0081/17 and by the Slovak Research and Development Agency under the contract no. APVV-15-0377.

Author details

Andrea Stašková[1*], Radomíra Nemcová[2], Stanislav Lauko[2] and Andrej Jenča[1]

1 Pavol Jozef Šafárik University in Košice, Slovakia

2 University of Veterinary Medicine and Pharmacy in Košice, Slovakia

*Address all correspondence to: andrea.staskova@upjs.sk

References

[1] Philip N, Suneja B, Walsch LJ. Ecological approaches to dental caries prevention: Paradigm schift or shibboleth. Clinical Research. 2018;**52**:153-165. DOI: 10.1159/000484985

[2] Arweiler NB, Netuschil L. The oral microbiota. Advances in Experimental Medicine and Biology. 2016;**902**:45-60. DOI: 10.1007/978-3-319-31248-4_4

[3] Do T, Devine D, Marsch PD. Oral biolms: Molecular analysis, challenges, and future prospects in dental diagnostics. Clinical, Cosmetic and Investigational Dentistry. 2013;**5**:11-19. DOI: 10.2147/CCIDE.S31005

[4] Wilcox CR, Stuart B, Leaver H, Lown M, Willcox M, Moore M, et al. Effectiveness of the probiotic *Streptococcus salivarius* K12 for the treatment and/or prevention of sore throat: A systematic review. Plum Analytics. 2019;**25**:673-680. DOI: 10.1016/j.cmi.2018.12.031

[5] Rosenblatt R, Steinberg D, Mankuta D, Zini A. Acquired oral microflora of newborns during the first 48 hours of life. The Journal of Clinical Pediatric Dentistry. 2015;**39**:442-446. DOI: 10.17796/1053-4628-39.5.442

[6] Kleessen B, Bezirtzoglou E, Mättö J. Culture-based knowledge on biodiversity, development and stability of human gastrointestinal microflora. Microbial Ecology in Health and Disease. 2000;**12**:53-63. DOI: 10.1080/089106000750060305

[7] Siqueira JF, Rôcas IN. Diversity of endodontic microbiota revisited. Journal of Dental Research. 2009;**88**:969-981. DOI: 10.1177/0022034509346549

[8] Peterson SN, Snesrud E, Liu J, Ong AC, Kilian M, Schork N, et al. The dental plaque microbiome in health and disease. PLoS One. 2013;**8**:1-10. DOI: 10.1371/journal.pone.0058487

[9] Jakubovics NS. Saliva as the sole nutritional source in the development of multispecies communities in dental plaque. Micriobiology Spectrum. 2015;**3**:1-11. DOI: 10.1128/microbiolspec.MBP-0013-2014

[10] Moutsopoulos NM, Konkel JE. Tissue-specific immunity at the oral mucosal barrier. Trends in Immunology. 2018;**39**:276-287. DOI: 10.1016/j.it.2017.08.005

[11] Ništiar F. Ústna dutina ako charakteristický ekosystém. 2016. Available from: http://patfyz.medic.upjs.sk/SSTUDMAT/ORALPAFYEKOSYST.pdf

[12] Kõll P, Mändar R, Marcotte H, Leibur E, Mikelsaar M, Hammarström L. Characterizacion of oral lactobacilli as potential probiotics for oral health. Oral Microbiology and Immunology. 2008;**23**:139-147. DOI: 10.1111/j.1399-302X.2007.00402.x

[13] Hrebík M. Ochorenia ústnej dutiny. 2016. Available from: http://zdravoteka.sk/choroby/ochorenia-ustnej-dutiny/

[14] Ghabanchi J, Zibaei M, Afkar MD, Sarbazie AH. Prevalence of oral *Entamoeba gingivalis* and *Trichomonas tenax* in patients with periodontal disease and health population in shiraz, southern Iran. Indian Journal of Dental Research. 2010;**21**:89-91. DOI: 10.4103/0970-9290.62821

[15] Papaioannou W, Gizani S, Haffajee AD, Quirynen M, Mamai-Homata E, Papagiannoulis L. The microbiota on different oral surfaces in healthy children. Oral Microbiology and Immunology. 2009;**24**:183-189. DOI: 10.1111/j.1399-302X.2008.00493.x

[16] Zaura E, Nicu AE, Krom BP, Keijser BJ. Acquiring and maintaining a normal oral micriobiome: Current perspective. Frontiers in Cellular and Infection Micriobiology. 2014;**4**:85. DOI: 10.3389/fcimb.2014.00085

[17] Edlund A, Yang Y, Hall AP, Guo L, Lux R, He X, et al. An *in vitro* biofilm model system maintaining a highly reproducible species and metabolic diversity approaching that on the human oral microbiome. Micriobiome. 2013;**1**:25. DOI: 10.1186/2049-2618-1-25

[18] Stratmeyer MS, Diefenderfer KE, Leiendecker TM. Retrospective assessment of caries experience among US naval academy midshipmen. Journal of Public Health Dentistry. 2015;**76**: 47-55. DOI: 10.1111/jphd.12110

[19] Ellies M, Laskawi R. Diseases of the salivary glands in infants and adolescent. Head and Face Medicine. 2010;**6**:1. DOI: 10.1186/1746-160X-6-1

[20] Abíková T, Urbanová W. Slina a její význam. 1. Část. Stoma Team. 2013;**13**:71-74

[21] Broukal Z. Výživa a zubní kaz. 2006. Available from: http://zdravi.euro.cz/clanek/sestra/vyziva-a-zubni-kaz-274858

[22] Hellwig E, Kliment J, Attin T. Záchovná stomatologie a paradontologie. 1st ed. Praha: Grada; 2003. 331 p

[23] Hecová H, Merglová V, Stehlíková J, Chaloupka P. Výskyt *Streptococcus mutans* a stav orálního zdraví u těhotných žen. Česká stomatologie/Praktické zubní lékařství. 2010;**9**:69-74

[24] Krajčík Š. Princípy diagnostiky a terapii v geriatrii. 1st ed. Bratislava: Charis; 2008. 192 p

[25] Wotke J. Patologie orofaciální oblasti. 1st ed. Praha: Grada; 2001. 184 p

[26] Petrášová A, Dráčová J, Ondrašovičová J. Ošetrenie pacientov s rizikom vzniku fokálnej infekcie odontogénneho pôvodu. In: 1. Kongres biomedicíny v oromaxilofaciálnej oblasti; 8-10 October 2009; Košice. Košice: EQUILIBRIA, s.r.o.; 2009. pp. 119-122

[27] Kalvach Z, Zadák Z, Jirák R, Zavázalová H, Holmerová I, Weber P. Geriatrické syndrómy a geriatrický pacient. 1st ed. Praha: Grada; 2008. 336 p

[28] Peřinka L. Základy klinické endodoncie. 1st ed. Praha: Quintessenz; 2003. 288 p

[29] Maoyang L, Xuan S, Wang Z. Oral microbiota: A new view of body health. Food Science and Human Wellness. 2019;**8**:8-15. DOI: 10.1016/j.fshw.2018.12.001

[30] Mäkinen KK. Sugar alcohols, caries incidence, and remineralization od caries lesions: A literature review. International Journal of Dentistry. 2009;**2010**:23. DOI: 10.1155/2010/981072

[31] Stejskalová J. Konzervační zubní lékařství. 1st ed. Praha: Galén; 2003. 235 p

[32] Madárová Ľ. Klinická endodoncia. 1st ed. Košice: UPJŠ; 1996. 236 p

[33] Pazdera J. Základy ústní a čeľustní chirurgie. 3rd ed. Olomouc: Univerzita Palackého v Olomouci; 2013. 312 p

[34] Ďurovič E, Vodrážka J, Ďurovičová J, Vincze K. Choroby slizníc a ústnej dutiny. 1st ed. Prešov: Vydavateľstvo Michala Vaška; 2005. 367 p

[35] Jenkins WMM, Allan CJA, Collins WJN. Guide to Periodontics. 3rd ed. Oxford: Butterworth-Heinemann; 1994. 250 p

[36] Satko I, Stanko P, Švidraň J. Orálna a maxilofaciálna chirurgia. 2nd ed. Bratislava: Vydavateľstvo UK; 2008. 305 p

[37] Lindhe J, Karring T, Lang NP. Clinical Periodontology and Implant Dentistry. 4th ed. Oxford: Wiley-Blackwel; 2003. 1044 p. DOI: 10.1177/154405910308201117

[38] Julsgaard M, Christensen LA, Gibson PR, Gearry RB, Fallingborg J, Hvas CL, et al. Concentrations of adalimumab and infliximab in mothers and newborns, and effects on infection. Gastroenterology. 2016;**151**:110-119. DOI: 10.1053/j.gastro.2016.04.002

[39] Mutschelknauss RE. Praktická parodontológia: klinické postupy. 1st end. Praha: Quintessenz; 2002. 532 p

[40] Marsh PD. Dental plaque: Biological significance of a biofilm and community life-style. Journal of Clinical Periodontology. 2005;**32**:7-15. DOI: 10.1111/j.1600-051X.2005.00790.x

[41] Marsh PD, DO T, Beighton D, Devine DA. Influence of saliva on the oral microbiota. Periodontology. 2016;**2000, 70**:80-92. DOI: 10.1111/prd.12098

[42] Dorko E, Baranová Z, Jenča A, Kizek P, Pilipčinec E, Tkáčiková Ľ. Diabetes mellitus and candidiases. Folia Microbiologica. 2005;**50**:255. DOI: 10.1007/BF02931574

[43] Mohan H. Patológia. 1st ed. Bratislava: Balneotherma; 2011. 976 p

[44] Palmer CA, Kent R, Loo CY, Hughes CV, Stutius E, Pradhan N, et al. Diet and caries-associated bacteria in severe early childhood caries. Journal of Dental Research. 2010;**89**:1224-1229. DOI: 10.1177/0022034510376543

[45] Thylstrup A, Fejerskov O. Textbook of Clinical Cardiology. 2nd ed. Copenhagen: Munksgaard International Publishers; 1994. 424 p

[46] Krzyściak W, Jurcuak A, Kościelniak D, Bystrowska B, Skalniak A. The virulence of *Streptococcus mutans* and the ability to form biofilm. European Journal of Clinical Microbiology and Infectious Diseases. 2014;**33**:499-515. DOI: 10.1007/s10096-013-1993-7

[47] James SM, Tagg JR. The prevention of dental caries by BLIS-mediated inhibition of mutans streptococci. The New Zealand Dental Journal. 1991;**87**:80-83

[48] Dawes C, Pedersen AM, Villa A, Ekström J, Proctor GB, Vissink A, et al. The functions of human saliva: A review sponsored by the world worshop on oral medicine VI. Archives of Oral Biology. 2015;**60**:863-874. DOI: 10.1016/j.archoralbio.2015.03.004

[49] Laleman I, Teughels W. Probiotics in the dental practice: A review. Quintessence International. 2015;**46**:255-264. DOI: 10.3290/j.qi.a33182

[50] FAO/WHO. Guidelines for the Evalaution of Probiotics in Food, Joint, FAO/WHO Working Group Meeting; 30 April-1 May 2002; London Ontario, Canada; 2002

[51] Waleed M, Naser Y, Samah W, Ansam F, Belal R, Rasha A, et al. Assessing worldwide research activity on probiotics in pediatrics using Scopus database: 1994-2014. World Allergy Organization Journal. 2016;**9**:25. DOI: 10.1186/s40413-016-0116-1

[52] Yang S-C, Lin C-H, Sung CT, Fang J-Y. Antibacterial acitvities of bacteriocins: Application in foods and pharmaceuticals. Frontiers in Microbiology. 2014;**5**:241. DOI: 10.3389/fmicb.2014.00241

[53] Allaker RP, Ian Douglas CW. Non-conventional therapeutics for oral infections. Virulence. 2015;**6**:196-207. DOI: 10.4161/21505594.2014.983783

[54] Persson GR. Immune responses and vaccination against periodontal infections. Journal of Clinical

Periodontology. 2005;**32**:39-53. DOI: 10.1111/j.1600-051X.2005.00800.x

[55] Kumar PS. Oral microbiota and systemic disease. Anaerobe. 2013;**24**:90-93. DOI: 10.1016/j.anaerobe.2013.09.010

[56] Devine A, Marsh PD. Prospects for the development of probiotics and prebiotics for oral applications. Journal of Oral Microbiology. 2009;**1**. DOI: 10.3402/jom.v1i0.1949

[57] Teughels W, Loozen G, Quirynen M. Do probiotics offer opportunities to manipulate the periodontal oral microbiota? Journal of Clinical Periodontology. 2011;**11**:159-177. DOI: 10.1111/j.1600-051X.2010.01665.x

[58] Stašková A, Lauko S, Englerová K, Gancarčíková S, Nemcová R, Maďar M. Porovnanie inhibičných účinkov vybraných izolátov *Streptococcus salivarius* a komerčne dostupného *Streptococcus salivarius* (K12) na vybrané potenciálne patogény dentálneho biofilmu. Čo nového v mikrobiológii. Konferencia mladých mikrobiológov; 15-18 March 2018; Demänovská dolina. Bratislava-Praha: Československá spoločnosť mikrobiologická; 2018. p. 40

[59] Teughels W, Durukan A, Ozcelik O, Pauwels M, Quirynen M, Haytac MC. Clinical and microbiological effects of *Lactobacillus reuteri* probiotics in the treatment of chronic periodontitis: A randomized placebo-controlled study. Journal of Clinical Periodontology. 2013;**40**:1025-1035. DOI: 10.1111/jcpe.12155

[60] Bonifait L, Grenier D, Chandad F. Probiotics for oral health: Myth of reality? Journal of the Canadian Dental Association. 2009;**75**:585-590

[61] Byers HL, Tarelli E, Homer KA, Beighton D. Isolation and characterisation of sialidase from a strain of *Streptococcus oralis*. Journal of Medical Microbiology. 2000;**49**:235-244. DOI: 10.1099/0022-1317-49-3-235

[62] Hicks J, Garcia-Godoy F, Flaitz C. Biological factors in dental caries: Role of remineralization and fluoride in the dynamic process of demineralization and remineralization (part 3). The Journal of Clinical Pediatric Dentistry. 2000;**28**:203-214. DOI: 10.17796/jcpd.28.3.w0610427l746j34n

[63] Štefanovič J, Hanzen J. Mikroorganizmy človeka v zdraví a chorobe. 1st ed. Bratislava: HPL Servis; 2012. 190 p

[64] Kuchta M, Halušková V, Kaletová V, Gombošová K, Bálintová E. Možnosti preventívneho a terapeutického využitia probiotík v pediatrii. Pediatria. 2006;**1**:275-280

[65] Ondriová I, Fertaľová T, Magurová D. Klinické využitie probiotík v prevencii a liečbe porúch tráviaceho traktu. Klinické farmakologie a farmacie. 2015;**29**:116-118

[66] Endres JR, Qureshi I, Farber T, Hauswirth J, Hirka G, Pasics I, et al. One-year chronic oral toxicity with combined reproduction toxicity study of a novel probiotic, *Bacillus coagulans*, as a food ingredient. Food and Chemical Toxicology. 2011;**49**:1174-1182. DOI: 10.1016/j.fct.2011.02.012

[67] Ahrne S, Johansson Hagslatt ML. Effect of lactobacilli on paracellular permeability in the gut. Nutrients. 2011;**3**:104-117. DOI: 10.3390/nu3010104

[68] Karpa KD. Probiotics for *Clostridium difficile* diarrhea: Putting it into perspective. The Annals of Pharmacotherapy. 2007;**41**:1284-1287. DOI: 10.1345/aph.1K228

[69] Shi LH, Balakrishnan K, Thiagarajah K, Ismail NIM, Yin OS. Beneficial properties

of probiotics. Tropical Life Sciences Research. 2016;**27**:73-90. DOI: 10.21315/tlsr2016.27.2.6

[70] Broadley KJ, Akhtar Anwar M, Herbert AA, Fehler M, Jones EM, Devies WE, et al. Effects of dietary amines on the gut and its vasculature. The British Journal of Nutrition. 2009;**101**:1645-1652. DOI: 10.1017/S0007114508123431

[71] Juneja VK, Sofos JN. Pathogens and Toxins in Foods: Challenges and Interventions. 1st ed. Vancouver: ASM Press; 2010. 500 p. DOI: 10.1128/9781555815936

[72] Matín-Muňoz MF, Fortuni M, Caminoa M, Belver T, Quirce S, Caballero T. Anaphylactic reaction to probiotics. Cow's milk and hen's egg allergens in probiotic compounds. Pediatric Allergy and Immunology. 2012;**23**:778-784. DOI: 10.1111/j.1399-3038.2012.01338.x

Section 4

Ecology of Biofilms

Chapter 18

The Importance of Biofilms to the Fate and Effects of Microplastics

John A. Glaser

Abstract

Microplastics are global pollutants in water media ranging from drinking water to freshwater streams to oceanic pollutant gyres. Besides the obvious appearance involving a scattered presence in the environmental landscape, microplastics are ubiquitous across modern society in products, food, and beginning to have strong economic effects too. Ingestion of microplastics is virtually unavoidable for each of us as we consume food, breathe air, or drink liquids. For example, beer has been found to be contaminated with plastic materials having the dimensions of micro- and nanoparticles. In the environment, the formation of biofilms on microplastics is widely observed and this can significantly alter properties important to environmental and human health. Significant research has been conducted on the role of biofilms in the fate and effect of microplastics on environmental and human health, with a general message to avoid contact with microplastics in the environment until more complete strategies for cleanup are developed.

Keywords: biofilms, fate and effects, microplastics, pathogenic human threats, pollutants, toxicity

1. Introduction

Plastic derived from the Greek *plasticos* refers to synthetic carbonaceous polymers that exhibit the desired degree of physical flexibility required for molding. During the past 60 years, the product of organic polymer production exploded to virtually all nooks and crannies across the globe [1]. In 2020, global plastic production is composed of a few well-known polymers used in a wide range of products having differing compositions and properties. Current plastic polymer production levels exceed 320 million metric tons (Mt). This surpassed production in the previous decade when significant production capacities were idled [2]. Massive plastic pollution in the world's oceans is estimated to exceed 5 trillion pieces of plastic with a mass of 250,000 Mt [3].

Carbon-based commercialized polymeric materials having desirable physical and chemical properties constitute a wide range of applications. Plastics have been part of the broad range of commercial materials entering the global economy since 1950. The mass production of virgin polymers has been estimated at 8300 Mt. for the period from 1950 to 2015 [4]. Global consumption of plastics continues at a rate of roughly 311 Mt. per year with 90% derived from a petroleum origin and has become a major worldwide solid waste problem. Plastic packaging enhancements have changed the composition of solid waste to where the plastic fraction exceeds 10% in 2005 [5]. In the plastic recycle flow, packaging plastics are poorly recycled. The bulk of plastic waste is disposed in landfills and the natural environment which may exceed 12,000 Mt. of plastic waste by 2050 if current production and waste management trends continue unabated [2].

Macroplastics or the polymers from which they are constructed have been recognized as valuable materials composed of repeating units and applicable to many material design requirements [6]. Each repeating unit of a polymer is referred to as the "-mer" with "polymer" denoting a chemical composed of many repeating units. Plastics are unique materials having the benefits of being light weight, versatile, having reasonably long service lives, and attractive cost. Across the land and seas, the accumulation of plastic litter found in natural environments looms as a global issue [7]. Potential negative impacts to wildlife, human health, and the economy offer strong incentives to thoroughly explore our approach to the sustainable use of plastics [8].

2. Plastics in the environment

Easily observed plastic pollution is often referred to as macroplastics which have dimensions greater than 1 mm. Smaller plastic particles are referred as micro- or nanoparticle. The aspects of long-term pollution and human health effects have been issues of social concern in recent times [9]. The wanton dispersal of plastic film bags and drink bottles mar our global landscape, waterways, and oceans/seas. Plastics apparent resistance to degradation elongates their residence time in the environment. Environmental processes can contribute to the debris by activating degradation pathways which lead to the conversion of macroplastics to smaller dimension plastic materials [10]. Plastics can carry with them pollutants such as plasticizers, antioxidants, and other persistent organic pollutants **Table 1** [11–15]. Human health concerns have been focused on the monomeric components, additives, and certain combinations of the chemical employed in the synthesis of a plastic [16].

Characteristic	Behavior
Density	Determines the vertical water column position
Crystallinity	Controls susceptibility to photochemical oxidation
Extent of oxidation	Chemical composition determines the ease of oxidation and weathering
Biodegradability	Contributes to the general structural deterioration of microplastics through biological means
Monomer residual	Potential source of toxicity and small molecule pollutants
Transport properties	Affinity for hydrophobic chemicals and metals
Polymer additives	Highly variable depending on polymer composition and application of polymer
Surface properties	Important to aggregate formation and biofouling

Table 1.
Characteristics influencing microplastic behavior.

3. Microplastics

The chemical composition of the major plastics provides some basic understanding of their environmental behavior (**Table 2**) [17]. The physical dimensions of plastic particles are classified by size class which refers to the particle's largest dimension that is important to the design of analytical collection protocols used in sampling microplastics sensitive to particle shape [18, 19]. The term microplastics refers to anthropogenic polymer materials having the dimensions of less than 5 mm (0.2 inch) occurring as

Polymer category	Specific gravity	Water column movement	Degree of crystallinity %
Polyethylene (PE)	0.91–0.94	Float	
Low density LDPE	"	"	45–55
High density HDPE	"	"	70–80
Polypropylene (PP)	0.90–0.92	"	
Atactic PP	"	"	~0
Isotactic PP	"	"	70–80
Polystyrene (expanded) (PS)	0.01–1.05	"	
Seawater	~1.02		
Polystyrene	1.04–1.09	Sink	
Polyvinyl chloride (PVC)	1.16–1.30	"	
Polyamide	1.13–1.15	"	35–45
Polyethylene terephthalate (PET)	1.34–1.39	"	30–40
Polyester resin + glass fibers	>1.35	"	
Cellulose acetate	1.22–1.24	"	

Table 2.
Plastic properties important to the fate and effects of microplastics.

plastic pollution in the environment [20]. Smaller particles referred to as nanoplastics are becoming an issue of growing concern that falls into the size range of 10–1,000 nm [21]. The consensus definition and categorization of plastic debris are yet to be achieved. Uneven size classes are employed for sampling for microplastics to represent random size classes, and even material composition is a matter of debate [22, 23].

3.1 Definition

Microplastic specifications can be found in two broad categories, primary and secondary [24]. Primary microplastics are manufactured particles that are characterized as microbeads, nurdles, and fibers in size dimensions of 5 mm or smaller. Any interception technology must be equipped with appropriately sized filters to remove the particles from contaminated environmental media. Secondary microplastics are formed from larger plastics or macroplastics through the effects of weathering and physical deterioration in the environment. Weathering by photochemical oxidation, UV rays, and wind and wave action leads to the fragmentation of macroplastics to form microplastics. Aquatic plastic debris can be organized by size as mega (>1 m)-, macro (<1 m)-, meso (<2.5 cm)-, micro (<5 mm)-, and nano (<1 μm)-dimensions [25]. A recently proposed size schema separates microplastics in marine environments into the following categories: nano (1–1000 nm)-, micro (1–1000 μm)-, meso (1–10 mm)-, and macroplastics (≥1 cm). Size schemes are proposed to address the sampling problems encountered in the field, but these schemes are lacking since it is difficult to provide a microplastic sample that is spatially representative of a specific environmental space [26–29].

3.2 Composition

Chemical composition and environmental impacts of microplastic samples differ broadly (**Table 2**). Microplastic composition reflects the use and disposal of the most popular macroplastics such as the polyolefins [polypropylene (PP) and polyethylene (PE)], polyvinyl chloride (PVC), polyurethane (PU), polyethylene terephthalate (PET), polystyrene (PS), and polycarbonate (PC). The composition of this list represents a large fraction of plastic use and global plastic production [2]. The high molecular weight of most plastic polymers renders them biochemically inert initially and hence have an inherent low toxicity due to lack of water solubility [30]. Many polymer compositions can contain small concentrations of unpolymerized monomer [31]. Monomers can be toxic and carcinogenic as in the case of styrene or vinyl chloride [32]. Problematic plastics such as PVC, PU, PS, and PC can contain toxic monomers or additives. Additives can include fillers, plasticizers, coloring agents, antimicrobials, flame retardants, and other material property modifiers [33]. These materials represent a source of health risks for humans and other species [34].

3.3 Origin

Microplastics can be produced directly for use as raw materials in the fabrication of larger items. Environmental processes are known to form microplastic particles through mechanical destruction of macroplastic materials such as automobile tires disintegrating during wear and use [35]. As ingredients of abrasive, cleaning, and cosmetic products, microplastics have been manufactured as articles of commerce [36]. Microplastics were found to form during material wear of macroplastics by industrial processes and via physical breakdown of macroplastics [35, 36]. Their abundance and in situ effects of the environment have not been well quantified due in part to the random composition of particles of non-uniform shapes which are difficult to assess by representative samples [37]. The abundance of micro-, meso-, and macroplastics floating in the marine environment has been estimated from aggregated data derived from a host of surveys [38]. An estimate of global plastic pollution identifies at least 5.25 trillion plastic pieces of plastics, and most of its composition is microplastics [39]. Plastic marine debris (PMD) surveys suggested estimates of the total burden could be at least an order of magnitude lower than what has been observed in the environment [40]. A concern for a missing debris component has been interpreted as losses to deep sea and sediment sinks as prominent components to marine plastic fate [41].

3.4 Analytical protocols

An understanding of microplastic pollution requires the use of proper and clear terminology for use in the design of data collection and supporting analytical protocols, enhanced coordination of strategic design for research directions, and most importantly a consensus development of mitigation management practices tailored to the global problem solution [42]. Composition, dimensions, and shape of plastic debris can be defined explicitly to properly design sampling protocols and conduct the requisite analytical determinations (biological, chemical, and physical) using a wide array of techniques ranging from microscopy to different forms of spectroscopy [43]. Physicochemical properties (polymer composition, solid state, solubility) are employed as standards accompanying size, shape, color, and origin for categorical identification [44].

Standardized quantification and analysis procedures designed to analyze microplastics are critical to the design and data collection for comparative research studies [45]. Microplastics have high surface area solids and should be described in consensus terms [46]. The surface area of environmentally sampled microplastics was found to be a very important descriptor along with an accurate parameter to describe plastic size coupled with a description of plastic quantity per spatial area. As widespread contaminants, microplastics can be found in virtually all environmental partitions [47]. Features such as spatial information, contamination sources, fate, and environmental concentration are difficult to assemble and the variety of analytical procedures currently in use hinders a timely and proficient gathering of information [48]. Methods currently used to sample and detect microplastics are under review which is aimed to identify flaws in design and suggest alternatives [49]. Analytical protocols must be designed to include bulk sample collection, particle separation, digestion, identification and quantification, and mitigation of cross-contamination in the form of transportable and consensus tools. This enhanced ability to sample and analyze microplastics enables the use of more representative samples and helps enhance the determination of the sample features mentioned previously. Incorporation of these features provides an enhanced ability to sample and analyze microplastics leading to the utilization of more representative samples attuned to the sample features required for the formulation of standard methods. The inclusion of new and novel analytical methodology can assist the chemical, biological, and physical characterization of samples [50].

3.5 Concerns

Without the proper knowledge of the environmental behavior of microplastics, we are incapable of solving the growing problem of microplastic management as applied to reducing the problem dimensions and human health risk. The necessary knowledge rests on properly designed research efforts and the use of harmonized and consensus analytical tools employed in the data gathering. What parameters for quantifying microplastics are available at a status that permits the comparison of field results acceptable to the general research community?

4. Biofilms

A consortium of microorganisms composed of cells adhering to a surface is called a biofilm [51, 52]. The physical setting for cells to adhere to a surface occurs through the intermediacy of extracellular polymeric substances (EPS) which forms a slimy extracellular matrix **Figure 1** [53]. Microbial cells in the biofilm produce the EPS which are composites of extracellular polysaccharides, proteins, lipids, and DNA [54, 55]. The cellular agglomeration of biofilms forms a three-dimensional structure as a community that offers significant protection against the forces levied by the environment [56].

4.1 Structure

Microbial cells composing a biofilm are distinct from the planktonic cells of the same organism, which are single-cell organisms that are free to float or swim in an aquatic medium [57]. Biofilm structures are formed in response to a variety of different factors enabling biofilm development [58, 59]. Surface recognition is important to specific or nonspecific attachment sites, toxic materials, or antibiotics, and nutritional stress may complicate biofilm growth **Figure 2** [60]. A cell that switches to the biofilm mode of growth undergoes a shift of observable behavior of the bacteria resulting from the interaction of its genotype with the environment that is required of a microbial cell in the transition from planktonic to sessile growth in the regulation genes of the biofilm. A biofilm can mimic a hydrogel, a three-dimensional (3D) network of hydrophilic polymers complex containing a

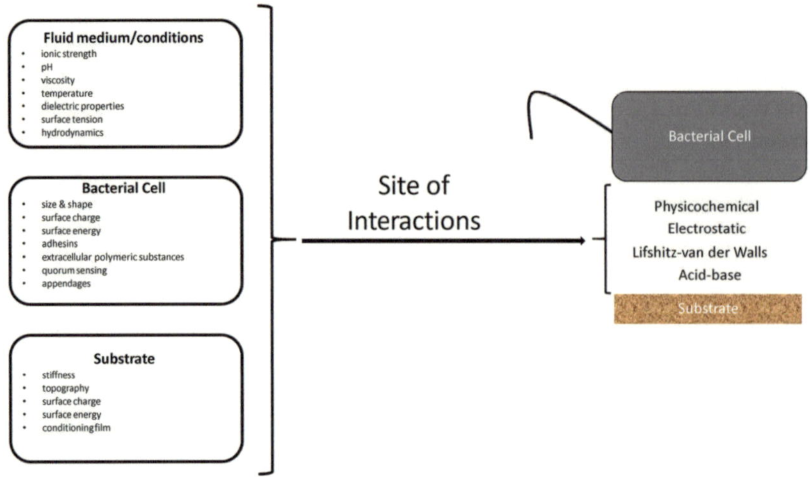

Figure 1.
Site of biofilm interactions.

large quantity of water, which retains its structure through chemical or physical cross-linking polymer chains [61]. Biofilm formation can lead to the formation of a coordinated functional microbial community. The bacteria composing a biofilm can share nutrients due to their proximity in the biofilm and protection from harmful factors of the environment. Biofilms usually begin to form when a free-swimming bacterium attaches to a surface [62].

Colonization of a surface requires a significant transition from the free-living planktonic existence in the bulk aquatic phase to a surface-attached state. A biofilm life cycle is portrayed in **Figure 2** [63]. This process is initiated by the reversible adhesion of a few single cells to a surface leading to a reversible attachment where weakly attached cells are sloughed to the bulk medium, or irreversible attachment where interactions of the cells and a surface are reinforced [64]. Irreversibly attached cells at a surface continue to agglomerate to form microcolonies through cellular division and can proceed to form a mature biofilm when the conditions support growth [65]. As the biofilm matures, factors that will prevent sustainable growth can be triggered by limited nutrients supply or lowered oxygen concentrations may reverse biofilm formation through the dispersal of cells from the biofilm to the bulk aquatic phase. Released cells may attach to a new surface [66]. For single-cell adhesion, three factors leading to single-cell adhesion require attention: the chemical and physical composition of the aquatic environment, the solid surface, and the transitioning microbiota [67].

4.2 Characteristics

Microorganisms form from attached phase growth structures (biofilms) or multicellular microbial communities by transitioning from planktonic (freely-swimming) biota to components of a complex, surface-attached community (**Figure 1**). These communities of adhering microorganisms in the form of biofilms provide protection to the microbes participating in its development. The process begins with planktonic microorganism encountering a surface where some adsorb followed by surface release to final attachment by the secretion of exopolysaccharides which act as an adhesive for the growing biofilm (**Figure 2**) [68]. Switching from a planktonic existence to an attached-life state (sessile) requires a complex

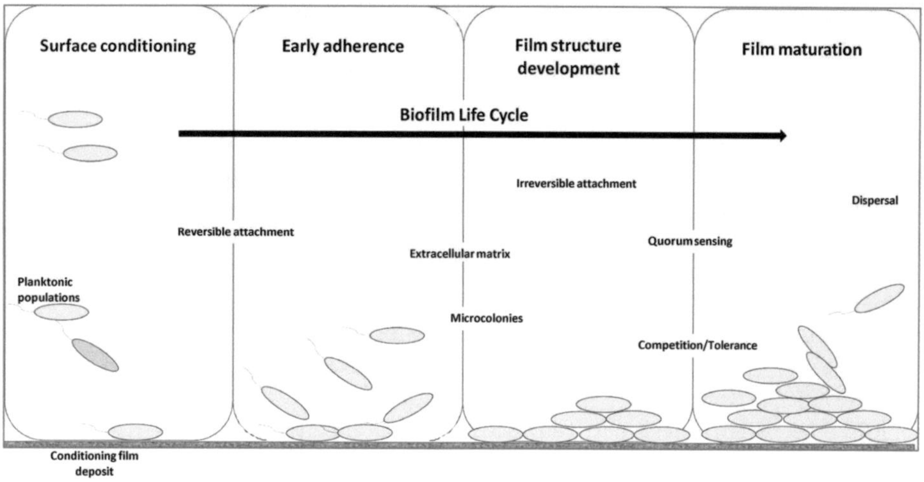

Figure 2.
Biofilm life cycle.

process composed of several factors derived from biological, chemical, and physical properties of the environment, the surface, and the bacterial cell (**Figure 2**) [69]. Initial weak, reversible interactions between a bacterium and a surface lead to irreversible adhesion. New phenotypic characteristics are exhibited by the bacteria of a biofilm in response to environmental signals. Initial cell-polymer surface interactions, biofilm maturation, and the return to the planktonic mode of growth have regulatory circuits and genetic elements controlling these diverse functions. Studies have been conducted to explore the genetic basis of biofilm development with the development of new insights. Compositionally, these films have been found to be a single microbial species or multiple microbial species with attachment to a range of biotic and abiotic surfaces [70]. Mixed-species biofilms are generally encountered in most environments. With proper nutrient and carbon substrate provided, biofilms can grow to massive sizes. A biofilm can achieve large film structures that may be sensitive to physical forces such as agitation. Such energy regimes can lead to biofilm detachment. An example of biofilm attachment and utility can be found in the wastewater treatment sector where large polypropylene disks are rotated through industrial or agriculture wastewater and then exposed to the atmosphere to treat pollutants through the intermediacy of cultured biofilms attached to the rotating polypropylene disk.

4.3 Plastic colonization and plastisphere communities

Plastic's role in freshwater and marine systems is poorly understood from many perspectives especially microbiology. Microscopic scrutiny and next-generation sequencing of PMD from locations in the North Atlantic were used to characterize attached microbial communities. A microbial community having a high degree of diversity was identified as the "Plastisphere" from the pitting of the debris surface which suggested bacterial shapes engaged in the utilization of the polymer by enzymatic means [71]. Opportunistic pathogens were observed as specific members of the genus *Vibrio* [72, 73]. Attached plastisphere communities were found to be distinct from surrounding surface water, suggesting that PMD could be a novel ecological habitat in the open ocean. Most natural floating marine substrates have shorter half-lives than PMD which is enhanced by a hydrophobic surface that assists microbial colonization and biofilm formation. The adhesion of individual bacteria to a surface-initiated biofilm formation is supported by a collection of factors arising from initial adhesion to the growth of a mature biofilm [74].

Bacteria communicate with one another using chemical signal molecules [75]. This process, termed quorum sensing, allows bacteria to monitor the environment to adjust community behavior at a population-wide scale in response to community changes in the number and species present [76]. The information conveyed by these molecules works to synchronize activities for a wide group of cells. This cell-to-cell communication is used by bacteria to coordinate population density-dependent changes in behavior. Quorum sensing involves the production of and response to diffusible or secreted signals, which can vary substantially across different types of bacteria and important to the first stage of encounter between a bacterium and a solid surface [77].

Initial bacterial adhesion to a surface, bacterial mass transport, the role of substratum surface properties in initial adhesion and the transition from reversible to irreversible adhesion have been analyzed through a physiochemical lens to yield great insight. Surface thermodynamics and Derjaguin Landau Verwey Overbeek analyses can describe bacterial support using smooth, inert colloidal particles to estimate bacterial cells. A depiction of initial bacterial adhesion to surface-programmed biofilm growth was found to have four major stages: bacterial mass transport towards a surface, reversible bacterial adhesion, conversion to irreversible adhesion, cell wall deformation, and associated developing properties [78]. The production of EPS can be surface-programmed [79]. Initial bacterial adhesion to surfaces and biofilm growth at the solid surface is driven by aspects of physico-chemistry [80].

Bacterial adhesion is important to the fate and transport of plastics in aquatic environments. There has been no systematic investigation of bacterial adhesion to different types of plastics. A limited evaluation of short-term and long-term adhesion for different types of bacteria and four types of plastics, polyethylene (PE), polypropylene (PP), polyethylene terephthalate (PET), and polyvinyl chloride (PVC), was conducted [81]. The target physicochemical factors of surface charge, hydrophobicity/hydrophilicity, roughness, and plastic hardness were characterized. Surface hardness of the plastics was identified as a major factor dominating the adhesion of bacteria onto plastic surfaces in contrast to the other factors [82]. There were significant differences in bacterial cell adhesion for the types of plastics. The different plastic types influenced the bacterial adhesion due to intrinsic surface properties in both short- and long-term studies [83]. Generally, surface roughness, topography, surface free energy, surface charge, electrostatic interactions, and surface hydrophobicity are anticipated to be important to the process of biofilm attachment [84].

5. Environmental effects and fate

A complex network of interactions existing among the physical, chemical, and biological aspects of microplastics in an aquatic environment is shown in **Figure 3** [85]. The microplastic interfaces with pollutant chemicals and biofilms. In this system the plastic surface can be composed of pollutant chemical, biofilm, or biofilm contaminated with pollutant. With time the interactions of microorganisms and microplastics modify pollutant characteristics establishing how and why cells attach to plastic particles. The complexity of the relationship between plastic particles and microorganism attachment relies on factors influencing community development of biofilm and physical characteristics of microplastic particles.

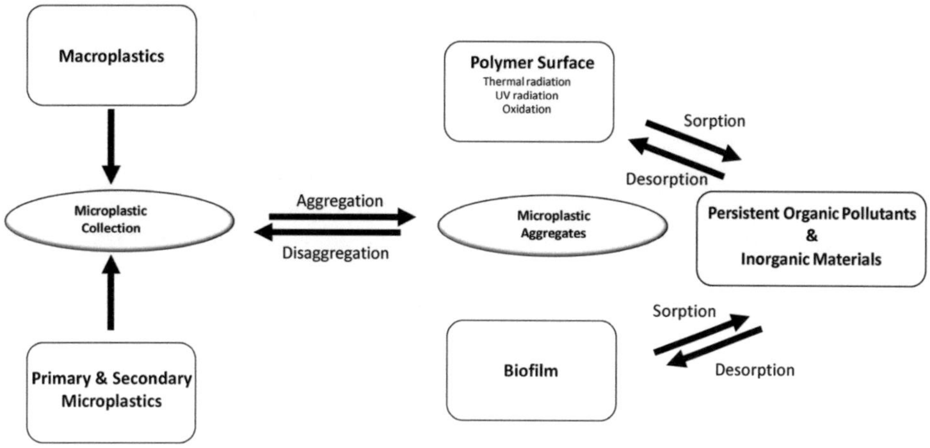

Figure 3.
Microplastic formation and environmental degradation.

5.1 Sorption

Plastics contaminating aquatic environments have been shown to hold various pollutant chemicals arising from the plastic manufacture and environmental pollution. Understanding of sorption and desorption of chemicals by plastics is pivotal to the evaluation of plastics, and their role is important to the environmental dynamics of these chemicals and as a vector of pollution and human health concerns [86]. The chemicals can be of inorganic and organic composition. Environmental microplastic pollution is an assembly of effects found in freshwater and marine conditions relating the complex interrelationship of physical processes, pollutant chemicals, and biota in the formation of biofilms. Sorption of chemicals and microbes to microplastic surfaces involves sorption of chemicals and biota directly to the plastic surface that may or may not be covered with other pollutant chemicals or biofilm. Direct sorption of chemicals to the plastic or biofilm covered plastic surface may exhibit different effects.

The sorption of neutral chemicals to solids from a water phase requires partitioning of freely floating or partially dissolved organic chemical moieties from an aqueous phase to a plastic surface [87, 88]. Factors affecting the partitioning process are the magnitude of the sorption coefficient, temperature, pH, and other coexisting organic and inorganic constituents present in the water phase [89]. The environmental partitioning process is seldom if ever at equilibrium and non-equilibrium conditions describe the general status of environmental conditions. Stagnant or quiescent conditions in the environment may come the closest to equilibrium partitioning conditions. Non-equilibrium conditions in environmental aquatic systems arise from turbulent conditions ranging from flows through broken or incomplete flow paths found in freshwater streams, sea wave action, and wind and turbulent weather-related phenomena. Sorption properties are also related to phenomena such as the chemical/physical properties of the solid, the extent of physical degradation, biodegradation, and agglomerating processes such as biofouling [90].

Sorption is a physical process of the environment where chemicals are transferred from a fluid phase such as water and air to a solid phase [91]. The term "sorption" collectively refers to both absorption and adsorption which are components of the sorption process. Molecular penetration of a chemical and association within a solid phase matrix defines absorption [92]. Whereas, adsorption refers to a process where molecules are confined at the interface between fluid and solid phases as an adherent physical form [93]. Sorption is directly related to properties of the solid, a chemical, and the surface-to-volume ratio of the solid which for microplastic particles is quite large [94]. Apart from surface area, plastics exhibit a range of properties and dimensions, implying the relevance of absorption and adsorption to understanding the importance to the understanding of microplastics' fate and effect. Physisorption or physical sorption occurs from noncovalent intermolecular interactions such as van der Waals interactions. The interaction forces of solids and chemicals though the noncovalent interactions and their combinations and physisorption are usually reversible. Generally, the sorption of materials and chemicals to environmental solids is by physisorption.

5.2 Chemicals

Microplastics can sorb and accumulate both organic and inorganic contaminants detrimental to humans and ecosystem life when released to organisms that may ingest them [95]. Sorption is a major determinant for bioavailability and contributes to the effects of combined exposure to chemicals and microplastics related to the toxicity and bioaccumulation in humans and ecosystem flora. Neutral charged areas of the microplastic surface offer attractive settings for deposition of chemicals due to attractive hydrophobic forces. This is in contrast with hydrophilic or charged compounds that are attracted to the negative-charged areas on the microplastic surface through electrostatic interactions and aquatic media characteristics [94, 96]. Organic chemicals associated with microplastic debris are typically in the semi-volatile or non-volatile categories such as polychlorinated biphenyls and some organic pesticides [97, 98]. Inorganic chemical species are generally ionic. Fuel chemicals and other higher-boiling constituents can be found in the microplastic debris [88, 99–103]. Weathering can be significantly changed the composition containing volatile compounds.

Sorption evaluations can identify the chemicals with higher affinity to microplastics under a variety of environmental conditions. Bench scale sorption studies permit the evaluation of the mass balance for a specific chemical or chemical mixtures. The distribution of chemicals in an environment contaminated with microplastics can be estimated from experimentally determined sorption capacities. Toxicity parallels sorption data, but greater sorption to microplastics does not necessarily lead to higher toxicity or bioaccumulation of a pollutant chemical.

5.3 Buoyancy and aggregation

Biofilm formation at the surface of microplastics may lead to density changes of particles that alter the specific gravity for the mass of microplastic debris [104]. Mineral detritus when incorporated in microplastic debris will increase the density which leads to sinking. Biofilm distribution and bioavailability are expected to be adjusted in response to the buoyancy of microplastics [105]. Biofouling causes changes in the buoyancy of microplastics and, with increasing specific gravity, leads to descension in the water column to a depth of comparable density. Microplastic sampling in the water column can lead to an underestimation of quantities since turbulence leads to vertical mixing.

Aggregate debris formation can be enhanced by biofilm formation on microplastic surfaces commonly expected in situations where diverse bacterial communities colonize the microplastic surfaces. Aggregation has been confirmed by experiment as a factor leading to the apparent removal of microplastics from the surface layer of the marine ecosystems [106].

Microplastics aggregate rapidly with biogenic particles found in the marine environment [107]. The incorporation of organic material is accelerated through gross aggregate formation. It is anticipated that natural aggregation dynamics will influence particle size distribution and the export rates of organic matter which may mirror the similar processes of freshwater and marine ecosystems.

5.4 Plastic biodegradation

Significant abiotic and biotic conditions exist to show that plastics are vulnerable to these forces found in the environment. Plastic weathering contributes to structural defects and size reduction but incomplete decay. Chemical and physical degradation processes contribute to the overall weathering process. Plastics are composed of a wide variety of chemical structure features that degrade in a spectrum of kinetics under biotic and abiotic conditions. Biodegradation of plastics under aerobic conditions forms new products during the degradation path leading potentially to mineralization forming process end-products such as CO_2, H_2O, or CH_4 depending on the terminal electron acceptor [108]. Oxygen is the terminal electron acceptor for the aerobic degradation process. Aerobic conditions lead to the formation of CO_2 and H_2O in addition to the cellular biomass of microorganisms during the degradation of the plastic forms. When sulfidogenic conditions are encountered, plastic biodegradation can lead to the formation of CO_2 and H_2O. Polymer degradation accomplished under anaerobic conditions produces organic acids, H_2O, CO_2, and CH_4. The aerobic process has been found to be more efficient than anaerobic conditions. The anaerobic process produces less energy due to the absence of O_2, serving the electron acceptor, which is more efficient in comparison to CO_2 and SO_4^{-2} [109]. The exposed surface of plastics is where the initial effects of biodegradation are encountered. The biodegradation rate is directly related to the composition of the plastic. The increase of microbial-colonized surface area leads to faster biodegradation rates assuming all other environmental conditions to be equal [110]. Microorganisms can break organic chemicals into simpler chemical forms through biochemical transformation. Plastic biodegradation is a process in which any change in the polymer structure occurs through the structure altering action of microbial enzymes leading to plastic property changes in the form of molecular weight reduction, mechanical strength changes, and surface properties. A more complete understanding of plastic daughter products of environmental degradation is required to more thoroughly understand the effectiveness of environmental plastic degradation.

5.4.1 Human health and pathogenicity

A wide spectrum of pathogenic microorganisms exists and some form biofilms with microplastics in aquatic environments [111]. Freshwater ecosystem analysis has the formation of biofilms on microplastic substrates by a selected grouping of human pathogens utilizing high-throughput sequencing of 16S rRNA that had distinctive community structures [112]. Opportunistic human pathogens such as *Pseudomonas monteillii*, *Pseudomonas mendocina*, and a plant pathogen *Pseudomonas syringae* were detected forming a microplastic biofilm. The opportunistic pathogens were enriched in a biofilm, and the microplastic biofilm exhibited a unique microbial community structure. Distinctive antibiotic resistance genes were detected in the microplastic biofilm. It appears that microplastic surfaces are novel microbial niches and may serve as a vector for antibiotic resistance genetic traits and pathogens in freshwater bodies, engendering environmental risk and exerting adverse impacts on human health [113].

Vibrios are Gram-negative-curved bacilli naturally occuring in marine, estuarine, and freshwater systems [114]. A group of factors has been shown to drive certain microorganisms' virulence in *in vivo* studies, and some are fitness factors in the environment [115]. Factors associated with virulence, nutrient acquisition, competition, survival in unfavorable biotic and abiotic conditions, and attachment and colonization were found to be in the group [116]. As human and animal pathogens, it is important to understand virulence factors, attachment factors, regulatory factors, and antimicrobial resistance factors, which have been characterized for their importance to the organism's fitness apart from its external environment. Virulence and fitness factors were designated and characterized for the three main human pathogens *Vibrio cholerae*, *V. parahaemolyticus*, and *Vibrio vulnificus*.

Bacterial fitness depends on the ability to colonize and grow in hosts, avoid immunological inhibition, and be transmitted to a new host [117]. Established virulence factors can be considered fitness factors, as these factors render the organisms more fit under specific circumstances. Mobile components such as pathogenicity islands carry genes that strengthen the fitness of *Vibrios* even when not producing a toxic effect in a host [118]. Elevated mutation rates can also facilitate evolution of bacteria, making it possible to survive under a wide array of environmental conditions [119].

The three-dimensional complex communities of microbes found in biofilms form on both organic and inorganic substrates that render bacteria more protected from environmental stressors [112]. Biofilms have been demonstrated and characterized for *V. vulnificus*, *V. parahaemolyticus*, *V. cholerae*, *V. fischeri*, *V. harveyi*, and *V. anguillarum* [120].

Pathogen fitness factors and virulence factors produce similar effects in different environments [121]. In unfavorable environments, microbial survival requires factors supporting attachment and colonization such as polysaccharide synthesis, secretion, colonization, motility, toxicity, and genetic regulation. Accompanying these factors may be additive and synergistic effects important to active colonization of biotic and abiotic substrates.

6. Conclusions

The global society's concern over microplastics is directly related to its persistence and potential adverse effects on aquatic biota. In aquatic environments, microorganisms can colonize surfaces by forming adherent biofilms. Biofilm's role in the fate and effects of microplastic has not been completely delineated since active research is aimed to fill copious information gaps. The physical interactions of plastic surfaces and their microbial colonizers is becoming more functionally integrated in the understanding of the effects of microplastic weathering, vertical transport in the water column, and processes of sorption and contaminant release [122]. Biofilm-plastic interactions are recognized for their influence on the fate and effects of microplastics through modification of a particle's physical and chemical properties.

The use of proper and clear terminology for the design of data collection and supporting analytical protocols is necessary for the collection of representative data which is important to the strategic design of research directions based on consensus data development [42]. The necessary analytical determinations (biological, chemical, and physical) developed from a wide array of current and developing technologies offer answers to questions concerning the details of microplastics in the environment. Spatial information, contamination sources, fate, and environmental concentrations are necessary to a timely and proficient gathering of information [48]. New and novel analytical methodology designed to assist the chemical, biological, and physical characterization of samples is welcomed [50].

An understanding of surface biofouling of submerged surfaces is important to decipher surface colonization processes relative to of the behavior of plastic in the environment [123–126]. An enhanced understanding of biofilm formation on submerged surfaces is required to develop a more complete pictures of microbial colonization and the basic processes involved in biofilm formations. Biofilm-plastic interactions important to hydrodynamic processes, such as vertical transport, require the use of environmentally representative biofilm.

The effect of biofilm formation and its connection to the kinetics of chemical partitioning required additional scrutiny [127, 128]. The complexity of surfaces available to sorption processes needs attention to discover the relative importance of the multiple surface adsorption of organic and inorganic pollutant chemicals. The importance of surface topography to the sorption process requires further research. Mechanisms to explain toxic chemical transport by microplastics employing established biofilm contaminated with heavy metals and organic chemicals will be very helpful.

Microbial effects specific to the ability of biofilm-forming microorganisms on a microplastic surface in contact with aqueous media are important to the development of biofilms and their control. Human pathogens such as strains of *Vibrio* spp. have been isolated in formed biofilm on microplastics. The pathogen-populated biofilms must be scrutinized for their possible role in the transmission of materials that could be lethal [129].

Studies are available suggesting that biofilms on microplastics do not present a threat over biofilms on naturally occurring surfaces [130, 131]. The pathogenic populated biofilms are viewed as having no new adverse effect on human food supplies. Since we are still an early state of learning with the environmental effects of microplastics, it is incumbent that we continue to scrutinize biofilm effects and their relation to human health and the health of aquatic ecosystems [132].

This chapter has focused on the question of a role for bacterial biofilms to the environmental effects attributable to microplastics. The importance of biofilms to plastics and their degradation is becoming more completely revealed through continuing focused research effort. The alacrity with which biofilms form on plastic in the environment is functionally connected to ambient conditions and the weather effects to which the plastic has been subjected. Microplastics and adherent biofilms provide potential vector mechanisms to assist the transport of pollutant chemicals and pathogens to a wide area of the aquatic environment.

Conflict of interest

No conflict of interest is known or expected.

Disclaimer

The findings and conclusions in this chapter have not been formally disseminated by the United States Environmental Protection Agency and should not be construed to represent any agency determination or policy.

Author details

John A. Glaser
Center for Environmental Solutions and Emergency Response, Office of Research and Development, United States Environmental Protection Agency, Cincinnati, Ohio, United States

*Address all correspondence to: glaser.john@epa.gov

References

[1] Massy J. A Little Book about BIG Chemistry: The Story of Man-Made Polymers. Cham: Springer International Publishing. 2017. DOI: 10.1007/978-3-319-54831-9

[2] Geyer R, Jambeck JR, Law KL. Production, use, and fate of all plastics ever made. Science Advances. 2017;**3**(7):e1700782. DOI: 10.1126/sciadv.1700782

[3] Eriksen M, Lebreton LC, Carson HS, Thiel M, Moore CJ, Borerro JC, et al. Plastic pollution in the world's oceans: More than 5 trillion plastic pieces weighing over 250,000 tons afloat at sea. PLOS One. 2014;**9**(12):e111913. DOI: 10.1371/journal.pone.0111913

[4] Jambeck JR, Geyer R, Wilcox C, Siegler TR, Perryman M, Andrady A, et al. Plastic waste inputs from land into the ocean. Science. 2015;**347**(6223):768-771. DOI: 10.1126/science.1260352

[5] Hopewell J, Dvorak R, Kosior E. Plastics recycling: Challenges and opportunities. Philosophical Transactions of the Royal Society B: Biological Sciences. 2009;**364**(1526):2115-2126. DOI: 10.1098/rstb.2008.0311

[6] Brinson HF, Brinson LC. Characteristics, applications and properties of polymers. In: Polymer Engineering Science and Viscoelasticity. Boston, MA: Springer; 2015. pp. 57-100. DOI: 10.1007/978-1-4899-7485-3_3

[7] Schwarz AE, Ligthart TN, Boukris E, Van Harmelen T. Sources, transport, and accumulation of different types of plastic litter in aquatic environments: A review study. Marine Pollution Bulletin. 2019;**143**:92-100. DOI: 10.1016/j.marpolbul.2019.04.029

[8] Haider TP, Völker C, Kramm J, Landfester K, Wurm FR. Plastics of the future? The impact of biodegradable polymers on the environment and on society. Angewandte Chemie, International Edition. 2019;**58**(1):50-62. DOI: 10.1002/anie.201805766

[9] Halden RU. Plastics and health risks. Annual Review of Public Health. 2010;**31**:179-194. DOI: 10.1146/annurev.publhealth.012809.103714

[10] Min K, Cuiffi JD, Mathers RT. Ranking environmental degradation trends of plastic marine debris based on physical properties and molecular structure. Nature Communications. 2020;**11**:727. DOI: 10.1038/s41467-020-14538-z1

[11] Bergmann M, Gutow L, Klages M, editors. Marine Anthropogenic Litter. Cham: Springer Open; 2015. ISBN: 978-3-319-16509-7

[12] Law KL. Plastics in the marine environment. Annual Review of Marine Science. 2017;**9**:205-229. DOI: 10.1146/annurev-marine-010816-060409

[13] Andrady AL. Microplastics in the marine environment. Marine Pollution Bulletin. 2011;**62**(8):1596-1605. DOI: 10.1016/j.marpolbul.2011.05.030

[14] Ivleva NP, Wiesheu AC, Niessner R. Microplastic in aquatic ecosystems. Angewandte Chemie, International Edition. 2017;**56**(7):1720-1739. DOI: 10.1002/anie.201606957

[15] Worm B, Lotze HK, Jubinville I, Wilcox C, Jambeck J. Plastic as a persistent marine pollutant. Annual Review of Environment and Resources. 2017;**42**:1-26. DOI: 10.1146/annurev-environ-102016-060700

[16] Sharma S, Chatterjee S. Microplastic pollution, a threat to marine ecosystem and human health: A short review. Environmental Science and Pollution

Research. 2017;**24**(27):21530-21547. DOI: 10.1007/s11356-017-9910-8

[17] Andrady AL, editor. Plastics and the Environment. New York: John Wiley & Sons; 20 Feb 2003. ISBN: 0-471-09520-6

[18] Frias JP, Nash R. Microplastics: Finding a consensus on the definition. Marine Pollution Bulletin. 2019;**138**:145-147. DOI: 10.1016/j.marpolbul.2018.11.022

[19] Hartmann NB, Hüffer T, Thompson RC, Hassellöv M, Verschoor A, Daugaard AE, et al. Are we speaking the same language? Recommendations for a definition and categorization framework for plastic debris. Environmental Science & Technology. 2019;**53**(3):1039-1047. DOI: 10.1021/acs.est.8b05297

[20] GESAMP. Sources, fate and effects of microplastics in the marine environment: A global assessment. In: Kershaw PJ, editor. IMO/FAO/UNESCO-IOC/UNIDO/WMO/IAEA/UN/UNEP/UNDP Joint Group of Experts on the Scientific Aspects of Marine Environmental Protection. GESAMP Reports and Studies Series 90. London: International Maritime Organization; 2015. p. 96

[21] Gigault J, Ter Halle A, Baudrimont M, Pascal PY, Gauffre F, Phi TL, et al. Current opinion: What is a nanoplastic? Environmental Pollution. 2018;**235**:1030-1034. DOI: 10.1016/j.envpol.2018.01.024

[22] Filella M. Questions of size and numbers in environmental research on microplastics: Methodological and conceptual aspects. Environment and Chemistry. 2015;**12**:527-538. DOI: 10.1071/EN15012

[23] Kooi M, Koelmans AA. Simplifying microplastic via continuous probability distributions for size, shape, and density. Environmental Science & Technology Letters. 2019;**6**:551-557. DOI: 10.1021/acs.estlett.9b00379

[24] Cole M, Lindeque P, Halsband C, Galloway TS. Microplastics as contaminants in the marine environment: A review. Marine Pollution Bulletin. 2011;**62**(12):2588-2597. DOI: 10.1016/j.marpolbul.2011.09.025

[25] Kershaw PJ, Rochman CM. Sources, fate and effects of microplastics in the marine environment: Part 2 of a global assessment. Reports and studies. In: IMO/FAO/Unesco-IOC/WMO/IAEA/UN/UNEP Joint Group of Experts on the Scientific Aspects of Marine Environmental Protection (GESAMP) Eng No. 93. 2015. ISSN: 1020-4873

[26] Underwood AJ, Chapman MG, Browne MA. Some problems and practicalities in design and interpretation of samples of microplastic waste. Analytical Methods. 2017;**9**(9):1332-1345. DOI: 10.1039/C6AY02641A

[27] Covernton GA, Pearce CM, Gurney-Smith HJ, Chastain SG, Ross PS, Dower JF, et al. Size and shape matter: A preliminary analysis of microplastic sampling technique in seawater studies with implications for ecological risk assessment. Science of the Total Environment. 2019;**667**:124-132. DOI: 10.1016/j.scitotenv.2019.02.346

[28] Cutroneo L, Reboa A, Besio G, Borgogno F, Canesi L, Canuto S, et al. Microplastics in seawater: Sampling strategies, laboratory methodologies, and identification techniques applied to port environment. Environmental Science and Pollution Research. 2020;**27**:8938-8952. DOI: 10.1007/s11356-020-07783-8

[29] Ryan PG, Suaria G, Perold V, Pierucci A, Bornman TG, Aliani S. Sampling microfibres at the sea surface: The effects of mesh size, sample volume

and water depth. Environmental Pollution. 2020;**258**:113413. DOI: 10.1016/j.envpol.2019.113413

[30] Galloway TS. Micro-and nano-plastics and human health. In: Marine Anthropogenic Litter. Cham: Springer; 2015. pp. 343-366. DOI: 10.1007/978-3-319-16510-3_13

[31] Teuten EL, Saquing JM, Knappe DR, Barlaz MA, Jonsson S, Björn A, et al. Transport and release of chemicals from plastics to the environment and to wildlife. Philosophical Transactions of the Royal Society B: Biological Sciences. 2009;**364**(1526):2027-2045. DOI: 10.1098/rstb.2008.0284

[32] Sheftel VO. Handbook of Toxic Properties of Monomers and Additives. Boca Raton, FL: CRC Press; 1995. ISBN-13: 978-1566700757

[33] Bolgar M, Hubball J, Groeger J, Meronek S. Handbook for the Chemical Analysis of Plastic and Polymer Additives. Boca Raton, FL: CRC Press; 2015. ISBN-13: 978-1439860748

[34] Hahladakis JN, Velis CA, Weber R, Iacovidou E, Purnell P. An overview of chemical additives present in plastics: Migration, release, fate and environmental impact during their use, disposal and recycling. Journal of Hazardous Materials. 2018;**344**:179-199. DOI: 10.1016/j.jhazmat.2017.10.014

[35] Browne MA. Sources and pathways of microplastics to habitats. In: Marine Anthropogenic Litter. Cham: Springer; 2015. pp. 229-244. DOI: 10.1007/978-3-319-16510-3_9

[36] Anderson AG, Grose J, Pahl S, Thompson RC, Wyles KJ. Microplastics in personal care products: Exploring perceptions of environmentalists, beauticians and students. Marine Pollution Bulletin. 2016;**113**(1-2):454-460. DOI: 10.1016/j.marpolbul.2016.10.048

[37] Silva AB, Bastos AS, Justino CI, da Costa JP, Duarte AC, Rocha-Santos TA. Microplastics in the environment: Challenges in analytical chemistry—A review. Analytica Chimica Acta. 2018;**1017**:1-9. DOI: 10.1016/j.aca.2018.02.043

[38] Shahul Hamid F, Bhatti MS, Anuar N, Anuar N, Mohan P, Periathamby A. Worldwide distribution and abundance of microplastic: How dire is the situation? Waste Management & Research. 2018;**36**(10):873-897. DOI: 10.1177/0734242X18785730

[39] Costa MF, da Costa JP, Duarte AC. Sampling of micro (nano) plastics in environmental compartments: How to define standard procedures? Current Opinion in Environmental Science & Health. 2018;**1**:36-40. DOI: 10.1016/j.coesh.2017.10.001

[40] Hale RC, Seeley ME, La Guardia MJ, Mai L, Zeng EY. A global perspective on microplastics. Journal of Geophysical Research, Oceans. 2020;**125**(1). DOI: 10.1029/2018JC014719

[41] Thompson RC, Olsen Y, Mitchell RP, Davis A, Rowland SJ, John AW, et al. Lost at sea: Where is all the plastic? Science. 2004;**304**(5672):838-838. DOI: 10.1126/science.1094559

[42] Loder MGJ, Gerdts G. Methodology used for the detection and identification of microplastics—A critical appraisal. In: Bergmann M, Gutow L, Klages M, editors. Marine Anthropogenic Litter. Cham: Springer International Publishing; 2015. pp. 201e227. DOI: 10.1007/978-3-319-16510-3_8

[43] Prata JC, da Costa JP, Duarte AC, Rocha-Santos T. Methods for sampling and detection of microplastics in water and sediment: A critical review. TrAC, Trends in Analytical Chemistry. 2019;**110**:150-159. DOI: 10.1016/j.trac.2018.10.029

[44] Renner G, Schmidt TC, Schram J. Analytical methodologies for monitoring micro (nano) plastics: Which are fit for purpose? Current Opinion in Environmental Science & Health. 2018;**1**:55-61. DOI: 10.1016/j.coesh.2017.11.001

[45] Klein S, Dimzon IK, Eubeler J, Knepper TP. Analysis, occurrence, and degradation of microplastics in the aqueous environment. In: Freshwater Microplastics. Cham: Springer; 2018. pp. 51-67. DOI: 10.1007/978-3-319-61615-5_3

[46] Primpke S, Christiansen SH, Cowger W, De Frond H, Deshpande A, Fischer M, et al. EXPRESS: Critical assessment of analytical methods for the harmonized and cost efficient analysis of Microplastics. Applied Spectroscopy. 2020;**6**:0003702820921465. DOI: 10.1177/0003702820921465

[47] Nguyen B, Claveau-Mallet D, Hernandez LM, Xu EG, Farner JM, Tufenkji N. Separation and analysis of microplastics and nanoplastics in complex environmental samples. Accounts of Chemical Research. 2019;**52**(4):858-866. DOI: 10.1021/acs.accounts.8b00602

[48] Masura J et al. Laboratory Methods for the Analysis of Microplastics in the Marine Environment: Recommendations for Quantifying Synthetic Particles in Waters and Sediments. NOAA Technical Memorandum NOS-OR&R-48, Silver Spring, MD; 2015

[49] Koelmans AA, Nor NH, Hermsen E, Kooi M, Mintenig SM, De France J. Microplastics in freshwaters and drinking water: Critical review and assessment of data quality. Water Research. 2019;**155**:410-422. DOI: 10.1016/j.watres.2019.02.054

[50] Zarfl C. Promising techniques and open challenges for microplastic identification and quantification in environmental matrices. Analytical and Bioanalytical Chemistry. 2019;**411**:3743-3756. DOI: 10.1007/s00216-019-01763-9

[51] Flemming HC, Wingender J. The biofilm matrix. Nature Reviews. Microbiology. 2010;**8**(9):623. DOI: 10.1038/nrmicro2415

[52] Hobley L, Harkins C, MacPhee CE, Stanley-Wall NR. Giving structure to the biofilm matrix: An overview of individual strategies and emerging common themes. FEMS Microbiology Reviews. 2015;**39**(5):649-669. DOI: 10.1093/femsre/fuv015

[53] Dragoš A, Kovács ÁT. The peculiar functions of the bacterial extracellular matrix. Trends in Microbiology. 2017;**25**(4):257-266. DOI: 10.1016/j.tim.2016.12.010

[54] Dragoš A, Kiesewalter H, Martin M, Hsu CY, Hartmann R, Wechsler T, et al. Division of labor during biofilm matrix production. Current Biology. 2018;**28**(12):1903-1913. DOI: 10.1016/j.cub.2018.04.046

[55] Flemming HC. EPS—Then and now. Microorganisms. 2016;**4**(4):41. DOI: 10.3390/microorganisms4040041

[56] Hartmann R, Singh PK, Pearce P, Mok R, Song B, Díaz-Pascual F, et al. Emergence of three-dimensional order and structure in growing biofilms. Nature Physics. 2019;**15**(3):251. DOI: 10.1038/s41567-018-0356-9

[57] Dang H, Lovell CR. Microbial surface colonization and biofilm development in marine environments. Microbiology and Molecular Biology Reviews. 2016;**80**(1):91-138. DOI: 10.1128/MMBR.00037-15

[58] Harrison JP, Hoellein TJ, Sapp M, Tagg AS, Ju-Nam Y, Ojeda JJ. Microplastic-associated biofilms: A comparison of freshwater

and marine environments. In: Freshwater Microplastics. Cham: Springer; 2018. pp. 181-201. DOI: 10.1007/978-3-319-61615-5_9

[59] Oberbeckmann S, Löder MG, Labrenz M. Marine microplastic-associated biofilms—A review. Environment and Chemistry. 2015;**12**(5):551-562. DOI: 10.1071/EN15069

[60] Kataky R, Knowles E. Biofilm formation on abiotic surfaces and their redox activity. Current Opinion in Electrochemistry. 2018;**12**:121-128. DOI: 10.1016/j.coelec.2018.07.007

[61] Kandemir N, Vollmer W, Jakubovics NS, Chen J. Mechanical interactions between bacteria and hydrogels. Scientific Reports. 2018;**8**(1):1-1. DOI: 10.1038/s41598-018-29269-x

[62] Lobelle D, Cunliffe M. Early microbial biofilm formation on marine plastic debris. Marine Pollution Bulletin. 2011;**62**(1):197-200. DOI: 10.1016/j.marpolbul.2010.10.013

[63] Grinberg M, Orevi T, Kashtan N. Bacterial surface colonization, preferential attachment and fitness under periodic stress. PLOS Computational Biology. 2019;**15**(3):e1006815. DOI: 10.1371/journal.pcbi.1006815

[64] Miao L, Wang P, Hou J, Yao Y, Liu Z, Liu S, et al. Distinct community structure and microbial functions of biofilms colonizing microplastics. Science of the Total Environment. 2019;**650**:2395-2402. DOI: 10.1016/j.scitotenv.2018.09.378

[65] Armbruster CR, Parsek MR. New insight into the early stages of biofilm formation. Proceedings of the National Academy of Sciences. 2018;**115**(17):4317-4319. DOI: 10.1073/pnas.1804084115

[66] McDougald D, Rice SA, Barraud N, Steinberg PD, Kjelleberg S. Should we stay or should we go: Mechanisms and ecological consequences for biofilm dispersal. Nature Reviews. Microbiology. 2012;**10**(1):39. DOI: 10.1038/nrmicro2695

[67] Garrett TR, Bhakoo M, Zhang Z. Bacterial adhesion and biofilms on surfaces. Progress in Natural Science. 2008;**18**(9):1049-1056. DOI: 10.1016/j.pnsc.2008.04.001

[68] Lembre P, Lorentz C, Di Martino P. Exopolysaccharides of the biofilm matrix: A complex biophysical world. In: The Complex World of Polysaccharides. Rijeka, Croatia: InTech; 2012. pp. 371-392. DOI: 10.5772/51213

[69] Achinas S, Charalampogiannis N, Euverink GJ. A brief recap of microbial adhesion and biofilms. Applied Sciences. 2019;**9**(14):2801. DOI: 10.3390/app9142801

[70] Flemming HC, Wuertz S. Bacteria and archaea on earth and their abundance in biofilms. Nature Reviews. Microbiology. 2019;**17**(4):247-260. DOI: 10.1038/s41579-019-0158-9

[71] De Tender C, Schlundt C, Devriese LI, Mincer TJ, Zettler ER, Amaral-Zettler LA. A review of microscopy and comparative molecular-based methods to characterize “Plastisphere” communities. Analytical Methods. 2017;**9**(14):2132-2143. DOI: 10.1039/C7AY00260B

[72] Zettler ER, Mincer TJ, Amaral-Zettler LA. Life in the “plastisphere”: Microbial communities on plastic marine debris. Environmental Science & Technology. 2013;**47**(13):7137-7146. DOI: 10.1021/es401288x

[73] Jacquin J, Cheng J, Odobel C, CONAN P, Pujo-pay M, Jean-Francois G. Microbial ecotoxicology

of marine plastic debris: A review on colonization and biodegradation by the 'plastisphere'. Frontiers in Microbiology. 2019;**10**:865. DOI: 10.3389/fmicb. 2019.00865

[74] Yuan Y, Hays MP, Hardwidge PR, Kim J. Surface characteristics influencing bacterial adhesion to polymeric substrates. RSC Advances. 2017;**7**(23):14254-14261. DOI: 10.1039/ C7RA01571B

[75] Waters CM, Bassler BL. Quorum sensing: Cell-to-cell communication in bacteria. Annual Review of Cell and Developmental Biology. 2005;**21**:319-346. DOI: 10.1146/annurev. cellbio.21.012704.131001

[76] Abisado RG, Benomar S, Klaus JR, Dandekar AA, Chandler JR. Bacterial quorum sensing and microbial community interactions. MBio. 2018;**9**(3):e02331-e02317. DOI: 10.1128/ mBio.02331-17

[77] Whiteley M, Diggle SP, Greenberg EP. Bacterial quorum sensing: The progress and promise of an emerging research area. Nature. 2017;**551**(7680):313-320. DOI: 10.1038/ nature24624

[78] Carniello V, Peterson BW, van der Mei HC, Busscher HJ. Physico-chemistry from initial bacterial adhesion to surface-programmed biofilm growth. Advances in Colloid and Interface Science. 2018;**261**:1-14. DOI: 10.1016/j.cis.2018.10.005

[79] Di Martino P. Extracellular polymeric substances, a key element in understanding biofilm phenotype. AIMS Microbiology. 2018;**4**(2):274. DOI: 10.3934/microbiol.2018.2.274

[80] Renner LD, Weibel DB. Physicochemical regulation of biofilm formation. MRS Bulletin. 2011;**36**(5):347-355.a2. DOI: 10.1557/ mrs.2011.65

[81] Cai L, Wu D, Xia J, Shi H, Kim H. Influence of physicochemical surface properties on the adhesion of bacteria onto four types of plastics. Science of the Total Environment. 2019;**671**:1101-1107. DOI: 10.1016/j. scitotenv.2019.03.434

[82] Elbourne A, Chapman J, Gelmi A, Cozzolino D, Crawford RJ, Truong VK. Bacterial-nanostructure interactions: The role of cell elasticity and adhesion forces. Journal of Colloid and Interface Science. 2019;**546**:192-210. DOI: 10.1016/j.jcis.2019.03.050

[83] Pinto M, Langer TM, Hüffer T, Hofmann T, Herndl GJ. The composition of bacterial communities associated with plastic biofilms differs between different polymers and stages of biofilm succession. PLOS One. 2019;**14**(6). DOI: 10.1371/journal.pone.0217165

[84] Persat A, Nadell CD, Kim MK, Ingremeau F, Siryaporn A, Drescher K, et al. The mechanical world of bacteria. Cell. 2015;**161**(5):988-997. DOI: 10.1016/j.cell.2015.05.005

[85] Engler RE. The complex interaction between marine debris and toxic chemicals in the ocean. Environmental Science & Technology. 2012;**46**(22):12302-12315. DOI: 10.1021/ es3027105

[86] Endo S, Koelmans AA. Sorption of hydrophobic organic compounds to plastics in the marine environment: Equilibrium. In: Hazardous Chemicals Associated with Plastics in the Marine Environment. Cham: Springer; 2016. pp. 185-204. DOI: 10.1007/698_2016_11. ISBN: 978-3-319-95566-7

[87] Delle Site A. Factors affecting sorption of organic compounds in natural sorbent/water systems and sorption coefficients for selected pollutants. A review. Journal of Physical and Chemical Reference Data. 2001;**30**(1):187-439. DOI: 10.1063/1.1347984

[88] Yu F, Yang C, Zhu Z, Bai X, Ma J. Adsorption behavior of organic pollutants and metals on micro/nanoplastics in the aquatic environment. Science of the Total Environment. 2019;**694**:133643. DOI: 10.1016/j.scitotenv.2019.133643

[89] Barrer RM. Specificity in physical sorption. Journal of Colloid and Interface Science. 1966;**21**(4):415-434. DOI: 10.1016/0095-8522(66)90007-9

[90] Schwarzenbach RP, Gschwend PM, Imboden DM. Environmental Organic Chemistry. John Wiley & Sons; 2016. ISBN-13: 978-1118767238

[91] Karapanagioti HK, Werner D. Sorption of hydrophobic organic compounds to plastics in the marine environment: Sorption and desorption kinetics. In: Hazardous Chemicals Associated with Plastics in the Marine Environment. Cham: Springer; 2018. pp. 205-219. DOI: 10.1007/698_2018_256

[92] Gouin T, Roche N, Lohmann R, Hodges G. A thermodynamic approach for assessing the environmental exposure of chemicals absorbed to microplastic. Environmental Science & Technology. 2011;**45**(4):1466-1472. DOI: 10.1021/es1032025

[93] Guo X, Wang J. The chemical behaviors of microplastics in marine environment: A review. Marine Pollution Bulletin. 2019;**142**:1-4. DOI: 10.1016/j.marpolbul.2019.03.019

[94] Fred-Ahmadu OH, Bhagwat G, Oluyoye I, Benson NU, Ayejuyo OO, Palanisami T. Interaction of chemical contaminants with microplastics: Principles and perspectives. Science of the Total Environment. 2019;**16**:135978. DOI: 10.1016/j.scitotenv.2019.135978

[95] Alimi OS, Farner Budarz J, Hernandez LM, Tufenkji N. Microplastics and nanoplastics in aquatic environments: Aggregation, deposition, and enhanced contaminant transport. Environmental Science & Technology. 2018;**52**(4):1704-1724. DOI: 10.1021/acs.est.7b05559

[96] Rochman CM. The complex mixture, fate and toxicity of chemicals associated with plastic debris in the marine environment. In: Marine Anthropogenic Litter. Cham: Springer; 2015. pp. 117-140. DOI: 10.1007/978-3-319-16510-3_5

[97] Koelmans AA, Bakir A, Burton GA, Janssen CR. Microplastic as a vector for chemicals in the aquatic environment: Critical review and model-supported reinterpretation of empirical studies. Environmental Science & Technology. 2016;**50**(7):3315-3326. DOI: 10.1021/acs.est.5b06069

[98] Tourinho PS, Kočí V, Loureiro S, van Gestel CA. Partitioning of chemical contaminants to microplastics: Sorption mechanisms, environmental distribution and effects on toxicity and bioaccumulation. Environmental Pollution. 2019;**252**(Part B):1246-1256. DOI: 10.1016/j.envpol.2019.06.030

[99] Rochman CM, Hentschel BT, Teh SJ. Long-term sorption of metals is similar among plastic types: Implications for plastic debris in aquatic environments. PLOS One. 2014;**9**(1):e85433. DOI: 10.1371/journal.pone.0085433

[100] Richard H, Carpenter EJ, Komada T, Palmer PT, Rochman CM. Biofilm facilitates metal accumulation onto microplastics in estuarine waters. Science of the Total Environment. 2019;**683**:600-608. DOI: 10.1016/j.scitotenv.2019.04.331

[101] Massos A, Turner A. Cadmium, lead and bromine in beached microplastics. Environmental Pollution. 2017;**227**:139-145. DOI: 10.1016/j.envpol.2017.04.034

[102] Johansen MP, Cresswell T, Davis J, Howard DL, Howell NR,

Prentice E. Biofilm-enhanced adsorption of strong and weak cations onto different microplastic sample types: Use of spectroscopy, microscopy and radiotracer methods. Water Research. 2019;**158**:392-400. DOI: 10.1016/j.watres.2019.04.029

[103] Yang Y, Liu G, Song W, Ye C, Lin H, Li Z, et al. Plastics in the marine environment are reservoirs for antibiotic and metal resistance genes. Environment International. 2019;**123**:79-86. DOI: 10.1016/j.envint.2018.11.061

[104] Fletcher M. Bacterial biofilms and biofouling. Current Opinion in Biotechnology. 1994;**5**(3):302-306. DOI: 10.1016/0958-1669(94)90033-7

[105] Chen X, Xiong X, Jiang X, Shi H, Wu C. Sinking of floating plastic debris caused by biofilm development in a freshwater lake. Chemosphere. 2019;**222**:856-864. DOI: 10.1016/j.chemosphere.2019.02.015

[106] Kaiser D, Estelmann A, Kowalski N, Glockzin M, Waniek JJ. Sinking velocity of sub-millimeter microplastic. Marine Pollution Bulletin. 2019;**139**:214e220. DOI: 10.1016/j.marpolbul.2018.12.035

[107] Michels J, Stippkugel A, Lenz M, Wirtz K, Engel A. Rapid aggregation of biofilm-covered microplastics with marine biogenic particles. Proceedings of the Royal Society B. 2018;**285**(1885):20181203. DOI: 10.1098/rspb.2018.1203

[108] Lucas N, Bienaime C, Belloy C, Queneudec M, Silvestre F, Nava-Saucedo JE. Polymer biodegradation: Mechanisms and estimation techniques–a review. Chemosphere. 2008;**73**(4):429-442. DOI: 10.1016/j.chemosphere.2008.06.064

[109] Gottschalk G. Bacterial Metabolism. Netherlands: Springer Science & Business Media; 2012. ISBN-13: 978-0387961538

[110] Panikov NS. Microbial Growth Kinetics. Springer Science & Business Media; 1995. ISBN-13: 978-0387913728

[111] Kirstein IV, Kirmizi S, Wichels A, Garin-Fernandez A, Erler R, Martin L, et al. Dangerous hitchhikers? Evidence for potentially pathogenic *Vibrio* spp. on microplastic particles. Marine Environmental Research. 2016;**120**:1-8. DOI: 10.1016/j.marenvres.2016.07.004

[112] Wu X, Pan J, Li M, Li Y, Bartlam M, Wang Y. Selective enrichment of bacterial pathogens by microplastic biofilm. Water Research. 2019;**165**:114979. DOI: 10.1016/j.watres.2019.114979

[113] Imran M, Das KR, Naik MM. Co-selection of multi-antibiotic resistance in bacterial pathogens in metal and microplastic contaminated environments: An emerging health threat. Chemosphere. 2018;**17**:846-857. DOI: 10.1016/j.chemosphere.2018.10.114

[114] Teschler JK, Zamorano-Sánchez D, Utada AS, Warner CJ, Wong GC, Linington RG, et al. Living in the matrix: Assembly and control of *Vibrio cholerae* biofilms. Nature Reviews. Microbiology. 2015;**13**(5):255. DOI: 10.1038/nrmicro3433

[115] Johnson CN. Fitness factors in *vibrios*: A mini-review. Microbial Ecology. 2013;**65**(4):826-851. DOI: 10.1007/s00248-012-0168-x

[116] Takemura AF, Chien DM, Polz MF. Associations and dynamics of *Vibrionaceae* in the environment, from the genus to the population level. Frontiers in Microbiology. 2014;**5**:38. DOI: 10.3389/fmicb.2014.00038

[117] Pope CF, McHugh TD, Gillespie SH. Methods to determine fitness in bacteria. In: Antibiotic

Resistance Protocols. Humana Press; 2010. pp. 113-121. DOI: 10.1007/978-1-60327-279-7_9

[118] Johnson CN. Influence of environmental factors on *Vibrio* spp. in coastal ecosystems. Microbiology Spectrum. 2015;**3**(3). DOI: 10.1128/microbiolspec.VE-0008-2014

[119] Klein S, Pipes S, Lovell CR. Occurrence and significance of pathogenicity and fitness islands in environmental *vibrios*. AMB Express. 2018;**8**(1):1-7. DOI: 10.1186/s13568-018-0704-2

[120] Ceccarelli D, Colwell RR. *Vibrio* ecology, pathogenesis, and evolution. Frontiers in Microbiology. 2014;**5**:256. DOI: 10.3389/fmicb.2014.00256

[121] Kroukamp O, Bester E, Wolfaardt GM. Biofilms: Besieged cities or thriving ports? In: Hurst C, editor. The Structure and Function of Aquatic Microbial Communities. Cham: Springer; 2019. pp. 53-90. DOI: 10.1007/978-3-030-16775-2_3. ISBN: 978-3-030-16773-8

[122] Hartmann NB, Rist S, Bodin J, Jensen LH, Schmidt SN, Mayer P, et al. Microplastics as vectors for environmental contaminants: Exploring sorption, desorption, and transfer to biota. Integrated Environmental Assessment and Management. 2017;**13**(3):488-493. DOI: 10.1002/ieam.1904

[123] Horton AA, Dixon SJ. Microplastics: An introduction to environmental transport processes. Wiley Interdisciplinary Reviews Water. 2018;**5**(2):e1268. DOI: 10.1002/wat2.1268

[124] Van Sebille E, Aliani S, Law KL, Maximenko N, Alsina JM, Bagaev A, et al. The physical oceanography of the transport of floating marine debris. Environmental Research Letters. 2020;**15**(2):023003. DOI: 10.1088/1748-9326/ab6d7d

[125] Khatmullina L, Chubarenko I. Transport of marine microplastic particles: Why is it so difficult to predict? Anthropocene Coasts. 2019;**2**(1):293-305. DOI: 10.1139/anc-2018-0024

[126] Waldschläger K, Schüttrumpf H. Effects of particle properties on the settling and rise velocities of Microplastics in freshwater under laboratory conditions. Environmental Science & Technology. 2019:1958-1966. DOI: 10.1021/acs.est.8b06794

[127] Hall-Stoodley L, Costerton JW, Stoodley P. Bacterial biofilms: From the natural environment to infectious diseases. Nature Reviews. Microbiology. 2004;**2**(2):95-108. DOI: 10.1038/nrmicro821

[128] Rummel CD, Jahnke A, Gorokhova E, Kühnel D, Schmitt-Jansen M. Impacts of biofilm formation on the fate and potential effects of microplastic in the aquatic environment. Environmental Science & Technology Letters. 2017;**4**(7):258-267. DOI: 10.1021/acs.estlett.7b00164

[129] Verla AW, Enyoh CE, Verla EN, Nwarnorh KO. Microplastic–toxic chemical interaction: A review study on quantified levels, mechanism and implication. SN Applied Sciences. 2019;**1**(11):1400. DOI: 10.1007/s42452-019-1352-0

[130] SAPEA, Science Advice for Policy by European Academies. A Scientific Perspective on Microplastics in Nature and Society. Berlin: SAPEA; 2019. 176 p. DOI: 10.26356/microplastics

[131] Oberbeckmann S, Kreikemeyer B, Labrenz M. Environmental factors support the formation of specific bacterial assemblages on microplastics. Frontiers in Microbiology. 2018;**8**:2709. DOI: 10.3389/fmicb.2017.02709

[132] Koelmans AA, Besseling E, Foekema E, Kooi M, Mintenig S, Ossendorp BC, et al. Risks of plastic debris: Unravelling fact, opinion, perception, and belief. Environmental Science & Technology. 2017;**51**(20): 11513-11519. DOI: 10.1021/acs.est.7b02219

Chapter 19

Extending an Eco-Evolutionary Understanding of Biofilm-Formation at the Air-Liquid Interface to Community Biofilms

Robyn Jerdan, Olga Iungin, Olena V. Moshynets, Geert Potters and Andrew J. Spiers

Abstract

Growing bacterial populations diversify to produce a number of competing lineages. In the *Pseudomonas fluorescens* SBW25 model system, Wrinkly Spreader mutant lineages, capable of colonising the air-liquid interface of static microcosms by biofilm-formation, rapidly appear in diversifying populations with a fitness advantage over the ancestral wild-type strain. Similarly, a biofilm is rapidly produced by a community containing many biofilm-competent members, and selection by serial transfer of biofilm samples across microcosms results in a gradually changing community structure. Both the adaptive radiation producing Wrinkly Spreaders and the succession of biofilm communities in these static microcosms can be understood through evolutionary ecology in which ecological interactions and evolutionary processes are combined. Such eco-evolutionary dynamics are especially important for bacteria, as rapid growth, high population densities and strong selection in the context of infections can lead to fast changes in disease progression and resistance phenotypes, while similar changes in community function may also affect many microbially mediated biotechnological and industrial processes. Evolutionary ecology provides an understanding of why bacterial biofilms are so prevalent and why they are such a successful colonisation strategy, and it can be directly linked to molecular analyses to understand the importance of pathways and responses involved in biofilm-formation.

Keywords: adaptive radiation, air-liquid (A-L) interface biofilms, evolutionary ecology, experimental evolution, fitness, microcosms, oxygen gradients, *Pseudomonas*, Wrinkly Spreaders

1. Introduction

Our research interests have focussed on air-liquid (A-L) interface biofilm-formation by the model pseudomonad *Pseudomonas fluorescens* SBW25 and the adaptive Wrinkly Spreader in experimental microcosms (see our recent reviews [1, 2]), and we have recently begun to extend our investigations into biofilm-formation by communities dominated by similar biofilm-competent pseudomonads. Our research has also developed from a molecular biology perspective [1] towards a more evolutionary and ecological understanding [2] of why biofilms are such a successful colonisation strategy used by bacteria in a wide variety of environments.

In contrast to our changing perspectives, we realise that although biofilm research is interdisciplinary, it appears dominated by molecular biologists working with medically relevant model species with a focus on a mechanistic understanding of biofilm-formation which has remained unchanged from that of the early biofilm pioneers [3, 4]. However, contemporary biofilm research includes a wide range of other disciplines, including evolutionary ecology which provides a framework for understanding how the cooperation needed between bacterial cells to produce biofilms is established and maintained, how bacteria diversify and adapt within these structures, and how biofilm communities respond to changing environmental conditions.

We note that although biofilm reviews addressing evolutionary ecology are published regularly, evolutionary ecology content is negatively correlated with molecular biology and medical content in those reviews with a wider focus.[1] This should be of concern, as any mechanistic understanding of biofilms lacking an evolutionary ecological element will not be able to evaluate the importance of these structures nor make long-term predictions about persistence or function in a wide range of medical, biotechnological and industrial contexts. Furthermore, these negative correlations suggest that the medical molecular microbiology community is ignoring or is unaware of the contributions evolutionary ecology could make towards understanding and mitigating the impact of biofilm-associated disease.

2. Importance of an eco-evolutionary perspective in biofilm research

Evolutionary ecology seeks to understand how ecological interactions can affect selection and adaptation and the consequences of evolutionary change [5–7]. These interactions occur within and between populations, as well as with the environment, and ecological processes involving these interactions explain community dynamics and succession. In contrast, evolutionary processes are usually considered as driving lineages through time, and when subject to selection can result in adaptive changes and ultimately speciation (we use the term lineage here to include mutations, alleles and genotypes, individuals and mutants, and species, all of which can be followed through time and across generations to investigate ecological interactions or evolution). However, ecological and evolutionary changes are directly linked and can occur on the same time-scale [8, 9]. Such eco-evolutionary dynamics are especially important in bacterial populations and communities, where growth rates and numbers are high and selective pressures can be extreme, leading to the rapid fixation of adaptive mutations and striking changes in phenotype or community function.

Evolution research should not therefore be limited to examining fossils or contemporary ecosystems but can be undertaken over relatively short time-scales

[1] We have assessed changing interests in biofilm research by undertaking a simple content analysis of open access reviews published between 2000 and 2004 and 2014–2019 listed by Google Scholar and PubMed on 10 October 2019 ($n = 40$), scoring each for medical (M), molecular biology (MB), and evolutionary or ecological (EE) content. No significant differences were seen in each content type between dates (Wilcoxon, $P < 0.05$) or between contents for each date (Kruskal-Wallis, $P < 0.05$). In early publications we found a significant correlation between M & MB (Spearman $\rho = -0.83$, $P < 0.0001$), but not between M & EE ($P = 0.12$) or MB & EE ($P = 0.96$). In recent publications there were significant correlations between M & EE ($\rho = -0.74$, $P = 0.0002$) and MB & EE ($\rho = -0.43$, $P = 0.06$), but not between M & MB ($P = 0.49$).

in experimental evolution studies using microbial populations and microcosms [10–20]. In particular, the ease with which bacterial populations can be cultured, short generation times and large population sizes which allow mutations to accumulate (diversification) and be identified, the ability to freeze isolates indefinitely, and undertake genetic analyses, make bacteria an ideal model to explore aspects of evolutionary ecology.

Two significant eco-evolutionary processes are particularly relevant to biofilm research. The first of these are ecological interactions which help assemble, stabilise or change community structure [21–23] (community change is often referred to as succession). The main two-way interactions between members of a community are mutualism, commensalism, competition and predation. Cooperation, one example of mutualism where both partners benefit, is usually considered an intraspecific or within-lineage interaction, though it can also occur between closely related lineages or lineages with very similar phenotypes as in the case of community biofilms. External forcing such as physical disturbance can alter ecological interactions (**Figure 1a**) and the impact of this can be measured in terms of system stability and productivity, and possibly even by a change in function. Evolutionary processes, including selection, speciation, drift and dispersal also effect community composition and diversity [21, 23, 24].

The second significant eco-evolutionary process relevant to biofilm research is adaptive radiation [5], the evolution of diversity through random mutation and selection (**Figure 1b**), which in the context of bacteria, can happen very rapidly within a few generations. Developing populations accumulate mutations or diversify, and those mutants with a fitness advantage over their competitors can be considered successful or adaptive. Although evolution is normally thought of as the slow accumulation of mutations with small additive effects on fitness, bacterial microcosms are usually dominated by the first adaptive lineage to appear or by adaptive lineages which appear early on in the process of diversification [14].

Adaptive lineages often make use of new ecological opportunities with key innovations that allow them to interact with the environment in a fundamentally different way [5, 25, 26]. Ecological interactions also occur between lineages and will result in the fixation or loss of particular mutations. These interactions clearly link community change and adaptive radiation, as they help determine the importance of novel ability, such as biofilm-formation, brought in by immigration or key innovation resulting from mutation. In terms of the cooperation required for

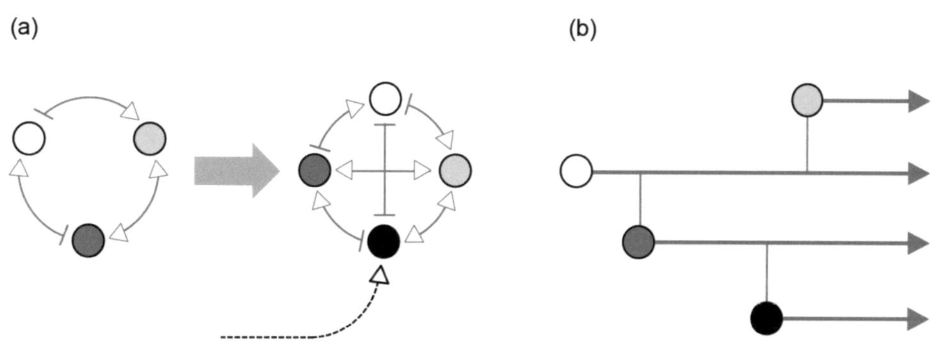

Figure 1.
Eco-evolutionary processes involve ecological interactions and adaptive radiation. Basic ecological interactions determine community dynamics which can change over time, for example, by the immigration (dashed line) of a new member with a novel ability (arrows and bars linking nodes represent positive and negative interactions between community members, respectively) (a). Adaptive mutations occurring in diversifying populations established by a common ancestor can lead to new lineages with key innovations which then compete with other lineages (vertical lines represent mutations giving rise to new lineages) (b).

biofilm-formation, kin selection ensures that cost of construction, often considered in terms of public goods or common pool resources such as the extracellular polymeric substances (EPS) which provide the main structural element of biofilms, is spread across all members who then all share in the benefits [15, 27].

Cooperation is further stabilized in biofilms by spatial separation of producers and cheaters who do not contribute to the cost of construction and a reduction of distance over which the benefits of cooperation act [19]. It is important to note that where external forcing or selection occurs, or where there is an ecological opportunity, community structures will change and lineages continue to adapt, until the theoretical end-point of evolution in a community known as an evolutionary stable community is achieved [28].

3. The SBW25 model system

Pseudomonas fluorescens SBW25 was originally isolated from the sugar beet (*Beta vulgaris* subsp. *vulgaris*) phyllosphere and has been used in experimental evolution studies where the appearance of mutant lineages with altered colony morphologies (these are sometimes referred to as morphotypes or morphs) in diversifying populations have allowed the dynamics of diversification and the fitness of adaptive mutations to be readily investigated [1, 2, 13, 14, 19, 29]. In this system, competitive trade-offs between lineages result in negative frequency-dependent selection and indicate that the major driver of adaptive radiation is competition for limited resources [29].

Fitness, a measure of an individual's reproductive success, is determined at the population level in such microcosms. For bacteria, fitness is readily assessed by comparing the maximum growth rate (v_{Max}) of one population with that of a second, reference population. Simple growth rate comparisons are typically used to infer the success of mutations for which enzymatic or regulatory changes are being investigated, but a more meaningful ecological comparison can be made by growing the two populations together, allowing them to interact with one another and to compete for limiting resources. Competitive fitness [14, 30, 31][2] can be readily determined using co-cultures if the two populations produce different colony morphologies allowing viable number counts to be made on agar plates, or if they can be labelled using fluorescent markers, to allow more rapid enumeration with automated cell counters.

The evolutionary consequences of ecological processes are readily studied using microcosms. They provide defined environments for bacterial growth, and because they are reproducible, treatments can be replicated, experiments are repeatable, and selective pressures can be changed by altering resources or inocula. Nonetheless, the use of microcosms in evolution studies faces some criticisms, including the fact that they are unnatural and very simple environments, and that these studies are essentially contrived [15]. However, although populations may be founded in these synthetic environments, evolutionary and ecological dynamics are interpreted in terms of recent evolutionary history which may span 10–60,000

[2] The competitive fitness (W) of one population (A) compared to a reference population (B) is determined as the ratio of Malthusian parameters (m_A/m_B) where $m = ln$ [final numbers/initial numbers] for each population over the period of the assay [30] (m is scaled here for generation time using ln as a correcting factor [31]). When $W_{A,B}$ is greater than one, A has a competitive advantage over B (and B is at a disadvantage), when $W_{A,B}$ is equal to one, the two populations are neutral, and when $W_{A,B}$ is less than one, A is at a disadvantage (and B has a competitive advantage). As W might be dependent on the initial ratios of the two competing populations, it can show a frequency-dependent response. The selection coefficient (s) is also often used as a measure of survival and success ($s = 1 - W$).

generations in 1 day–25 years (for example, in our system which is described in the following sections, and Lenski's long-term evolution experiment (LTEE) [32]) during which the populations adapt to these environments. Microcosms are not used to replicate the complexity of natural environments but are rather models in which key factors involved in the process of adaptive radiation can be tested [15]. These studies are of course contrived, in the sense that they are designed and in some cases the outcomes are inevitable, but the value of such an approach is that they can be initiated at any point along the evolutionary process and are not limited by the initial diversity or time (for example, the fitness of an adaptive lineage or mutant compared to the ancestral strain can be immediately explored by using genetic manipulation to produce the mutation without having to wait until it appears naturally) [15].

In liquid cultures, wild-type SBW25 populations diversify as random mutations occur, dividing the initially homogeneous or isogenic population into a number of related but diversified lineages. One re-occurring lineage frequently found in static microcosms was the Wrinkly Spreader (WS) mutant class, named after the wrinkled and flat colonies produced on agar plates which are readily distinguished from the smooth and rounded colonies produced by wild-type SBW25 (**Figure 2a**) [29] (quantitative aspects of the WS phenotype are referred to as wrinkleality [1, 35]). WS mutants are further distinguished by an altered niche preference in static microcosms, where they form a robust and well-attached physically cohesive-class biofilm [36] at the air-liquid (A-L) interface, rather than growing throughout the liquid column like wild-type SBW25 (**Figures 2b** and **3b**) [29] (A-L interface biofilms are sometimes referred to as a pellicle [37]).

Wrinkly Spreaders are considered to be adaptive (evolved) lineages because they have a competitive fitness advantage over their ancestor, wild-type SBW25, which does not normally form biofilms in static microcosms [29, 38]. However, in shaking microcosms WS mutants are disadvantaged because they cannot form biofilms [38] and on agar plates the WS phenotype is genetically unstable [39]. Biofilm-formation by Wrinkly Spreaders and SBW25 [40] is neither unusual nor peculiar, as many other soil, plant and water-associated pseudomonads form A-L interface biofilms in static microcosms under the same conditions [36].

Figure 2.
Ancestral SBW25 and adaptive Wrinkly Spreaders. Wild-type SBW25 and Wrinkly Spreader colonies are readily identified on agar plates (a). In static microcosms (b), wild-type SBW25 grows throughout the liquid column (left microcosm) and the Wrinkly Spreader forms a robust biofilm at the A-L interface (right microcosm). These microcosms are 28–30 ml glass vials containing 6 ml growth medium; they are incubated with shaking which provides a homogeneous and unstructured environment with good aeration, or statically which leads to a heterogeneous and structured environment dominated by an O_2 gradient [29, 33]. When tipped out, the WS biofilm retains shape (c) demonstrating just how robust these structures are (see ***Figure 4*** *for more biofilm images). Photographs: A. Spiers.*

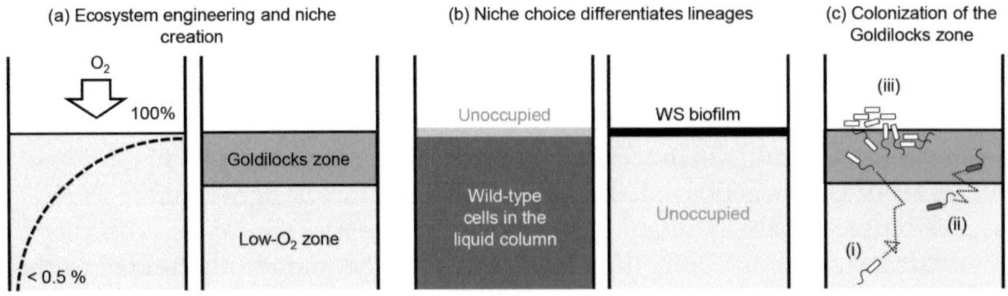

Figure 3.
The success of the Wrinkly Spreader in static microcosms can be understood from an evolutionary ecological perspective. The ecosystem engineering of the initial wild-type SBW25 colonists produces an O_2 gradient (dotted curve) which creates an O_2-rich upper zone (the Goldilocks zone) and a lower depleted zone (a). Wild-type SBW25 and Wrinkly Spreaders show different niche preferences with the WS colonising the top of the Goldilocks zone at the A-L interface (b). The WS biofilm-forming strategy is a more efficient use of resources than constant aerotaxis (swimming) to counter Brownian motion, microcurrents and vibrations which would move cells away from the optimal growth zone (c) (cell tracks indicate (i) aerotaxis towards the goldilocks zone and (ii) displacement from this region; WS biofilms (iii) are formed at the A-L interface).

The distinctive WS colony morphology allowed an investigation of the genes required for biofilm-formation, as mini-transposon mutants of the archetypal WS with wild-type-like colony morphologies were also defective in biofilm-formation [38]. This approach identified the cellulose biosynthesis (*wss*) operon required for the production of partially acetylated cellulose which was the primary biofilm matrix or EPS [38, 41]. However, the WS colony morphology and biofilm also involves an additional EPS, poly-β-1,6-N-acetyl-D-glucosamine (PNAG), as well as lipopolysaccharide (LPS), and interactions between cellulose, PNAG, LPS, and cells are required to maintain biofilm strength and integrity [42, 43]. Mini-transposon analysis also identified a chemotaxis-like (*wsp*) operon with a diguanylate cyclase (DGC) response regulator [38, 44–46]. Subsequent sequence analysis of this operon from the archetypal WS determined the presence of a single nucleotide mutation changing one amino acid residue in the methylesterase subunit [45] which acts as a negative regulatory component of the system. This results in the over-activation of the DCG, leading to increased c-*di*-GMP levels and the activation of the cellulose synthase complex. Mutations in other Wsp subunits, regulators and DGCs activated the WS phenotype in a series of independently isolated mutants [35, 43, 45, 47–49].

This understanding of the underlying molecular biology of the WS phenotype allowed a mechanistic link to be made between adaptive mutation and fitness [45] and demonstrated how easily perturbations c-*di*-GMP homeostasis could result in a key innovation through the activation of a system normally repressed in wild-type SBW25 [1, 2]. The relative ease of recovering WS lineages from diversifying populations of wild-type SBW25, demonstrating a change in niche preference and determining the competitive fitness advantage compared to the ancestral strain, also makes the SBW25 system a model for demonstrating evolution in laboratory classes [50, 51].

The microcosm system has therefore since been used to examine how wild-type colonists modify their environment [33], cells access the A-L interface [52], different environmental conditions drive WS evolution, phenotype and fitness [35, 53], and whether quorum regulation might be involved in biofilm-formation [54]. In the following subsections, we describe how the ecosystem engineering of the colonists provides the ecological opportunity and creates the niche for adaptive WS lineages and explain why biofilm-formation is the better strategy for colonizing this new niche.

4. Ecosystem engineering, ecological opportunity and niche creation

Sterile static microcosms have a uniform O_2 distribution throughout the liquid column. However, after inoculation the metabolic activity of wild-type SBW25 cells rapidly produces a steep O_2 gradient, with less than 0.1% normal levels of dissolved O_2 below 1 mm after 5 h [33]. The ecosystem engineering by these early colonists is driven by O_2 uptake levels which exceed the O_2 flux from the air above into the liquid column, and as a result the initially spatially homogeneous and unstructured environment is divided into an upper high-O_2 zone and a lower O_2-depleted zone (**Figure 3a**). The transition between the two zones is arbitrary but reflects a significant change in growth by wild-type SBW25. Further growth makes the O_2 gradient even more extreme, with less than 1% O_2 found below the top 200 μm layer of the liquid column after 5 days [33].

This depletion of O_2 is an example on a bacterial scale of the social dilemma known as the tragedy of the commons. In this, O_2 is a shared and limiting resource known as the commons, and if used selfishly and without restraint by members of the community it will be depleted and eventually destroyed [55]. Despite the growing difference between high and low-O_2 zones, wild-type SBW25 cells remain distributed throughout the liquid column though there is an appreciable accumulation of cells at the top [52]. Growth rates will be higher in this region which we have described as the Goldilocks zone[3] of optimal growth [2, 53], rather than lower down, as growth is limited by O_2 availability rather than by nutrient levels in this microcosm system [33, 53].

The ecosystem engineering of the initial colonists is also an example of niche creation (niche construction or biogenic habitat formation) [19, 56], as the high-O_2 zone now represents an ecological opportunity [5, 25, 26] for any adaptive lineage capable of colonizing this region more successfully than the initial colonists. Adaptive radiation and niche creation are inter-linked [5, 19, 25, 26, 57], and in this system the high-O_2 zone is colonized primarily by the Wrinkly Spreaders by biofilm-formation at the A-L interface (**Figure 3b** and **c**). Single-cell confocal Raman spectroscopy has demonstrated that WS cells recovered from within the biofilm have the same spectral profile as those grown under high-O_2 conditions, while cells recovered from the liquid column below the biofilm are more similar to those grown under low-O_2 conditions [58].

WS cells under high-O_2 conditions also grow faster than those under low O_2-conditions [33]. However, although WS cells do not grow faster than wild-type SBW25 cells under high O_2-conditions [33], their rapid domination of the A-L interface and subsequent population growth displaces the wild-type colonists from this region in a process known as niche exclusion. WS growth at the A-L interface further reduces O_2 flux into the lower parts of the liquid column in a density dependent manner, effectively limiting the growth of any non-biofilm-forming competitor and WS biofilms have more impact on niche divergence as populations lacking WS produce shallower O_2 gradients [59].

As the WS biofilm population increases, the division between the high and low-O_2 zones also moves up into the biofilm [33], allowing further niche differentiation within the biofilm structure itself. Substantial fitness variation has been observed

[3] 'Goldilocks and the Three Bears', written by Robert Southey, is a tale about a girl called Goldilocks who enters the home of a family of bears while they are away. She tests their chairs, beds and breakfast porridge, always choosing the one most favourable for her, before eventually being chased away when the bears return. The 'Goldilocks zone' is also used to refer to the habitable zone around a star where the temperature is just right for liquid water to exist on an orbiting planet. Here we use the term, stricto sensu, to mean the A-L interface plus the high-O_2 zone immediately below it.

between independently isolated WS [38, 43, 47, 49], suggesting multiple lineages may develop in these populations and compete with one another as Red Queens [60][4] and further competition occurs with resident cheater lineages which no longer produce cellulose [61] and do not contribute to biofilm-formation [62–64].

Fluorescent microscopy suggests WS cells are most active near the top surface of the biofilm [33] and electron micrographs show that it is a very porous structure [65] (**Figure 4**). It is possible that continuous growth near the top progressively limits the growth of cells lower down in a manner known as the Ancestors' inhibition effect [61], though this can also be interpreted as altruistic behaviour by cells which push their descendants up into better O_2 conditions and help suffocate neighbouring competitors [19, 61]. Spatial separation caused by the clumping of WS cells

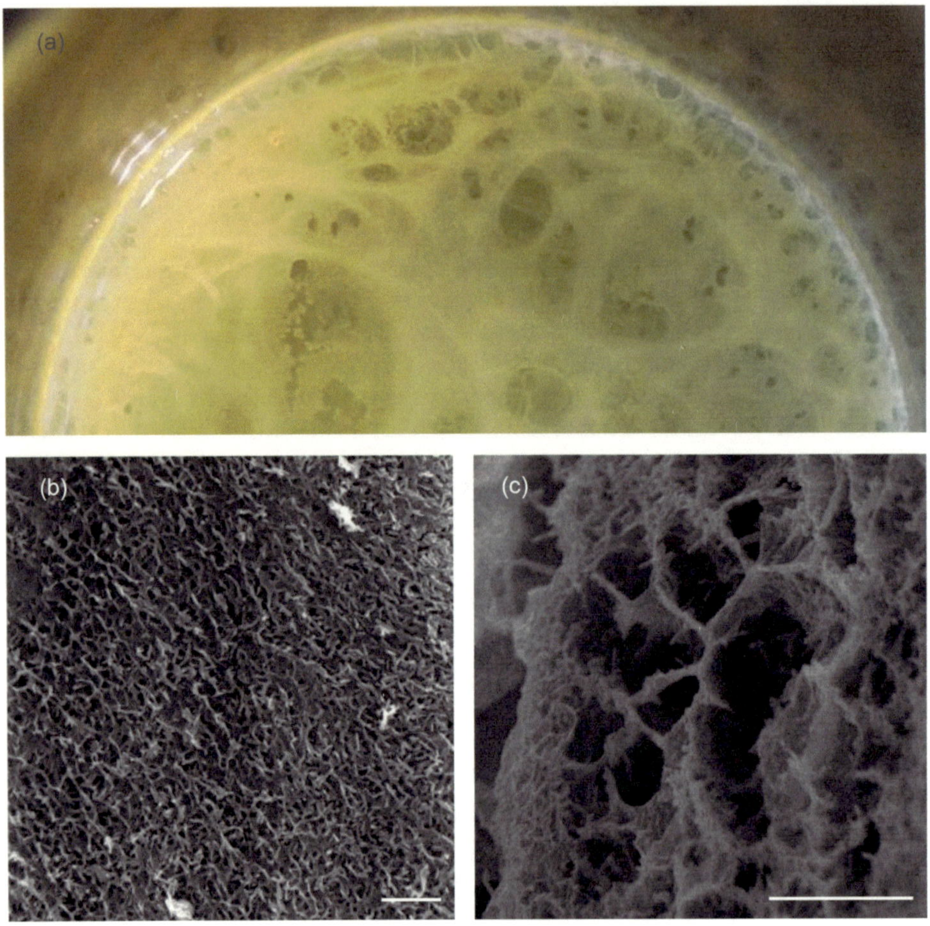

Figure 4.
The Wrinkly Spreader biofilm is a complex structure with voids and fibres apparent at different levels of magnification. Shown are views of biofilms in situ from above (a) and by electron microscopy (b and c) (scale bars represent 10 μm; the mean wild-type SBW25 bacterial body length is 3 μm [34] and individual cells are just visible in (c)). Photographs: (a) A. Spiers, (b and c) O. Moshynets.

[4] The Red Queen is a character in 'Through the looking-glass, and what Alice found there', written by Lewis Carrol. In the Red Queen's race, she and Alice were constantly running yet remained in the same spot. The Red Queen has been adopted as an evolutionary hypothesis which states that lineages must constantly adapt and evolve in order to compete successfully against others which are adapting and to a constantly changing environment. (The Red Queen should not to be confused with the Queen of Hearts who appears in an earlier story by Lewis Carrol.)

by the production of cellulose and the exclusion of cheaters, plus the continued development of the O_2 gradient within developing biofilms [33] which limits the distance over which the benefits of cooperation act, will help stabilize cooperation in biofilms [19] and allow kin selection to provide a competitive advantage to WS lineages. Biofilm development, including increasing depth and total biomass, as well as lineage and total population levels, ultimately ends with system failure when it rips and sinks to the bottom of the microcosm vial [2].

5. Biofilm-formation is the best strategy for colonising the high-O_2 zone

Aerobic motile bacteria such as SBW25 could gain access to the high-O_2 zone by aerotaxis [66], using flagella-mediated swimming motility and following the O_2 gradient up towards the A-L interface. Aerotaxis could also be used to maintain position against the physical displacement of cells caused by random diffusion, micro-currents and random knocks and vibrations occurring in microcosms during incubation. Although SBW25 is known to be capable of swimming, swarming and twitching motilities, we only recently demonstrated that wild-type and WS cells are aerotaxic [52] and that the average swimming velocity [34] is sufficient to overcome the negative effects of random diffusion on cell localization [52].

However, random diffusion still has a significant effect on maintaining position in the high-O_2 zone, and we were able to demonstrate this using modified

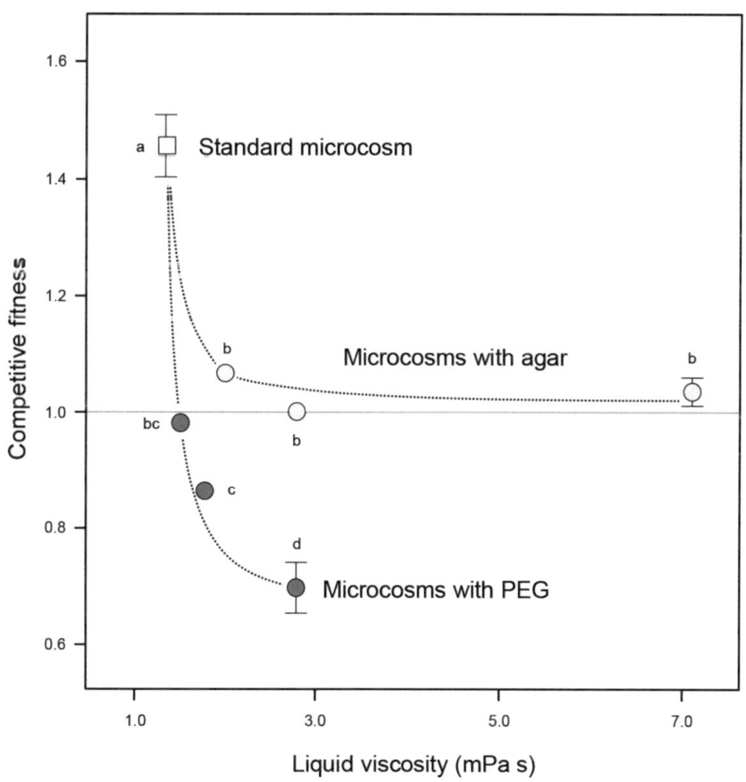

Figure 5.
WS fitness decreases with increasing liquid viscosity. Agar (light grey circles) and polyethylene glycol (dark grey circles) were used to increase the viscosity of standard microcosms (white square) and the competitive fitness of the archetypal WS determined in comparison with wild-type SBW25 under Fe-limited conditions where it cannot form a biofilm. Means are shown with standard errors. Dotted lines suggest trends and differences between means were investigated by Tukey-Kramer HSD; means sharing the same letters are not significantly different (α = 0.05). Data are replotted from [52] (Supplementary Information).

microcosms in which we had added low concentrations of agar or polyethylene glycol to increase viscosity, as diffusion is inversely dependent on liquid viscosity [52]. Both wild-type and WS cell localization improved with increasing viscosity and, furthermore, WS competitive fitness was found to decrease with increasing viscosity (**Figure 5**) [52]. This indicates that WS biofilm-formation is a better strategy allowing the colonization of the high-O_2 zone and more specifically, of the A-L interface, than constant aerotaxis.

We argue that the need to remain in place at the top of the liquid column efficiently in order to make use of greater O_2 availability is the fundamental explanation for the success of A-L interface biofilm-formation in static microcosms by motile aerobic such as the pseudomonads [36] where growth is limited by O_2-availability rather than by nutrients [53]. The success is determined by a cost-benefit trade-off, in which resource costs required for biofilm-formation by the community or constant aerotaxis by individual cells are balanced against population gains.

6. Biofilms are not equivalent structures or of equal value

Although biofilm-formation has been extensively investigated for a wide range of model bacteria, SBW25 is the only strain for which multiple A-L interface biofilms with qualitatively different phenotypes have been reported. Wild-type SBW25 produces a cellulose-based but fragile and poorly attached 'viscous mass' (VM) [36] biofilm when induced by exogenous Fe [40], and a genetically modified strain over-expressing the *wss* operon produces a similarly fragile biofilm [38]. In addition to the Wrinkly Spreaders, Fuzzy Spreaders have also been recovered from diversifying populations of SBW25 in static microcosms [29]. Though these were initially thought to be adaptive mutants which grew at the bottom of static microcosms and were adapted to anoxic conditions, they have subsequently been shown to produce fragile and short-lived A-L interface biofilms in which cells aggregate because of altered LPS expression [67]. A range of other biofilm-forming mutants have also evolved from genetically manipulated strains of SBW25, including the CBFS and PWS mutants which utilise PNAG as the primary biofilm matrix [43, 68].

WS and WS-like phenotypes are often caused by loss-of-function mutations affecting negative regulators, less frequently by promoter activation or gene-fusion mutations, and finally by rare mutations resulting in intragenic gain-of-function [47]. In general, these biofilm-forming lineages have a fitness advantage compared to non-biofilm-forming competitors [38, 43, 47, 49]. However, possible negative pleiotropy and epistasis effects [11, 14] might contribute to a lower-than-expected fitness advantage in some cases, and the accumulation of additional mutations not associated with the WS phenotype may also have a negative effect on fitness in a process known as Muller's ratchet [14].

In order to better understand the links between WS mutation, phenotype, and fitness, it has been necessary to develop quantitative assays to describe WS biofilms and an experimental approach to test the effect of physical disturbance on biofilm-formation and fitness. Variations in WS phenotype or wrinkleality [1, 35], including microcosm growth, biofilm strength and attachment levels, can be determined using a combined biofilm assay [69] that can quantitatively differentiate WS isolates recovered from different environments, whilst careful use of orbital shakers can provide intermediate levels of disturbance between static and shaking conditions.

Using this approach, we can differentiate CBFS, VM and WS biofilms on the basis of competitive fitness compared to a non-biofilm-forming strain. Under static conditions CBFS fitness is greater that either VM or WS biofilms, suggesting that the CBFS biofilm is the most cost-effective solution to colonising the A-L interface.

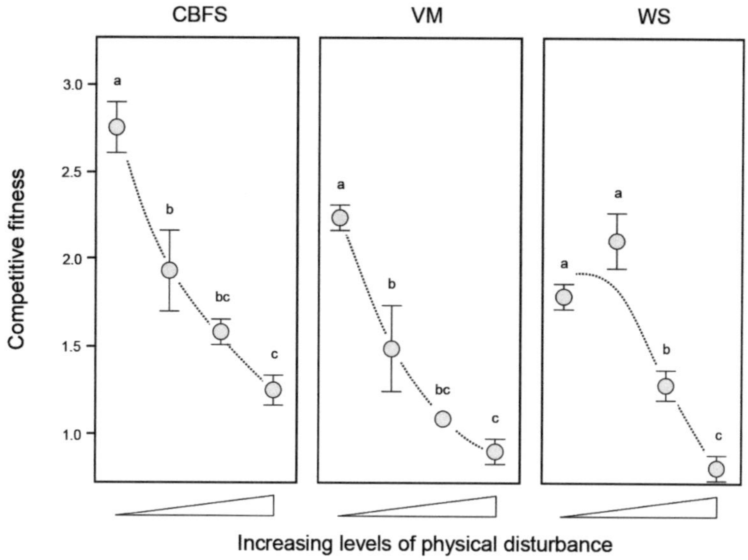

Figure 6.
The adaptive advantage of biofilms is dependent on levels of physical disturbance. Competitive fitness assays were used to assess the adaptive advantage of CBFS, VM and WS biofilms compared to a non-biofilm-forming competitor across a range of levels of physical disturbance from static to shaken conditions. Means are shown with standard errors. Dotted lines suggest trends and differences between means were investigated by Tukey-Kramer HSD; means sharing the same letters are not significantly different (α = 0.05). Data and analyses will be reported in full elsewhere (A. Koza and A. Spiers).

However, as the level of physical disturbance increases, CBFS and VM biofilms fail and sink before the more resilient WS biofilms. As a result, their fitness decreases before WS fitness (**Figure 6**). At maximum levels of disturbance where no biofilm can form, VM and WS fitness is lower than CBFS fitness. This suggests that the VM and WS phenotypes which continue to produce cellulose but cannot form biofilms are more costly than the CBFS phenotype which does not utilise this particular EPS. We noted that in these microcosms CBFS aggregates accumulated on the vial walls above the liquid line. Stranded cells may have better access to O_2 than those remaining in the liquid column and this may further increase competitive fitness.

We are also able to differentiate CBFS, VM and WS biofilms on the basis of structure and rheology, which, when combined with our fitness analyses, suggests that the CBFS biofilm is the most cost-effective structure allowing the colonisation of the A-L interface. It falls between the more costly and over-engineered WS and barely adequate VM biofilms and provides a greater fitness advantage because the levels of physical disturbance static microcosms are subject to will neither increase, which might favour the WS biofilm, nor fall, which may favour the VM biofilm.[5]

7. Community biofilm-formation in static microcosms

As the evolutionary dynamics of diversifying SBW25 populations and the fitness advantages of biofilm-formation in static microcosms are increasingly well understood, we have begun to consider the drivers of biofilm-formation in community-based multi-species biofilms [70–72]. Communities artificially established in microcosms from mixed inocula are particularly interesting as strong

[5] This 'neither too much nor too little' evaluation suits Red Queens who choose to compete for the occupation of the Goldilocks zone in our microcosms.

selection would be expected to play a role in community assembly with a rapid loss of redundant members who do not contribute to the new system. There are simple organising principles in microbial communities especially where competitive interactions are dominant [73, 74]. However, with the exception of WS-like biofilms initiated through mutation, cell-to-cell communication is thought to co-ordinate biofilm-formation and ensure that all members contribute to the cost of production without cheating [15, 27]. As a result, biofilm-formation is seen largely as a cooperative undertaking by closely related lineages, yet this appears to conflict with the view that competitive interactions generally dominate microbial communities.

In order to investigate the relative importance of cooperation and competition in community biofilms, we have developed a model system using soil-wash inocula which include biofilm-competent pseudomonads [36] and our static microcosms in which O_2 is the growth-limiting factor. This typically resulted in very fragile and poorly attached VM-like biofilms within 2–3 days with substantial growth also occurring throughout the liquid column. Preliminary trails suggested that growth levels were sensitive to different media and aeration conditions, and treatment with antibiotics, copper and perchlorate had differential effects on growth, biofilm strength and attachment levels, demonstrating that different selective pressures could alter community productivity and biofilm-formation.

We then undertook a serial transfer experiment selecting for biofilm-formation by transferring biofilm samples using a wire-loop across a series of 10 microcosms for a total of 60 days of incubation. Under such selection, we expected to see replicate communities dominated by robust WS-like biofilms and a decrease in strain diversity as non-biofilm-formers and uncompetitive strains were lost. We also expect to see a significant reduction in the number of bacteria growing below the biofilm in the liquid column, as competition for access to O_2 should drive ecological change and result in more 'effective' biofilm-formation.

However, replicate communities continued to produce weak biofilms despite their physically cohesive appearance [36], suggesting that the selective pressure for biofilm-formation was not particularly strong. Nonetheless, a significant loss of diversity was observed, and an analysis of random isolates suggested that the proportion of biofilm-formers increased, and a phenotypic shift occurred between the initial and final selected communities (**Figure 7**), confirming that these communities were subject to selective pressure. Although we expected to see the selected communities dominated by one or a few 'super' biofilm-formers, they appeared to be dominated by a mix of lineages with very similar phenotypes. This is perhaps not surprising, as our preparation of the soil-wash inocula would have selected for fast-growing aerobic and biofilm-competent bacteria such as *Pseudomonas* spp. from the original soil community (environmental filtering within the soil would also have selected for related lineages and lineages with similar phenotypes). Such mixes may be stable, as the coexistence of related lineages and the coexistence of unrelated lineages with similar phenotypes, is possible because they may not exhibit significant levels of negative interactions and might even facilitate one another [75].

We also found significant levels of growth in the liquid column below the biofilms, suggesting that lineages were colonising the A-L interface and low-O_2 region from the biofilm transfer samples and that migration was occurring between these two zones. It is possible that biofilm-competent lineages might avoid competition at the A-L interface by choosing a less competitive niche lower down the liquid column in a biochemical trade-off [76] in which lower growth rates resulting from O_2-limitation are balanced by the cost of biofilm-formation which would have been required at the A-L interface.

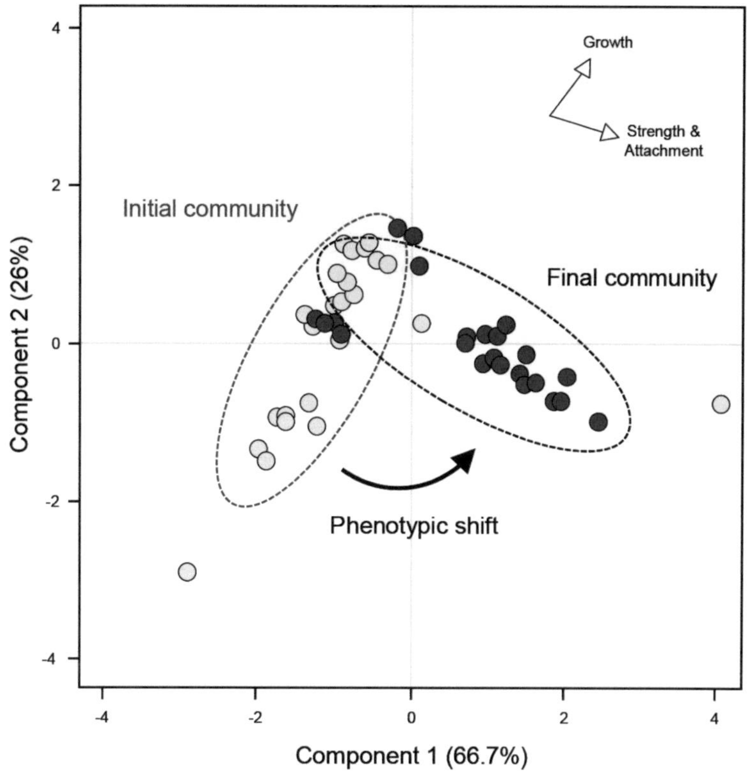

Figure 7.
Serial transfer of biofilm samples results in changes in biofilm characteristics of individual community members. Isolates were sampled from the initial (light grey circles) and final communities (dark grey circles) after serial transfer of biofilm material by wire loop across 10 microcosms over 60 days. Principal component analysis (PCA) of isolate biofilm characteristics, including total microcosm growth, biofilm strength and attachment levels, shows a phenotypic shift occurring between initial and final communities. Data and analyses will be reported in full elsewhere (R. Jerdan and A. Spiers).

Although this research is still on-going and will be published in full elsewhere, our current focus is to better understand the levels of competition occurring within the community biofilm and the role of the low-O_2 region in maintaining diversity in selected communities. A future goal is to investigate the dynamics of diversification of wild-type SBW25 in these communities in order to see how competition within the community biofilm effects WS evolution and fitness.

8. Conclusions

Biofilm research is interdisciplinary but is increasingly fragmented and polarised, with interest still dominated by molecular biologists working with medically relevant model species and a mechanistic focus on biofilm-formation. This perspective limits our understanding of more complex community-based biofilms, as ecological interactions and evolutionary processes play important roles in the development and success of these structures, with immigration and adaptive radiation introducing novel abilities or key innovations which may have a significant impact on community function. Biofilm research is now at the stage where an eco-evolutionary perspective should be included to produce a more comprehensive and holistic understanding of biofilms in a wide range of contexts, from model systems to biofilm-associated disease, biotechnology and industry.

Acknowledgements

We acknowledge Abertay University, Antwerp Maritime Academy, and the Institute of Molecular Biology and Genetics of the National Academy of Sciences of Ukraine, as well as past grants including BOF grant "Microbiologisch beïnvloede corrosie gecontroleerd door middel van een UV ballast water management system" from the University of Antwerp and Antwerp Maritime Academy, ERASMUS-Plus grant 2017-1-BE02-KA107-034651 and a Staff Mobility grant 2017-1-UK01-KA107-036400, NATO SPS grant 984834, and the International Exchange Programme of the Royal Society of Edinburgh, which have supported the biofilm-associated collaboration between G. Potters, O. Moshynets, and A. Spiers, and have most recently enabled visits by R. Jerdan to Kiev and of O. Iungin and O. Moshynets to Antwerp and Dundee.

Conflict of interest

The authors declare that there are no conflicts of interests.

Author details

Robyn Jerdan[1], Olga Iungin[2,3], Olena V. Moshynets[3], Geert Potters[4] and Andrew J. Spiers[1*]

1 School of Applied Sciences, Abertay University, Dundee, United Kingdom

2 Kyiv National University of Technologies and Design, Kiev, Ukraine

3 Institute of Molecular Biology and Genetics of the National Academy of Sciences of Ukraine, Kiev, Ukraine

4 Antwerp Maritime Academy and Department of Bioscience Engineering, University of Antwerp, Antwerp, Belgium

*Address all correspondence to: a.spiers@abertay.ac.uk

References

[1] Spiers AJ. A mechanistic explanation linking adaptive mutation, niche change and fitness advantage for the Wrinkly Spreader. International Journal of Evolutionary Biology. 2014;**2014**:10. Article ID: 675432

[2] Koza A, Kuśmierska A, McLaughlin K, Moshynets O, Spiers AJ. Adaptive radiation of *P. fluorescens* SBW25 in experimental microcosms provides an understanding of the evolutionary ecology and molecular biology of A-L interface biofilm-formation. FEMS Microbiology Letters. 2017;**364**:fnx109

[3] Costerton JW, Lewandowski Z, DeBeer D, Caldwell D, Korber D, James G. Biofilms, the customized microniche. Journal of Bacteriology. 1994;**176**:2137-2142

[4] Costerton JW, Lewandowski Z, Caldwell D, Korber D, Lappin-Scott HM. Microbial biofilms. Annual Review of Microbiology. 1995;**49**:711-745

[5] Schluter D. The Ecology of Adaptive Radiation. Oxford, UK: Oxford University Press; 2000

[6] Weber MG, Wagner CE, Best RJ, Harmon LJ, Matthews B. Evolution in a community context: On integrating ecological interactions and macroevolution. Trends in Ecology and Evolution. 2017;**32**:291-304

[7] Schoener TW. The newest synthesis: Understanding the interplay of evolutionary and ecological dynamics. Science. 2011;**331**:426-429

[8] Abrams PA. Modelling the adaptive dynamics of traits involved in inter- and intraspecific interactions: An assessment of three methods. Ecology Letters. 2001;**4**:166-175

[9] Pelletier F, Garant D, Hendry AP. Eco-evolutionary dynamics. Philosophical Transactions of the Royal Society B. 2009;**364**:1483-1489

[10] Dykhuizen DE. The potential for microorganisms and experimental studies in evolutionary biology. In: Bell MA, Futuyma DJ, Eanes WF, Levinton JS, editors. Evolution since Darwin. The First 150 Years. Sunderland, USA: Sinauer Associates; 2010. pp. 169-173

[11] Elena SF, Lenski RE. Evolution experiments with microorganisms: The dynamics and genetic bases of adaptation. Nature Reviews. Genetics. 2003;**4**:457-469

[12] Feldgarden M, Stoebel DM, Brisson D, Dykhuizen DE. Size doesn't matter: Microbial selection experiments address ecological phenomena. Ecology. 2003;**84**:1679-1687

[13] MacLean RC. Adaptive radiation in microbial microcosms. Journal of Evolutionary Biology. 2005;**18**:1376-1386

[14] Bell G. Selection. The Mechanism of Evolution. 2nd ed. Oxford: Oxford University Press; 2008. 553 p

[15] Buckling A, Maclean RC, Brockhurst MA, Colegrave N. The *Beagle* in a bottle. Nature. 2009;**457**: 824-829

[16] Kussell E. Evolution in microbes. Annual Review of Biophysics. 2013;**42**:493-514

[17] Adams J, Rosenzweig F. Experimental microbial evolution: History and conceptual underpinnings. Genomics. 2014;**104**:393-398

[18] Bailey SF, Bataillon T. Can the experimental evolution programme help us elucidate the genetic basis of adaptation in nature? Molecular Ecology. 2016;**25**:203-218

[19] Steenackers HP, Parijs I, Foster KR, Vanderleyden J. Experimental evolution in biofilm populations. FEMS Microbiology Reviews. 2016;**40**:373-397

[20] O'Malley M. The experimental study of bacterial evolution and its implications for the modern synthesis of evolutionary biology. Journal of the History of Biology. 2018;**51**:319-354

[21] Nemergut DR, Schmidt SK, Fukami T, O'Neill SP, Bilinski TM, Stanish LF, et al. Patterns and processes of microbial community assembly. Microbiology and Molecular Biology Reviews. 2013;**77**:342-356

[22] Pulsford SA, Lindenmayer DB, Driscoll DA. A succession of theories: Purging redundancy from disturbance theory. Biological Reviews. 2016;**91**:148-167

[23] Zhou J, Ning D. Stochastic community assembly: Does it matter in microbial ecology? Microbiology and Molecular Biology Reviews. 2017;**81**:e00002-e00017

[24] Vellend M. Conceptual synthesis in community ecology. The Quarterly Review of Biology. 2010;**85**:183-206

[25] Losos JB, Mahler DL. Adaptive radiation: The interaction of ecological opportunity, adaptation, and speciation. In: Bell MA, Futuyma DJ, Eanes WF, Levinton JS, editors. Evolution Since Darwin: The First 150 Years. Sunderland, USA: Sinauer Associates; 2010. pp. 381-420

[26] Yoder JB, Clancey E, Des Roches S, Eastman JM, Gentry L, Godsoe W, et al. Ecological opportunity and the origin of adaptive radiations. Journal of Evolutionary Biology. 2010;**23**: 1581-1596

[27] Nadell CD, Xavier JB, Foster KR. The sociobiology of biofilms. FEMS Microbiology Reviews. 2009;**33**:206-224

[28] Edwards KF, Kremer CT, Miller ET, Osmond MM, Litchman E, Klausmeier CA. Evolutionary stable communities: A framework for understanding the role of trait evolution in the maintenance of diversity. Ecology Letters. 2018;**21**:1853-1868

[29] Rainey PB, Travisano M. Adaptive radiation in a heterogeneous environment. Nature. 1998;**394**:69-72

[30] Lenski RE, Rose MR, Simpson SC, Tadler SC. Long-term experimental evolution in *Escherichia coli*. I. Adaptation and divergence during 2,000 generations. The American Naturalist. 1991;**138**:1315-1341

[31] Chevin L-M. On measuring selection in experimental evolution. Biology Letters. 2011;**7**:210-213

[32] Lenski RE. Experimental evolution and the dynamics of adaptation and genome evolution in microbial populations. The ISME Journal. 2017;**11**:2191-2194

[33] Koza A, Moshynets O, Otten W, Spiers AJ. Environmental modification and niche construction: Developing O_2 gradients drive the evolution of the Wrinkly Spreader. The ISME Journal. 2011;**5**:665-673

[34] Ping L, Birkenbeil J, Monajembashi S. Swimming behavior of the monotrichous bacterium *Pseudomonas fluorescens* SBW25. FEMS Microbiology Ecology. 2013;**86**:36-44

[35] Udall YC, Deeni Y, Hapca SM, Raikes D, Spiers AJ. The evolution of biofilm-forming Wrinkly Spreaders in static microcosms and drip-fed columns selects for subtle differences in wrinkleality and fitness. FEMS Microbiology Ecology. 2015;**91**:fiv057

[36] Ude S, Arnold DL, Moon CD, Timms-Wilson T, Spiers AJ. Biofilm formation and cellulose expression

among diverse environmental *Pseudomonas* isolates. Environmental Microbiology. 2006;**8**:1997-2011

[37] Moshynets OV, Spiers AJ. Viewing biofilms within the larger context of bacterial aggregations. In: Dhanasekaran D, Thajuddin N, editors. Microbial Biofilms—Importance and Applications. Rijeka: InTech Publishers; 2016

[38] Spiers AJ, Kahn SG, Travisano M, Bohannon J, Rainey PB. Adaptive divergence in *Pseudomonas fluorescens*. 1. Determinants of wrinkly spreader fitness and the cause of an evolutionary transition. Genetics. 2002;**161**:33-46

[39] Spiers AJ. Wrinkly-Spreader fitness in the two-dimensional agar plate microcosm: Maladaptation, compensation and ecological success. PLoS ONE. 2007;**2**(8):e740

[40] Koza A, Hallett PD, Moon CJ, Spiers AJ. Characterisation of a novel air-liquid interface biofilm of *Pseudomonas fluorescens* SBW25. Microbiology. 2009;**155**:1397-1406

[41] Spiers AJ, Bohannon J, Gehrig S, Rainey PB. Biofilm formation at the air-liquid interface by the *Pseudomonas fluorescens* SBW25 wrinkly spreader requires an acetylated form of cellulose. Molecular Microbiology. 2003;**50**:15-27

[42] Spiers AJ, Rainey PB. The *Pseudomonas fluorescens* SBW25 wrinkly spreader biofilm requires attachment factor, cellulose fibre and LPS interactions to maintain strength and integrity. Microbiology. 2005;**151**:2829-2839

[43] Lind PA, Farr AD, Rainey PB. Evolutionary convergence in experimental *Pseudomonas* populations. The ISME Journal. 2017;**11**:589-600

[44] Goymer P, Kahn SG, Malone JG, Gehrig SM, Spiers AJ, Rainey PB. Adaptive divergence in experimental populations of *Pseudomonas fluorescens*. II. Role of WspR in evolution and the development of the wrinkly spreader phenotype. Genetics. 2006;**173**:515-526

[45] Bantinaki E, Kassen R, Knight C, Robinson Z, Spiers AJ, Rainey PB. Adaptive divergence in experimental populations of *Pseudomonas fluorescens*. III. Mutational origins of wrinkly spreader diversity. Genetics. 2007;**176**: 441-453

[46] Malone JG, Williams R, Spiers AJ, Rainey PB. The structure-function relationship of WspR: A *Pseudomonas fluorescens* response-regulator with a GGDEF output domain. Microbiology. 2007;**153**:980-994

[47] Lind PA, Farr AD, Rainey PB. Experimental evolution reveals hidden diversity in evolutionary pathways. eLife. 2015;**4**:e07074

[48] McDonald MJ, Gehrig SM, Meintjes PL, Zhang X-X, Rainey PB. Adaptive divergence in experimental populations of *Pseudomonas fluorescens*. IV. Genetic constraints guide evolutionary trajectories in a parallel adaptive radiation. Genetics. 2009;**183**:1041-1053

[49] McDonald MJ, Cooper TF, Beaumont HJE, Rainey PB. The distribution of fitness effects of new beneficial mutations in *Pseudomonas fluorescens*. Biology Letters. 2011;**7**:98-100

[50] Green JH, Koza A, Moshynets O, Pajor R, Ritchie MR, Spiers AJ. Evolution in a test-tube: Rise of the Wrinkly Spreaders. Journal of Biological Education. 2011;**45**:54-59

[51] Spiers AJ. Getting Wrinkly Spreaders to demonstrate evolution in schools. Trends in Microbiology. 2014;**22**:301-303

[52] Jerdan R, Anna Kuśmierska A, Marija Petric M, Spiers AJ. Penetrating the air-liquid interface is the key to colonization and Wrinkly Spreader fitness. Microbiology. 2019;**165**:1061-1074

[53] Kuśmierska A, Spiers AJ. New insights into the effects of several environmental parameters on the relative fitness of a numerically dominant class of evolved niche specialist. International Journal of Evolutionary Biology. 2016;**2016**:10. Article ID: 4846565

[54] Moshynets OV, Foster D, Karakhim SA, McLaughlin K, Rogalsky SP, Rymar SY, et al. Examining c-di-GMP and possible quorum sensing regulation in *Pseudomonas fluorescens* SBW25: Links between intra- and inter-cellular regulation benefits community cooperative activities such as biofilm formation. Ukrainian Biochemical Journal. 2018;**90**:17-31

[55] Estrela S, Libby E, Van Cleve J, Débarre F, Deforet M, Harcombe WR, et al. Environmentally mediated social dilemmas. Trends in Ecology and Evolution. 2019;**34**:6-18

[56] Day RL, Laland KN, Odling-Smee J. Rethinking adaptation. The niche-construction perspective. Perspectives in Biology and Medicine. 2003;**46**:80-95

[57] Odling-Smee J, Erwin DH, Palkovacs EP, Feldman MW, Laland KN. Niche construction theory: A practical guide for ecologists. The Quarterly Review of Biology. 2013;**88**:3-28

[58] Huang WE, Ude S, Spiers AJ. *Pseudomonas fluorescens* SBW25 biofilm and planktonic cells have differentiable Raman spectral profiles. Microbial Ecology. 2007;**53**:471-474

[59] Loudon CM, Matthews B, Sevilgen DS, Ibelings BW. Experimental evidence that evolution by niche construction affects dissipative ecosystem dynamics. Evolutionary Ecology. 2016;**30**:221-234

[60] Liow LH, Van Valen L, Stenseth NC. Red Queen: From populations to taxa and communities. Trends in Ecology and Evolution. 2011;**26**:349-358

[61] Xavier JB, Foster KR. Cooperation and conflict in microbial biofilms. Proceedings of the National Academy of Sciences of the United States of America. 2007;**104**:876-881

[62] Rainey PB, Rainey K. Evolution of cooperation and conflict in experimental bacterial populations. Nature. 2003;**425**:72-74

[63] Brockhurst MA, Hochberg ME, Bell T, Buckling A. Character displacement promotes cooperation in bacterial biofilms. Current Biology. 2006;**16**:1-5

[64] Brockhurst MA, Colegrave N, Hodgson DJ, Buckling A. Niche occupation limits adaptive radiation in experimental microcosms. PLoS ONE. 2007;**2**:e193

[65] Spiers AJ, Deeni YY, Folorunso AO, Koza A, Moshynets O, Zawadzki K. Cellulose expression in *Pseudomonas fluorescens* SBW25 and other environmental pseudomonads. In: Van De Ven TGM, TGM GL, editors. Cellulose—Medical, Pharmaceutical and Electronic Applications. InTech Publishers: Rijeka; 2013

[66] Taylor BL, Zhulin IB, Johnson MS. Aerotaxis and other energy sensing behaviour in bacteria. Annual Review of Microbiology. 1999;**53**:103-128

[67] Ferguson GC, Bertels F, Rainey PB. Adaptive divergence in experimental populations of *Pseudomonas fluorescens*. V. Insight into the niche specialist fuzzy spreader

compels revision of the model *Pseudomonas* radiation. Genetics. 2013;**195**:1319-1335

[68] Gehrig S. Adaptation of *Pseudomonas fluorescens* SBW25 to the air-liquid interface: A study in evolutionary genetics [thesis]. Oxford, UK: University of Oxford; 2005

[69] Robertson M, Hapca SM, Moshynets O, Spiers AJ. Air-liquid interface biofilm formation by psychrotrophic pseudomonads recovered from spoilt meat. Antonie Van Leeuwenhoek. 2013;**103**:251-259

[70] Elias S, Banin E. Multi-species biofilms: Living with friendly neighbors. FEMS Microbiology Reviews. 2012;**36**:990-1004

[71] Røder HL, Sørensen SJ, Burmølle M. Studying bacterial multispecies biofilms: Where to start? Trends in Microbiology. 2016;**24**:503-513

[72] Tan CH, Lee KWK, Burmølle M, Kjelleberg S, Rice SA. All together now: Experimental multispecies biofilm model systems. Environmental Microbiology. 2017;**19**:42-53

[73] Vega NM, Gore J. Simple organizing principles in microbial communities. Current Opinion in Microbiology. 2018;**45**:195-202

[74] Foster KR, Bell T. Competition, not cooperation, dominates interactions among culturable microbial species. Current Biology. 2012;**22**:1845-1850

[75] Weber MG, Agrawal AA. Phylogeny, ecology, and the coupling of comparative and experimental approaches. Trends in Ecology and Evolution. 2012;**27**:394-403

[76] Giri S, Waschina S, Kaleta C, Kost C. Defining division of labor in microbial communities. Journal of Molecular Biology. 2019;**431**:4712-4731